Lecture Notes in Electrical Engineering

Volume 982

The book series *Lecture Notes in Electrical Engineering* (LNEE) publishes the latest developments in Electrical Engineering—quickly, informally and in high quality. While original research reported in proceedings and monographs has traditionally formed the core of LNEE, we also encourage authors to submit books devoted to supporting student education and professional training in the various fields and applications areas of electrical engineering. The series cover classical and emerging topics concerning:

- Communication Engineering, Information Theory and Networks
- Electronics Engineering and Microelectronics
- Signal, Image and Speech Processing
- Wireless and Mobile Communication
- Circuits and Systems
- Energy Systems, Power Electronics and Electrical Machines
- Electro-optical Engineering
- Instrumentation Engineering
- Avionics Engineering
- Control Systems
- Internet-of-Things and Cybersecurity
- Biomedical Devices, MEMS and NEMS

For general information about this book series, comments or suggestions, please contact leontina.dicecco@springer.com.

To submit a proposal or request further information, please contact the Publishing Editor in your country:

China

Jasmine Dou, Editor (jasmine.dou@springer.com)

India, Japan, Rest of Asia

Swati Meherishi, Editorial Director (Swati.Meherishi@springer.com)

Southeast Asia, Australia, New Zealand

Ramesh Nath Premnath, Editor (ramesh.premnath@springernature.com)

USA, Canada

Michael Luby, Senior Editor (michael.luby@springer.com)

All other Countries

Leontina Di Cecco, Senior Editor (leontina.dicecco@springer.com)

** This series is indexed by EI Compendex and Scopus databases. **

Rajeev Agrawal · Pabitra Mitra · Arindam Pal ·
Madhu Sharma Gaur
Editors

International Conference on IoT, Intelligent Computing and Security

Select Proceedings of IICS 2021

Editors
Rajeev Agrawal
GL Bajaj Institute of Technology
and Management
Greater Noida, India

Arindam Pal ⓘ
Commonwealth Scientific and Industrial
Research Organisation (CSIRO)
Canberra, ACT, Australia

Pabitra Mitra
Department of Computer Science
Engineering
Indian Institute of Technology
Kharagpur, India

Madhu Sharma Gaur ⓘ
GL Bajaj Institute of Technology
and Management
Greater Noida, India

ISSN 1876-1100   ISSN 1876-1119 (electronic)
Lecture Notes in Electrical Engineering
ISBN 978-981-19-8138-8   ISBN 978-981-19-8136-4 (eBook)
https://doi.org/10.1007/978-981-19-8136-4

This Springer imprint is published by the registered company Springer Nature Singapore Pte Ltd.
The registered company address is: 152 Beach Road, #21-01/04 Gateway East, Singapore 189721,
Singapore

Organizations

General Conference Chair

Prof. (Dr.) Rajeev Agrawal, GL Bajaj Institute of Technology and Management, Greater Noida, Uttar Pradesh, India

Program Chairs

Dr. Mohamed Elhoseny, Mansoura University, Egypt
Dr. Arindam Pal, Commonwealth Scientific and Industrial Research Organization (CSIRO), Australia
Dr. Maleq Khan, Texas A&M University–Kingsville, USA

Technical Chairs

Dr. Mamoun Alazab, Charles Darwin University, Australia
Dr. Selena He, Kennesaw State University, Georgia, USA
Dr. Chakchai So-In, Khon Kaen University, Thailand
Dr. Mohammad S. Alam, Texas A&M University-Kingsville
Dr. Dinesh Kumar Singh, GL Bajaj Institute of Technology and Management, Greater Noida

Organizing Secretariat

Dr. Shashank Awasthi, GL Bajaj Institute of Technology and Management, Greater Noida

Organizing Chairs

Dr. Sujay Deb, Indraprastha Institute of Information Technology, Delhi
Dr. Madhu Sharma Gaur, GL Bajaj Institute of Technology and Management, Greater Noida

Session Chair for Track 1

Dr. Brajesh Kumar, Mahatma Jyotiba Phule Rohilkhand University, Bareilly

Co-session Chair for Track 1

Dr. Divya Mishra, GL Bajaj Institute of Technology and Management, Greater Noida

Session Chair for Track 2

Dr. Krishan Kumar, National Institute of Technology, Srinagar
Dr. Deepak Punetha, Indian Institute of Technology, Bombay

Co-session Chair for Track 2

Dr. Rajiv Kumar, GL Bajaj Institute of Technology and Management, Greater Noida

Session Chair for Track 3

Dr. Nishant Kumar, Gurukula Kangri University, Haridwar
Dr. Abhinav Saxena, JSS Academy of Technical Education, Noida

Co-session Chair for Track 3

Dr. Astha Sharma, GL Bajaj Institute of Technology and Management, Greater Noida

Session Chair for Track 3

Dr. Surendiram B., National Institute of Technology, Puducherry
Dr. Emmanuel S. Pilli, Malaviya National Institute of Technology Jaipur

Conference Convener

Dr. Sanjeev Kumar, GL Bajaj Institute of Technology and Management, Greater Noida, Uttar Pradesh, India

Internal Advisory Committee

Dr. Shashank Awasthi, GL Bajaj Institute of Technology and Management, Greater Noida, Uttar Pradesh, India
Dr. Satyendra Sharma, GL Bajaj Institute of Technology and Management, Greater Noida, Uttar Pradesh, India
Dr. R. K. Mishra, GL Bajaj Institute of Technology and Management, Greater Noida, Uttar Pradesh, India
Dr. P. C. Vashist, GL Bajaj Institute of Technology and Management, Greater Noida, Uttar Pradesh, India
Dr. Sanjeev Pippal, GL Bajaj Institute of Technology and Management, Greater Noida, Uttar Pradesh, India
Dr. Mohit Bansal, GL Bajaj Institute of Technology and Management, Greater Noida, Uttar Pradesh, India
Dr. Vinod Yadav, GL Bajaj Institute of Technology and Management, Greater Noida, Uttar Pradesh, India
Dr. Sansar Singh Chauhan, GL Bajaj Institute of Technology and Management, Greater Noida, Uttar Pradesh, India
Dr. Prashant Mukherjee, GL Bajaj Institute of Technology and Management, Greater Noida, Uttar Pradesh, India

Internal Steering Committee

Dr. Rajiv Kumar, GL Bajaj Institute of Technology and Management, Greater Noida, Uttar Pradesh, India
Dr. Amrita Rai, GL Bajaj Institute of Technology and Management, Greater Noida, Uttar Pradesh, India
Dr. Upendra Dwivedi, GL Bajaj Institute of Technology and Management, Greater Noida, Uttar Pradesh, India
Dr. Astha Sharma, GL Bajaj Institute of Technology and Management, Greater Noida, Uttar Pradesh, India

Organizing Committee

Mr. Gaurav Bhaita, GL Bajaj Institute of Technology and Management, Greater Noida, Uttar Pradesh, India
Mr. Lalan Kumar, GL Bajaj Institute of Technology and Management, Greater Noida, Uttar Pradesh, India
Mr. Prem Sagar Sharma, GL Bajaj Institute of Technology and Management, Greater Noida, Uttar Pradesh, India
Ms. Anju Mishra, GL Bajaj Institute of Technology and Management, Greater Noida, Uttar Pradesh, India
Ms. Deepkiran, GL Bajaj Institute of Technology and Management, Greater Noida, Uttar Pradesh, India
Ms. Manjusa Gundale, GL Bajaj Institute of Technology and Management, Greater Noida, Uttar Pradesh, India
Ms. Aparna Sharma, GL Bajaj Institute of Technology and Management, Greater Noida, Uttar Pradesh, India
Ms. Priya Porwal, GL Bajaj Institute of Technology and Management, Greater Noida, Uttar Pradesh, India
Mr. Manish Kumar, GL Bajaj Institute of Technology and Management, Greater Noida, Uttar Pradesh, India
Mr. Anil Kumar Vats, GL Bajaj Institute of Technology and Management, Greater Noida, Uttar Pradesh, India
Mr. Virendra Kumar, GL Bajaj Institute of Technology and Management, Greater Noida, Uttar Pradesh, India
Mr. Bhavesh Kumar, GL Bajaj Institute of Technology and Management, Greater Noida, Uttar Pradesh, India
Mr. Vikram Singh, GL Bajaj Institute of Technology and Management, Greater Noida, Uttar Pradesh, India

Preface

With the emerging technologies, world is changing and technology intervention is bringing fast paradigm shift in the present era. The rapid development in IoT and intelligent computing technologies are happening day by day, and the things around us are becoming smart and intelligent. These new challenges posed and the ideas generated to solve problems need to be shared. However, this technology transformation is developing with higher security risk which requires discussion to bring awareness about necessary research in the field of intelligent computing and security.

This book presents selected proceedings of the International Conference on IoT, Intelligent Computing and Security: A Paradigm Shift. The Conference IICS-2021 was held on December 17–18, 2021, organized by the department of Master of Computer Applications (MCA), GL Bajaj Institute of Technology and Management, Greater Noida, Delhi NCR, India (affiliated to Dr. A. P. J. Abdul Kalam Technical University, Lucknow, Uttar Pradesh). The conference was conducted in hybrid mode (physical/online) mode due to COVID-19 pandemic.

The aim of IICS-2021 was to bring together vibrant stakeholders who share a passion for research, innovation, solution development partners, end users and our budding professionals around the world to deliberate upon the different challenging aspects and issues in the field of IoT, artificial intelligence, emerging computational intelligence and security solutions.

This two days' online international conference had participation from across the globe including, USA, Australia, France, Argentina, Russia, Nigeria. Norwegian and Bangladesh. The conference received 169 submissions, and review committee received 169 research papers out of which 53 papers were selected for final presentation after rigorous blind reviews involving more than 240 reviewers. The committees of the conference include more than 300 national/international committee chairs, advisory board members, technical program committee members, keynote speakers, presenters and experts from across the globe to share their views, innovations and accomplishments.

We are grateful to Respected Prof (Dr.) S. N. Singh, IIT Kanpur, Chief Guest, Dr. Satish K. Singh, IIIT Allahabad, Dr. Arti Noor, Senior Director, Center of Development in Advanced Computing (C-DAC) Noida, Uttar Pradesh, India, Guest of Honor for their gracious presence and motivation. We pay our heartfelt gratitude to Dr. Ram Kishore Agarwal, Chief Patrons, Shri Pankaj Agarwal, our Patron and our keynote speakers, Dr. Raj Jain, Professor of Computer Science and Engineering, Washington University in St. Louis, Dr. Carlos M. Travieso University of Las Palmas de Gran Canaria (ULPGC), Spain, Dr. K. C. Santosh University of South Dakota, USA, Mr. Daniel Lewis, CEO and Cofounder, Awen Collective (Industrial Cyber Security Software), University of Bristol, Pontypridd, Wales, UK, and Mr. Mitthan Meena, Founder and CEO at Microsec.AI Security for their valuable keynote address on day 1 and day 2.

Our sincerely thanks to all the national/international committee chairs, advisory board members, technical program committee members, keynote speakers, presenters and experts from across the globe and members of the organizing committee for their cooperation, hard work and support to make this conference successful.

We also thank Springer for publishing the proceedings in the Lecture Notes in Electrical Engineering (LNEE) series. Special thanks to all the authors and participants for their contributions in making an effective, successful and productive conference.

Greater Noida, India	Dr. Rajeev Agrawal
Kharagpur, India	Dr. Pabitra Mitra
Canberra, Australia	Dr. Arindam Pal
Greater Noida, India	Dr. Madhu Sharma Gaur
January 2022	

Keynote I

Title: Blockchains with AI for Security and Risk Management

Keynote Address

Dr. Raj Jain
Barbara J. and Jerome R. Cox, Jr., Professor of Computer Science and Engineering, Washington University in St. Louis

Abstract

Blockchains have found numerous applications in Fintech, supply chains and contracts because it is an ideal distributed consensus where all nodes agree on the validity of transactions in a block without needing a central trusted party. The consensus is binary—agree or disagree—true or false. In this era of big data, we need to move blockchains beyond data storage to provide knowledge. In the real world, there are many situations in which various participants may not fully agree, and their opinions may be probabilistic, leading to probabilistic agreements. In this talk, Prof. Jain will present his recent extensions using AI that allow blockchains to be used for group decisions that may not be binary. These extensions enable blockchains to be used for group decision making and risk management when the group sizes are large, and group members may want to remain anonymous. In particular, Prof. Jain will describe numerous use cases of this idea. Such situations frequently arise in network security and risky investments.

Biography

Raj Jain is currently the Barbara J. and Jerome R. Cox, Jr., Professor of Computer Science and Engineering at Washington University in St. Louis. Dr. Jain is Life Fellow of IEEE, Fellow of ACM, Fellow of AAAS and Recipient of the 2018 James B. Eads Award from St. Louis Academy of Science, 2017 ACM SIGCOMM Life-Time Achievement Award. Previously, he was one of the co-founders of Nayna Networks, Inc., Senior Consulting Engineer at Digital Equipment Corporation in Littleton, Mass, and then Professor of Computer and Information Sciences at Ohio State University in Columbus, Ohio. With 37,000+ citations, according to Google Scholar, he is one of the highly cited authors in computer science. Further information is at http://www.cse.wustl.edu/~jain/.

Keynote II

Title: Smart Affective States Identification for Neurodegenerative Diseases

Invited Speaker

Prof. Dr. Carlos M. Travieso-González
University of Las Palmas de Gran Canaria, Spain

Abstract

The use of image processing methods is a useful tool in order to extract information from persons for different application. In particular, soft biometrics application can be applied to detect race, age, gender, expression, etc. In this case, I have used that information to extract the grade of emotions. It can be an important and interesting indicator for medical doctors and to have more diagnostic evidences about a decision or criterion in neurodegenerative diseases.

There is that to difference the concept of expression and the grade of expression or arousal. For this kind of diseases, the important concept is the grade of expression because it says if the person loses or not that grade, with independence of the valence, positive or negative. The neurodegenerative disease is present on the arousal.

This kind of studies can represent new paradigms for the medicine and in great input from the technology field. The adding value of this kind of proposal is its low cost and easy use.

Biography

Carlos M. Travieso-González received the M.Sc. degree in 1997 in Telecommunication Engineering at Polytechnic University of Catalonia (UPC), Spain, and Ph.D. degree in 2002 at University of Las Palmas de Gran Canaria (ULPGC-Spain). He is Full Professor and Head of Signals and Communications Department at ULPGC. He belongs to ULPGC from 2001, teaching subjects on signal processing, pattern recognition and learning theory. His research lines are biometrics, biomedical signals and images, data mining, classification system, signal and image processing, machine learning and environmental intelligence. He has researched in more than 50 International and Spanish Research Projects, some of them as Head Researcher. He is Co-author of four books, Co-editor of 25 Proceedings Book and Guest Editor for eight JCR-ISI international journals and up to 24 chapters. He has over 460 papers published in international journals and conferences (83 of them indexed on JCR–ISI–Web of Science). He has published seven patents in Spanish Patent and Trademark Office. He has been Supervisor on nine Ph.D. theses (11 more are under supervision) and 130 master theses. He is Founder of The IEEE-IWOBI conference series and President of its Steering Committee, of The InnoEducaTIC conference series and of The APPIS conference series. He is Evaluator of project proposals for European Union (H2020 and Horizon Europe), Medical Research Council (MRC—UK), Spanish Government (ANECA), Research National Agency (ANR—France), DAAD (Germany), Argentinian Government and Colombian Institutions. He has been Reviewer in different indexed international journals (<70) and conferences (<240) since 2001. He is Member of IASTED Technical Committee on Image Processing from 2007 and Member of IASTED Technical Committee on Artificial Intelligence and Expert Systems from 2011. He will be APPIS 2020 General Chair and IEEE-IWOBI 2020 and was APPIS 2019 General Chair and IEEE-IWOBI 2019, IEEE-IWOBI 2018 General Chair, APPIS 2018 General Chair, InnoEducaTIC 2017 General Chair, IEEE-IWOBI 2017 General Chair, IEEE-IWOBI 2015 General Chair, InnoEducaTIC 2014 General Chair, IEEE-IWOBI 2014 General Chair, IEEE-INES 2013 General Chair, NoLISP 2011 General Chair, JRBP 2012 General Chair and IEEE-ICCST 2005 Co-chair. He is Associate Editor on Computational Intelligence and Neuroscience journal (Hindawi—Q1 JCR-ISI), Sensors (MDPI—Q1 JCR-ISI) and Entropy (MDPI—Q2 JCR-ISI). He was Vice-Dean from 2004 to 2010 in Higher Technical School of Telecommunication Engineers in ULPGC and Vice-Dean of Graduate and Postgraduate Studies from March 2013 to November 2017.

Keynote III

Title: Infectious DiseaseX: AI for Healthcare, How BigData is Big, and Explainabilty

Invited Speaker

Dr. K. C. Santosh
University of South Dakota, USA

Abstract

When we consider AI for healthcare, infectious disease outbreak is no exception. Three major topics: #AI4Healthcare, how #BigData is big (in medical imaging informatics) and #ExplainableAI will be discussed during the talk. The talk will begin with machine learning models that help in not only predicting but also detecting abnormalities due to infectious diseaseX such as pneumonia, TB and COVID-19. I will open my talk with infectious disease prediction models and unexploited data, where we will learn that predictive analytical tools are close to garbage-in garbage-out (at least for COVID-19). I will then cover multimodal learning and representation based on both shallow learning (handcrafted features) and deep learning (deep features) that typically apply on medical imaging tools. Like in computer vision, I will open an obvious question, how #BigData is big in addition to common techniques: data augmentation and transfer learning. Another crucial part of the talk is #ExplainableAI—I will discuss on where have we missed explainability? With all these facts, as most of models are limited to education and training, I will end up my talk with the statement "ML innovation should not limit to building models." What we need is #ExplainableAI in #Active Learning framework.

Contents

Security in Smart Computing Environment

Contemporary Computing Applications

Editors and Contributors

About the Editors

Dr. Rajeev Agrawal has worked as Director and Professor at G.L. Bajaj Institute of Technology and Management, Greater Noida, India. He has an illustrated experience of more than 27 years in teaching and research and holds a B.E. degree in Electronics Engineering and M.Tech. degree in System Engineering. He received Ph.D. in the area of Wireless Communication Channels from the School of Computer and System Sciences, JNU, New Delhi. He was visiting professor at Kennesaw State University, Georgia, USA, under a joint research project in the area of Remote Patient Monitoring and Medical Imaging. His research areas include planning and performance analysis of wireless networks and medical image analysis for automated diagnosis, performance analysis of fog, and edge networks. He has more than 70 publications in international journals and proceedings. He has been awarded by various state and national agencies for his contribution to research and academics. He is also serving as a reviewer for several reputed international journals and member of the editorial board for two international journals and an editor of a book by Elsevier in the area of Health Informatics and two lecture series by Springer.

Dr. Pabitra Mitra (Member, IEEE) received the B.Tech. degree in electrical engineering from IIT Kharagpur, Kharagpur, West Bengal, India, in 1996, and the Ph.D. degree from the Department of Computer Science and Engineering, Indian Statistical Institute, Kolkata, India, in 2005. Currently, he is working as a professor in the Department of Computer Science and Engineering, IIT Kharagpur. He has supervised eight Ph.D. students and 12 M.S. students (by research) on different issues related to AI and machine learning. His research interests are machine learning, data mining, pattern recognition, information retrieval, and image processing. As a part of publications, he has 40 journal articles, 7 chapters, 3 books, 100 conference proceedings, 3 reviews, and 2 editorials in his credit.

Dr. Arindam Pal is a senior research scientist at Data61 in Commonwealth Scientific and Industrial Research Organization (CSIRO). He is also a conjoint senior lecturer in the School of Computer Science and Engineering at UNSW Sydney. He has over 14 years of industrial research experience in companies like Microsoft, Yahoo!, Novell, CSIRO, Cognizant, and TCS Research. He has over 10 years of experience in Analytics and Machine Learning. He has published several papers at prestigious conferences such as ACM SIGIR, ACM CIKM, ACM JCDL, IEEE ICRA, and IEEE ICC. He has been granted 4 US patents and has filed 15 patents in various countries like the USA, European Union, and India. He has given more than 30 invited talks in Australia, India, USA, and Italy. He has managed and delivered high-quality software and impactful projects in Australia, the USA, and India. He has worked on evacuation planning, intelligent transportation, multi-robot task allocation, phishing detection, data privacy, and security. He earned his Ph.D. in Computer Science and Engineering from IIT Delhi. He works on the business and research problems of CSIRO and collaborates with faculty members of universities, both in Australia and abroad. He publishes academic papers in conferences and journals. He is a senior member of both ACM and IEEE.

Dr. Madhu Sharma Gaur (Communicating Editor) is a professor and head of the Department of Master of Computer Applications, GL Bajaj Institute of Technology and Management Greater Noida. She has received Ph.D. Degree in the area of Security and Trust Management in Mobile Pervasive Environment and serving as an academician, researcher, and computer application development architect for the last 20+ years. Her area of research interests includes IoT, AI/ML, security, and trust management. She has presented and published various research papers in Springer, IEEE, and Elsevier International Conferences and Journals of repute and also received the most cited paper by Springer *International Journal Human-centric Computing and Information Sciences*, in 2018. She has extensive exposure to software development, app development, and provides solutions for various industrial/social problems. She has also completed many industrial projects like visitor pass manager, payroll, stock maintenance, invoice and stock management, and order dispatch system.

Contributors

H. Aditya Pai Department of Computer Science & Engineering, Deemed to be University, Dehradun, India

Reshu Agarwal Amity Institute of Information Technology, Amity University, Noida, India

Jitendra Agrawal School of Information Technology, Rajiv Gandhi Proudyogiki Vishwavidyalaya, Bhopal, MP, India

Prashant Ahlawat School of Information Technology, Manipal University, Jaipur, India

S. Asha School of Computer Science and Engineering and Centre for Cyber Physical Systems, VIT University, Chennai, Tamil Nadu, India

V. S. Bakkialakshmi Department of Computer Science and Engineering, Hindustan Institute of Technology and Science, Chennai, India

Usha Batra School of Engineering & Sciences, D Goenka University, Gurgaon, India

Mohd. Belal Department of Computer Science, Aligarh Muslim University, Aligarh, Uttar Pradesh, India

Hanu Bhardwaj Department of Computer Science and Technology, Manav Rachna University, Faridabad, Haryana, India

Rishav Bhardwaj School of Electronics Engineering, VIT-AP Campus, Amrawati, India

Kamika Chaudhary Department of Computer Science, M.B. Government P.G. College, Haldwani, Nainital, India

Gaurav Choudhary DTU Compute, Technical University of Denmark, Lyngby, Denmark

A. Christy Jeba Malar Department of Information Technology, Sri Krishna College of Technology, Coimbatore, Tamilnadu, India

M. Deva Priya Department of Computer Science and Engineering, Sri Eshwar College of Engineering, Coimbatore, Tamilnadu, India

Javier Diaz Faculty of Informatics, Research Laboratory in New Information Technologies (LINTI), National University of La Plata, La Plata, Buenos Aires, Mexico

P. Divya Department of Computer Science and Engineering, Sri Krishna College of Technology, Coimbatore, Tamilnadu, India

Shivangi Diwan National Institute of Technology, Raipur, Chattisgarh, India

Zamam Farhat Aligarh Muslim University, Aligarh, India

Nishat Fatima Department of Computer Science and Engineering, Centre for Advanced Studies, Lucknow, Uttar Pradesh, India

Arti Gautam G L Bajaj Institute of Technology and Management, Greater Noida, India

Aayush Goel Bharati Vidyapeeth's College of Engineering, New Delhi, India

Harshit Goel Department of Computer Science and Engineering, Delhi Technological University, Delhi, India

Deepa Gupta Infosys Ltd., Bangalore, India

Lipika Gupta Chitkara University Institute of Engineering and Technology, Chitkara University, Punjab, India

Mukul Gupta G.L. Bajaj Institute of Management, Greater Noida, India

Rishi Gupta Manipal University Jaipur, Rajasthan, India

Sandeep Kumar Gupta AMET University, Chennai, India

M. S. Guru Prasad Department of Computer Science & Engineering, Deemed to be University, Dehradun, India

Ivana Harari Faculty of Informatics, Research Laboratory in New Information Technologies (LINTI), National University of La Plata, La Plata, Buenos Aires, Mexico

Tarun Jain Manipal University Jaipur, Rajasthan, India

Priya Jaiswal Department of Computer Application, M.C.M.T, Varanasi, UP, India

Sengathir Janakiraman Department of Information Technology, CVR College of Engineering, Hyderabad, Telangana, India

R. Kanmani Department of Electronics and Communication Engineering, SNS college of Technology, Coimbatore, Tamilnadu, India

Shylaja VinayKumar Karatangi G L Bajaj Institute of Technology and Management, Greater Noida, India

Rahul Katarya Department of Computer Science and Engineering, Delhi Technological University, Delhi, India

Shaminder Kaur Chitkara University Institute of Engineering and Technology, Chitkara University, Punjab, India

Rohit Kaushik G. L. Bajaj Institute of Technology and Management, Greater Noida, India

Abdullah Ahmad Khan Department of Computer Science, Aligarh Muslim University, Aligarh, Uttar Pradesh, India

Aijaz Khan VIT Bhopal University, Bhopal, Madhya Pradesh, India

Ankit Khatri Dr B R Ambedkar National Institute of Technology Jalandhar, Jalandhar, Punjab, India

Ravi Khatri Dr B R Ambedkar National Institute of Technology Jalandhar, Jalandhar, Punjab, India

K. Krishnaveni Department of Computer Science, Sri. S. Ramasamy Naidu MemorialCollege (Affiliated to Madurai Kamaraj University, Madurai), Virudhunagar District, Tamil Nadu, India

Surendra Kumar Shukla Department of Computer Science and Engineering, Graphic Era Deemed to be University, Dehradun, India

Abhishek Kumar Department of Computer Science & Engineering, Chaudhary Devi Lal University, Sirsa, Haryana, India

Rajeev Kumar School of Computer and Systems Sciences, Jawaharlal Nehru University, New Delhi, New Delhi, India

Rajesh Kumar Department of Electrical Engineering, JSS Academy of Technical Education, Noida, U.P, India

Sachin Kumar Department of Computer Application, V.B.S.P.U, Jaunpur, UP, India

Sushil Kumar School of Computer and Systems Science, Jawaharlal Nehru University, New Delhi, New Delhi, India

T. V. Vijay Kumar School of Computer and Systems Sciences, Jawaharlal Nehru University, New Delhi, India

Vinesh Kumar Phonics Group of Institutions, Roorkee, India

Priyanshi Kumari Department of Electrical Engineering, JSS Academy of Technical Education, Noida, U.P, India

Shilpi Kumari Department of Electrical Engineering, JSS Academy of Technical Education, Noida, U.P, India

Usha Kumari Department of Electronics and Communication, DCRUST Murthal, Murthal, India

Estela Macas International Ibero-American University—UNINI MX, Mexico City, Mexico

Rajashri Mahato Centre for Cyber Physical Systems, VIT University, Chennai, Tamil Nadu, India

A. Christy Jeba Malar Department of Information Technology, Sri Krishna College of Technology, Coimbatore, Tamilnadu, India

Preeti Malik Department of Computer Science and Engineering, Graphic Era University, Dehradun, India

Vikas Mittal School of Computer and Systems Sciences, Jawaharlal Nehru University, New Delhi, India

G. Nivedhitha Department of Computer Science and Engineering, Sri Krishna College of Technology, Coimbatore, Tamilnadu, India

Sarvottam Ola VIT University, Chennai, Tamil Nadu, India

S. Padmavathi Department of Computer Science and Engineering, Sri Krishna College of Technology, Coimbatore, Tamilnadu, India

Nitin Kumar Pal Government Polytechnic College Puranpur, Puranpur, U.P, India

Saurabh Pal Department of Computer Application, V.B.S.P.U, Jaunpur, UP, India

Bhaskar Pant Department of Computer Science and Engineering, Graphic Era Deemed to be University, Dehradun, India

Harshil Panwar Department of Computer Science and Engineering, Delhi Technological University, Delhi, India

Monika Parmar Chitkara University School of Engineering and Technology, Chitkara University, Solan, Himachal Pradesh, India

V. Praba Department of Computer Science, Sri. S. Ramasamy Naidu Memorial-College (Affiliated to Madurai Kamaraj University, Madurai), Virudhunagar District, Tamil Nadu, India

Om Prakash School of Computer and Systems Sciences, Jawaharlal Nehru University, New Delhi, New Delhi, India

Prashant Department of Electrical Engineering, JSS Academy of Technical Education, Noida, U.P, India

M. Deva Priya Department of Computer Science and Engineering, Sri Eshwar College of Engineering, Coimbatore, Tamilnadu, India

Jyoti Pruthi Department of Computer Science and Technology, Manav Rachna University, Faridabad, Haryana, India

Henry Quisnancela National University of La Plata, Faculty of Informatics, Research Laboratory in New Information Technologies (LINTI), Buenos Aires, Argentina

T. Raghunathan Department of Computer Science and Engineering, Sri Krishna College of Technology, Coimbatore, Tamilnadu, India

Jagdeep Rahul Department of Electronics and Communication Engineering, Rajiv Gandhi University, Doimukh, India

Amrita Rai G L Bajaj Institute of Technology and Management, Greater Noida, India

Vandana Rawat Department of Computer Science and Engineering, Graphic Era Deemed to be University, Dehradun, India

Mónica R. Romero Faculty of Informatics, Research Laboratory in New Information Technologies (LINTI), National University of La Plata, La Plata, Buenos Aires, Mexico

S. Saadhikha Shree Centre for Cyber Physical Systems, VIT University, Chennai, Tamil Nadu, India

Mridu Sahu National Institute of Technology, Raipur, Chattisgarh, India

Olena Sakovska Uman National University of Horticulture, Uman, Ukraine

Abhinav Saxena Department of Electrical Engineering, JSS Academy of Technical Education, Noida, U.P, India

Shishir Kumar Shandilya Vellore Institute of Technology, VIT Bhopal University, Bhopal, M.P, India

Amit Kumar Sharma Department of Electrical and Electronics Engineering, Galgotia College of Engineering and Technology, Greater Noida, U.P, India

Ashish K. Sharma Bajaj Institute of Technology, Wardha, Maharashtra, India

Dhananjay Sharma Department of Computer Science and Engineering, Delhi Technological University, Delhi, India

Durgesh M. Sharma Shri Ramdeobaba College of Engineering and Management, Nagpur, Maharashtra, India;
Vellore Institute of Technology, VIT Bhopal University, Bhopal, M.P, India

Hitesh Kumar Sharma School of Computer Science, University of Petroleum and Energy Studies, EnergyAcres, Dehradun, India

Lakhan Dev Sharma School of Electronics Engineering, VIT-AP Campus, Amrawati, India

Manoj Kumar Sharma School of Information Technology, Manipal University, Jaipur, India

Sandhya Sharma Chitkara University School of Engineering and Technology, Chitkara University, Solan, Himachal Pradesh, India

Surendra Kumar Shukla Department of Computer Science and Engineering, Graphic Era Deemed to Be University, Dehradun, India

Devesh Pratap Singh Department of Computer Science and Engineering, Graphic Era Deemed to be University, Dehradun, India

Jay Singh Department of Electrical and Electronics Engineering, G.L Bajaj Institute of Technology and Management, Greater Noida, U.P, India

Kiran Deep Singh Chitkara University Institute of Engineering and Technology, Chitkara University, Rajpura, Punjab, India

Neelam Singh Department of Computer Science and Engineering, Graphic Era Deemed to be University, Dehradun, India

Nitu Singh School of Information Technology, Rajiv Gandhi Proudyogiki Vishwavidyalaya, Bhopal, MP, India

Prabhdeep Singh Department of Computer Science & Engineering, Deemed to be University, Dehradun, India

Ram Sewak Singh Department of Electronics and Communication Engineering, Adama Science and Technology University, Adama, Ethiopia

Satya Singh Department of Computer Application, M.G.K.V.P, Varanasi, UP, India

Shipra Singh Infosys Ltd., Bangalore, India

Vikram Singh Department of Computer Science & Engineering, Chaudhary Devi Lal University, Sirsa, Haryana, India

Aditya Sinha Manipal University Jaipur, Rajasthan, India

T. Sudalaimuthu Department of Computer Science and Engineering, Hindustan Institute of Technology and Science, Chennai, India

Sudhakar School of Computer and Systems Science, Jawaharlal Nehru University, New Delhi, New Delhi, India

T. S. Pavith Surya Department of Information Technology, Sri Krishna College of Technology, Coimbatore, Tamilnadu, India

Mohammad Abdullah Tahir Aligarh Muslim University, Aligarh, India

Vishwas Tanwar Manipal University Jaipur, Rajasthan, India

Anupam Tiwari School of Engineering & Sciences, D Goenka University, Gurgaon, India

Umesh Kumar Tiwari Department of Computer Science and Engineering, Graphic Era Deemed to be University, Dehradun, India

Vikas Tripathi Department of Computer Science & Engineering, Deemed to be University, Dehradun, India

Ghufran Ullah Department of Computer Science, Aligarh Muslim University, Aligarh, Uttar Pradesh, India

B. Umamaheswari Department of Computer Science and Engineering, Amrita Vishwa Vidyapeetham, Amrita School of Engineering, Bengaluru, India

Shola Usharani School of Computer Science and Engineering, VIT University, Chennai, Tamil Nadu, India

Eesha Verma Amity Institute of Information Technology, Amity University, Noida, India

Mahima Verma Department of Electrical Engineering, JSS Academy of Technical Education, Noida, U.P, India

Subhask Kumar Verma AMET University, Chennai, India;
School of Business Management, Noida International University, Greater Noida, India

Rekha Yadav Department of Electronics and Communication, DCRUST Murthal, Murthal, India

Vrinda Yadav Department of Computer Science and Engineering, Centre for Advanced Studies, Lucknow, Uttar Pradesh, India

IoT and Intelligent Computing:
A Paradigm Shift

Internet of Medical Things Enabled by Permissioned Blockchain on Distributed Storage

Anupam Tiwari and Usha Batra

1 Introduction

COVID-19 pandemic has abruptly taken over the world in a matter of months at an expedited pace unthought-of. The pandemic has been able to change economies of nations, dissect into all growth and projections of development across the globe. All this has happened in exactly more or less the same way it happened 100 years back [1], i.e., 1918 Pandemic (H1N1 virus). The failures to identify infected patients, collate data, transparency and sharing real-time true analytics, real-time expedited availability of data, maintaining privacy and classification of data, etc. all have more or less remained in the same state today. Today, we are indeed different from what we were a century back, but even with so much of evolution in medicine and technologies, the state of failure is relatively same excluding few countries that promptly acted and retarded pandemic to an extent with enabled technologies. Even when we do have all the technologies around today, the reason why the global efforts have mostly gone in vain is the acute absence of integration and dearth of exploitation of technologies. While we still fight and try to tune a fine balance between privacy rights of individual and security of personnel information, there is an imminent need felt to not only improve the existing pandemic situation but also ascertain no such repeats take place.

A. Tiwari (✉) · U. Batra
School of Engineering & Sciences, D Goenka University, Gurgaon, India
e-mail: anupamtiwari@protonmail.com

© The Author(s), under exclusive license to Springer Nature Singapore Pte Ltd. 2023
R. Agrawal et al. (eds.), *International Conference on IoT, Intelligent Computing and Security*, Lecture Notes in Electrical Engineering 982,
https://doi.org/10.1007/978-981-19-8136-4_1

1.1 Problem Areas

Vide [2], there are five actionable arenas which need to be expeditiously attended and focused at global level by all nations and countries to retard any pandemics. These are briefly discussed below:

- **Unique personnel identity and sharing of health records**: When pandemics arrive, data suddenly becomes more important than ever. Questions are like where are the infected persons located currently, are they in motion or confined, what are the symptoms they showing, who all other persons are around? To answer all these kind questions, the need is availability of updated data in real time in a secured manner employing zero-knowledge-proof protocol for minimum invasion of effectuated personnel's privacy, between what is private and what is available to be shared in public domain needs to be defined.
- **Importance of supply chain and effected provenance**: The recent pandemic has put all healthcare-related supply chains amid high strain and demanding situations never foreseen. Potential impuissances of these supply chains have been exposed majorly like no transparency of data even to authorized agencies, slow access to information, data privacy and security and trust between participating nodes. COVID-19 has infracted the fictitious world which was relatively better equipped to counter any pandemic threats ever. As government issued emergency orders on usage for a COVID-19 vaccine vide the U.S. Food and Drug Administration [3], penetration of fake medicines, fake COVID-19 cyber cures, fraudulent corona virus tests, vaccines and treatments rose to new levels.
- **Incentive models**: As per the behavioral economics, any kind of incentives influences the behavior of humans. Pavlov's experiments laid the foundation of stimulus rewards. More so during pandemics, there will be many volunteers to help out all in respective capacities, there will be retired medical professionals willing to assist, medical professionals going beyond the call of duty to assist survival and fight pandemic, etc. Such value addition support during pandemics time needs to be incentivized beyond just a thankful service. This incentivization will facilitate a dedicated force which will work for a cause duly incentivized to each of the force participants and just not an expected thanklessness [4].
- **Finances and depletion of trust and interoperability**: Large supply chains during pandemics with multitudes of nodes bank upon cash exchanges intra-nodes suffer a setback since the fiat currency itself becomes a vulnerable source of spread among communities. Among geographic boundaries, factor of trust and interoperability of currency increases woven up in complex policies and trade protocol.
- **Verified and registered medical professionals**: Even before the pandemic safeguards and measures are in place for implementations, the most critical vacuum is felt in identifying validated and certified medical professionals around. The absence of centrally validated true data at national and international level of medical professionals has been majorly realized in COVID-19 pandemic. More

so, the absentia is there in spite of adequate medical professional present across, but without being known to demanding agencies.

The pandemic COVID-19 confirmed deaths, as per [5] stands at 5.1 million inside 21 months since onset in January 2020 and this figure could have been much lower, had there been adequate preparations. These preparations, in this paper context, do not imply adequate number of hospitals or ventilators as prelim measures since the authors feel these are reactive measures after the event. A resolved state of vulnerable arenas as discussed above could have played an essential role in retarding the pace of spread of pandemic. Blockchain can serve as a strong backend technology to resolve multiple challenges and aid in such circumstances.

1.2 Blockchain

Blockchain, the backend technology behind the widely known bitcoin cryptocurrency, has evolved itself today beyond the limited realms of finance. Blockchain technology today pertains not just to the mechanics widely known per se, Bitcoin, but has extended and evolved with much rich features than available in Bitcoin technology. After the introduction of Ethereum in 2014, the integration of blockchain has extended to all other domains of logistics, supply chains, banking, human resources, restaurants, procurements, defense, health care, etc. Ethereum further evolved with introduction of smart contracts deployment and possible multitude of application scenarios. Other prominent blockchain platforms include Ethereum blockchain, Hyperledger, Quoram, Multichain, IOTA, CORDA R3, etc. Few peculiar to medical-domain blockchain platforms include BurstIQ [6], Factom [7], Medicalchain [8], Robomed [9], Patientory [10], etc. to mention a few.

1.3 Distributed Storage

The need for distributed storage in a blockchain enabled medical Internet of Things (IoT) ecosystem arises simply attributed to the increasing storage requirements. It is estimated that entire electronic data that exists at any point of time doubles every two years. But at the same time, it is seen that one subset of the complete electronic data, i.e., healthcare data, surpasses the double approximates and is expected to be increasing with a compound annual growth rate of 36% through 2025 [11]. It is at this point, it is realized that a simple deployment of any blockchain platform might not be just the solution to resolve challenges. So while blockchain technology grants its peculiar characteristics to improve reliability, integrity and availability of data, the increasing velocity, veracity and volume of data would lead to storage handling issues as blocks continue to populate. Thus arises the need to also decentralize data [12] in a blockchain platform. Evolving distributed storage platforms that can be enabled with blockchain platforms include Sia [13], Storj [14], etc. Another platform has been used in our works, i.e., InterPlanetary File System (IPFS), yet another decentralized storage protocol for decentralized storage [15]. IPFS is a file sharing protocol enabled on a peer-to-peer (P2P) network that basically modifies the formal ways of sharing files across a network. The protocol is based on content-addressable storage that can retrieve a particular file based not on the location on the network ecosystem but based upon the content hash.

2   Schema Setup

A limited scale network with 6 nodes on Multichain blockchain platform was setup for simulating the proposed model vide Fig. 2 with hardware and software setup as seen in Table 1. Five nodes (Machines-2 to 6) were setup with different operating systems depicting spread out geo-locations. These nodes were loaded with Multi-chain 2 Enterprise GPLv3 [16] and were connected with the peer-to-peer network on permissioned blockchain and named *retard_pandemics*. Machine-1 was made the admin blockchain node. Figure 1 shows the major areas identified vide [2] as challenges to be resolved for retarding pace of pandemics, while the major parameters and setup of *retard_pandemics* blockchain are seen in Fig. 2. Further to these identified challenges, various emanating attributes responsible are seen. In the setup schema, these attributes are seen being appended in the *retard_pandemics* blockchain through streams feature of the Multichain blockchain. Thus for the setup, streams were created from different nodes as seen in Table 2. Figure 2 shows the overall schematic setup overview. Other related details are mentioned below:

- The blockchain is named *retard_pandemics*.
- Four streams UIDAI [17], *govern_finance*, *medsupply_chain* and *education_creds* were created.
- These streams generated unique transaction IDs. These transaction IDs were used for querying specifics in the *retard_pandemics* w.r.t data and linking transactions.
- All these streams were appended with sample data. The data was indexed inside the *retard_pandemics* in different blocks.
- Each transaction in the *retard_pandemics* blockchain generated a unique 64 bit transaction ID.
- Transaction in *retard_pandemics* refers to every kind of event. This event includes appending data, creating streams, granting permissions, creating addresses, etc.
- Transaction IDs so created for each event assist nodes to query peculiar events.
- Sample assets were created of value 2000 tokens at admin console with asset function, and 0.1 parts were created out of each [issue 1RhVJ4fK8JYDRt2R4PBhj12ChKoeCUYAEyiy1e asset1 2000 0.1].

For the simulation setup for IPFS, the file blk00000.dat [a collection of several raw blocks of *retard_pandemics* blockchain in one file] was chosen for upload to the

Table 1 Nodes established—hardware and application setup details

Machine name	IP configured	Hardware	Addresses	OS/Applications
Machine-1 Main blockchain admin console	190.168.10.20/24 160.10.20.10/24	Dell PowerEdge Server T30 Intel Xeon E3-1225 v5 (8 M Cache, 3.30 GHz) with 16 GB (2 × 8 GB) RAM and 1 TB SATA Hard Disk	1RhVJ4fK8JYDRt2R4PBhj12ChKoeCUYAEyiy1e	Ubuntu Server 20.04.1 LTS with Multichain 2 Enterprise GPLv3
Machine-2 (UIDAI)	190.168.10.21/24 160.10.20.11/24	Dell Inspiron 3880 10th Gen Intel Core i3 Desktop (8 GB RAM/1 TB HDD)	1JTC6GXSF9s2iADtkd6GkBU3dXShuqByQncU2Z	Ubuntu MATE 20.04 LTS (Setup with Multichain 2 Enterprise GPLv3)
Machine-3 (EHR)	190.168.10.22/24 160.10.20.12/24	Raspberry Pi 3 Model B 1 GB RAM 1.2 Ghz 64-Bit Quad Core Arm CPU	1QcVdpE2weNoJP3gsgxKxnHYF1BRcFAiVPCPwh	Raspberry Pi OS, Kernel version: 5.4 with Multichain 2 Enterprise GPLv3
Machine-4 (Supply chain provenance)	190.168.10.23/24 160.10.20.13/24		15CJVkCgCNTqMni7YsqcaJmSaM9gZMHV9vE2pm	
Machine-5 (Education and incentives)	190.168.10.24/24 160.10.20.14/24		1RLAPgXdVcMmAMfx57bqkde4dfLU4ms3kLbzFZ	
Machine-6 (Govt. finance schemes)	190.168.10.25/24 160.10.20.15/24		1Yov5JQUmFwHZoJTDPZk5YbSNxheShRBYJgLnS	

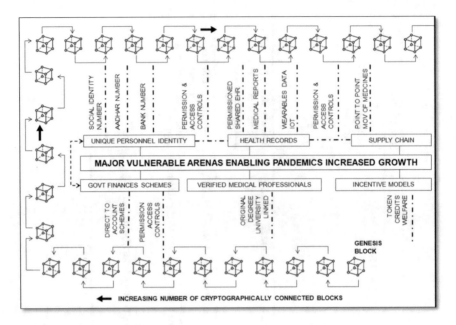

Fig. 1 Blockchain technology for resolving various challenges

IPFS gateway. IPFS stores its data, on a peer-to-peer network, which is split apart at different locations [24] and is retrievable by unique hashes. At the same time, it is vulnerable if this unique IPFS hash is known by anyone since it is stored as plain text. The files and data so stored vide IPFS protocol can be simply retrieved thus. Therefore to exploit IPFS protocol and also at the same time enable these files with secure communications on network, the sample file *blk00000.dat* was encrypted at the admin console and thereafter decrypted at an earmarked UIDAI node. Both the nodes involved in this encryption-decryption task were running the IPFS daemon. The file blk00000.dat (16777216 bytes) was encrypted with public key of UIDAI node and got converted as *blk00000.dat.enc*. The selection of this earmarked node UIDAI, in this setup, follows a round robin consensus wherein all nodes.

3   Results

- All data being indexed in the *retard_pandemics* is only visible to other nodes subject to stream subscription and grant of permissions (send, receive and connect). For example, data populated in stream UIDAI could only be visible to other streams including the admin console only after it subscribed the stream UIDAI and exclusive connect permissions were granted.

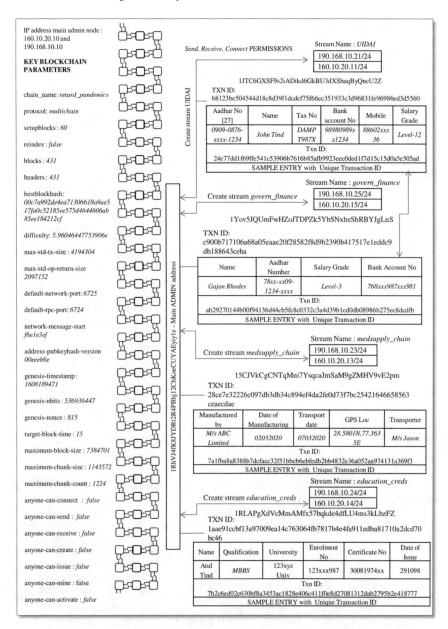

Fig. 2 Schematic architecture setup for proposed model

Table 2 Streaming details created in setup

Attribute	Stream	Remarks
Unique identity	*UIDAI* with stream reference number *84-265-10002*	This represents the unique social security numbers of country citizens like Aadhaar [17] India, ABRID [18] Brazil, NIA [19] Ghana, SSN [20] USA, NIN [21] UK, etc. These have been appended to *retard_pandemics* blockchain
Pandemic data	*pandemic_data* with stream reference number 339-266-19921	This stream appends pandemic data stats of countries [22]. The data has been considered as confidential and is appended as encrypted. The data is not visible to any node other than admin console, i.e., Machine-1. Brief schematic of flow of data from uploading node to access node is seen in Fig. 3 Data can also include vital parameters of health peculiar to citizens under scan through various modes of inputs like wearables connect to IoT or special government accredited APIs, random checks at shopping malls and public places or vide sensors like Rokid smart glasses [23]. These all parameters are interlinked with *uidai_number* stream on the *retard_pandemics* blockchain
Supply chain provenance	*medsupply_chain* with stream reference number 286-267-12371	The streams here refer to "point to point" appending of data as the medicines or stores move from the production premises to the last mile connectivity till pharmacy or the patient receives on the *retard_pandemics* blockchain
Medical education credentials and medical professionals	*education_creds* with stream reference number *151-267-44570*	These streams refer to authenticated appending of medical professional credentials by universities and institutes
Incentive models for medical professionals	*asset1* with asset reference number *252-267-57044*	These refer to issuances of tokens as incentives stimulus rewards. Policies can be adapted by government agencies for issuances

(continued)

Table 2 (continued)

Attribute	Stream	Remarks
Government finance schemes-direct credit	*govern_finance* with stream reference number *171-262-54921*	These streams refer to details of credits directed by governments to beneficiaries as per policies laid down. In the setup assumption here, a sample stream has been created to depict credit to below poverty line citizens Additionally, another stream *bank_account* will coordinate appending the bank details of beneficiary

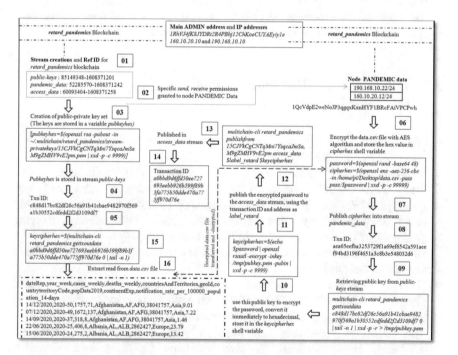

Fig. 3 Encryption-decryption operations for transfer of confidential data on blockchain

- All data being populated vide any streams in the *retard_pandemics* blockchain could populate only if exclusive receive permissions by the stream creator are granted.
- The node was able to see, index and connect only if exclusive send, connect, receive permissions are granted. Thus for each stream created in the

retard_pandemics blockchain, for data population from another node, exclusive permissions had to be granted for populating data, seeing transactions and connecting.

3.1   Distributed Storage with IPFS

- Block formation in the *retard_pandemics* blockchain files was seen located at admin console location under /home/pandemic_admin/.multichain/blocks as *blk00000.dat*.
- This blk00000.dat file was generated in the same directory/.multichain/blocks of all respective node users.
- blk00000.dat so generated at all nodes got the same md5sum as *d9bca30b952c445d2e29e3098a6f7b35* proving integrity of block data as a peculiar characteristic of *retard_pandemics* blockchain.

3.2   IPFS-Enabled Distributed Storage for Retard_pandemics

- blk00000.dat was encrypted at the admin console with the UIDAI public key and converted as blk00000.dat.enc.
- blk00000.dat.enc was added to the IPFS storage vide the IPFS "add" function and IPFS address generated as *QmRffqE3c7N3AwxPJx7F23DavnLVh43rfuGEohhbTvZ4Df*, while the same could be accessed by other nodes through any IPFS gateway querying *QmRffqE3c7N3AwxPJx7F23DavnLVh43rfuGEohhbTvZ4Df* also. Though all nodes have access to the file but since it is encrypted, none has the access to stored transactions inside.
- Hash function *md5sum* was checked as equal (*d9bca30b952c445d2e29e3098a6f7b35*) on both ends prior to encryption at admin console, Machine-1, and after decryption at UIDAI console, Machine-2.
- The extract of commands and work including encryption-decryption and validating above md5sum is seen available at IPFS hash *Qmanbe-HGXJuhF3WERF9P7uLQePuEApUy4K8D8s9iGy92qR*.
- The plain hex file *blk00000.dat* is available for exploring data blocks as generated in *retard_pandemics* atIPFS hash Qmf69KvNAGSuWbWwALtGzquSrzUP4FCJbdag8pnEWvGnyC.
- 100 asset tokens out of 2000 tokens generated were distributed to sample addresses while unique transactions IDs.

3.3 blk00000.dat File

The file was checked in hex editor, and following results are decoded from same:

- About 431 blocks were created in the *retard_pandemics* blockchain.
- All transaction IDs generated for various effected events of creating streams, granting permissions, storing data, decrypting data, etc. were visible in the blockchain in hex format.
- All transactions so created were also stored with their respective time stamps in epoch Unix format. This was converted into human-readable format, and information was validated with exact occurrences. For example, in Block 74, Fig. 4 shows time stamp of 4 bytes in epoch time stamp format as 79C0DC5F (swap endian converted to 5FDDB2FA), and time derived is GMT: Friday, December 18, 2020, 2:45:13 PM.

74th BLOCK EXTRACT

F6E1E3EF0102000003000000EEB44DBCB145992CC7BD28D75229D094B76087FFF8DB51C7A59B15B57CCBA
700C6EF21F2F5319DA416D0B04514869EFCC1DDCEC0423A8A1D948DA77594BFC7F379C0DC5FFFFF0020A9
020000020100000001000000000000000000000000000000000000000000000000000000000000FFFFFFFF0B014
90101062F503253482FFFFFFFFF010000000000000000726A4C6F53504B62473045022100E344AB4C386DDD5C20
D0D8D1665FD751602E1D194E6FF9180081BDC239D2CBFB0220094C69FDAA206F13D1E0E1410A6EFF09C020
C3C55DC58AE8041A21A6C7B7DE8D032102A65882CC65F1B3C3784AE0AC116732CFFAC06570F059C50770F4
CC3814FC59C1000000000010000000160553DED8669E91F83963D3C9351C36CF675CFCD1D9FD3C8184D5404E
53B12B8

- 4 bytes : Magic Byte: F6E1E3EF
- 4 bytes : Block size : 01020000 size of the block i.e. 513 bytes UINT32 - Little Endian (DCBA) means 1026 characters
- 4 bytes: Version number: 03000000
- 32 bytes : Previous Hash i.e hash of 73rd block
  EEB44DBCB145992CC7BD28D75229D094B76087FFF8DB51C7A59B15B57CCBA700
- 32 bytes : Merkle root
  C6EF21F2F5319DA416D0B04514869EFCC1DDCEC0423A8A1D948DA77594BFC7F3 swapped endian form is F3C7BF9475A78D941D8A3A42C0CEDDC1FC9E861445B0D016A49D31F5F221EFC6
- 4 bytes : Epoch Timestamp : 79C0DC5F(swap endian to 5FDDB2FA to get the time),time is GMT: Friday, December 18, 2020 2:45:13 PM
- 4 bytes : Difficulty : FFFF0020 depicts the difficulty/bits. Swapped endian is 2000FFFF, hex to dec we get 536936447
- 4 bytes : Nonce: A9020000 is the nonce converted to UINT32-LittleEndian (DCBA) as 681
- 01 byte : number of transaction, Hex to dec we get 02 means this block has two transactions
- 01000000 is the version
- 01 is the number of input
- 0000000000000000000000000000000000000000000000000000000000000000is the previous output
- FFFFFFFF is the sequence
- 0B is script length i.e 11 i.e next 22 bits are 01490101062F503253482F
- 01 is the number of outputs
- 0000000000000000 is the number of token created as reward(0 in our case)
- 32 bytes: Transaction ID:
  60553DED8669E91F83963D3C9351C36CF675CFCD1D9FD3C8184D5404E53B12B8.Swap endian of this is B8123BE504544D18C8D39F1DCDCF75F66CC351933C3D96831FE96986ED3D5560. This is transaction ID identified of creation of stream UIDAI as seen in Figure 2

Fig. 4 Block 74 internal information decoded from hex encoded data

- Block height 74 was decoded from the blockchain with all critical information stored and derived as seen in Fig. 4. The internal structure of the block is decoded and seen with byte details.
- Transaction ID *60553DED8669E91F83963D3C9351C36CF675CFCD1D9FD3 C8184D5404E53B12B8,* inside Block 74, is seen which is peculiar to creation of stream UIDAI as seen in Fig. 2. In the figure, we see that this transaction ID is generated vide creation of stream UIDAI and indexed inside the Block number 74 as *b8123be504544d18c8d39f1dcdcf75f66cc351933c3d96831fe96986ed3d5560* which is the swapped equivalent little endian of transaction ID visible in 74th block of *retard_pandemics* blockchain, i.e., *60553DED8669E91F83963D3C9351C36CF675CFCD1D9FD3C8184D5404E 53B12B8.*

4   Challenges and Discussions

In this paper, a model has been proposed with proof of concept that enables to expedite secure exchange of information enabled with fine-tuning of privacy aspects of EHR. The model can withstand low-scale datasets and perform as deemed. But with increasing number of nodes and high velocity of data pouring in, a number of challenges too needs to be resolved. Few of these challenges are identified as follows:

- **Blockchain platforms**: There are multiple numbers of evolving blockchain platforms which offer diverse functions peculiar to various domains. Hyperledger, Multichain, IOTA, CORDA, Ripple, Ethereum, IBM blockchain, etc. are few pertinent platforms to mention among many. While we intend retarding the pace of pandemics enabled by blockchain platforms at global level, the world needs to agree on a common platform to mechanize transactions between intra-devices. For the current works proposed, though the Multichain platform warrants no scalability limit, but no proofs exist of such huge datasets being stored on it.
- **Interoperability intra-blockchain platforms**: While the global consensus on choosing the right blockchain platform may get delayed or deem as difficult, interoperability between blockchain platforms might just resolve these challenges. Current works on to facilitate interoperability include [25–27], etc. An interoperable blockchain ecosystem will be conducive to expedited secure exchange of information with bare thread tuning of privacy on multiple blockchain platforms.
- **Missing standardization**: While almost all of the works in blockchain domain are being done in isolation, giving way to proprietary challenges, there is an imminent need felt for works in standardization at global level.
- **Identification of nodes and devices**: A real-time ecosystem of millions of devices doing transactions and information exchange would deem absolute identification of participation. Various works [28, 29] are currently on including IoT device identification via network flow-based fingerprinting and learning, DigiCert IoT, network traffic, etc. to mention a few pertinent ones. A simple compromise through

any existing vulnerability in devices firmware or interfacing APIs can be lethal in a live ecosystem of IoT.

- **Security and privacy issues in IPFS**: IPFS is an excellent way to host data by breaking it up and storing at geographically spread locations enabled on content-driven searches of unique hashes. One of the biggest challenges is to delete data once hosted. Few other distributed storage platforms include Swarm, Sia tech, Storj, MaidSafe, etc. All these platforms have unique functions to offer peculiar to user requirements. Encryption is currently not inbuilt to IPFS; therefore, the authors have used OpenSSL and RSA cryptography methods for encryption and exchanging information among nodes. Inbuilt encryption seems an inherent demand for establishing a distributed storage enabled for any blockchain platforms.

- **Big data**: Volume, veracity, variety, velocity and value of data are only going to increase in times to come with global medical data estimated to reach 175 zettabytes by 2025. The sheer volume of such data would deem rugged blockchain architectures on stable platforms.

- **Communication availability and location impact**: While the complete basis of the proposed model is based on the assumption that every node is connected with a good internet, it is also well known that as of January 2021, there are 4.66 billion active internet users worldwide, i.e., only 59.5% of the global population. Thus, ensuring internet connectivity to each global citizen remains a viable but difficult challenge [30].

5 Conclusion

This paper proposes a healthcare model focused on retarding growth of pandemics at global level enabled on a blockchain platform with distributed storage. It is imminent that pandemics cannot be controlled by a kill switch concept; they will happen irrespective till such time we reach a disease-less world state. Till such realizations are attained, we need to attempt retarding the pandemics by exploiting information systems enabled with new genre IT technologies, and blockchain offers one such way for realization. Though blockchain technology is still seen as an evolving technology, but the peculiar characteristics it offers to multiple domains reaffirm confidence of it being part in coming future. The paper has focused on areas including unique personnel identity of patients, sharing of EHR, supply chain and effected provenance, incentive models, finances and depletion of trust and interoperability and verified and registered medical professionals on distributed storage. Proof of concept working transactions have been established on a simulated Multichain permissioned blockchain *retard_pandemics*. Further to this, another identified challenge of storing huge datasets has been proposed to be resolved by associating IPFS with blockchain platform. The model establishes secure population of blocks on an IPFS by storing 431 blocks generated duly encrypted with *OpenSSL* methods.

The peculiarity of this work draws on the unique extension of blockchain-enabled medical IoT further with distributed storage, the need of which is imminent. Thus while other works in this domain have proposed multiple models enabling blockchain in medical IoT, there is definite lack of focus on storage of such data.

References

1. Smil V (2020) Pandemic memories and mortalities. IEEE Spectrum. https://spectrum.ieee.org/tech-history/heroic-failures/pandemic-memories-and-mortalities. Accessed 08 Nov 2021
2. Tapscott D (2020) Blockchain and pandemics report. Blockchain Research Institute. https://www.blockchainresearchinstitute.org/blockchain-and-pandemics/. Accessed on 12 Oct 2021
3. US Food and drug administration (2020) Beware of fraudulent coronavirus tests, vaccines and treatments. https://www.fda.gov/consumers/consumer-updates/beware-fraudulent-coronavirus-tests-vaccines-and-treatments Accessed on 21 Oct 2021
4. Barrera C, Hurder S (2021) "Cryptoeconomics: designing effective incentives and governance models for blockchain networks using insights from economics," foreword by Don Tapscott, Blockchain Research Institute, 13 Aug 2021
5. WHO Dashboard (2021) WHO coronavirus disease (COVID-19) dashboard. https://covid19.who.int/. Accessed on 19 Nov 2021
6. BurstIQ (2020) Research foundry: blockchain based healthcare data solutions. Apr 2020. [Online] Available: https://www.burstiq.com/. Accessed on 12 Oct 2021
7. Snow P (2019) Factom proprietary and confidential information. Factom. https://ewh.ieee.org/r5/central_texas/cn/presentations/Paul_Snow_Presentation_Jun_2019.pdf. Accessed on 01 Oct 2021
8. Albeyatti A (2020) Medicalchain 2.1. https://medicalchain.com/Medicalchain-Whitepaper-EN.pdf. Accessed on 29 Aug 2021
9. Kumar A, Krishnamurthi R, Nayyar A, Sharma K, Grover V, Hossain E (2020) A novel smart healthcare design, simulation, and implementation using healthcare 4.0 processes. IEEE Access 8:118433–118471. https://doi.org/10.1109/ACCESS.2020.3004790
10. Patientory. Patientory Inc. (2021) Forging the path to consumer-directed health through blockchain technology. https://patientory.com/wp-content/uploads/2019/11/Case_Study.pdf. Accessed on 20 Sept 2021
11. Kent J (2021) Big data to see explosive growth, challenging healthcare organizations. https://healthitanalytics.com/news/big-data-to-see-explosive-growth-challenging-healthcare-organizations. Accessed on 08 Sept 2021
12. Raman RK, Varshney LR (2018) Dynamic distributed storage for blockchains. In: 2018 IEEE international symposium on information theory (ISIT), Vail, CO, 2018, pp 2619–2623. https://doi.org/10.1109/ISIT.2018.8437335
13. Vorick D (2021) Sia: simple decentralized storage. https://sia.tech/sia.pdf. Accessed on 01 Sept 2021
14. Ricci J, Baggili I, Breitinger F (2019) Blockchain-based distributed cloud storage digital forensics: where's the beef? IEEE Secur Priv 17(1):34–42. https://doi.org/10.1109/MSEC.2018.2875877
15. Muralidharan S, Ko H (2019) An InterPlanetary File System (IPFS) based IoT framework. In: 2019 IEEE international conference on consumer electronics (ICCE), Las Vegas, NV, USA, 2019, pp 1–2. https://doi.org/10.1109/ICCE.2019.8662002
16. MultiChain 2.0 Inc. (2021) What's in multichain 2.0 enterprise. https://www.multichain.com/blog/2019/07/multichain-2-0-enterprise-demo/. Accessed on 05 Sept 2021
17. Unique Identification Authority of India (UIDAI). https://uidai.gov.in/. Accessed on 12 Sept 2021

18. The Brazilian Association of Companies in Digital Identification Technology (ABRID). http://abrid.org.br/. Accessed on 21 Oct 2021
19. National Identification Authority, Ghana. https://www.nia.gov.gh/. Accessed on 11 Oct 2021
20. Social Security Number and Card, USA. https://www.ssa.gov/ssnumber/. Accessed on 06 Oct 2021
21. National Insurance number, UK. https://www.gov.uk/apply-national-insurance-number. Accessed on 01 Oct 2021
22. Allen Institute For AI and 8 collaborators (2021) COVID-19 Open Research Dataset (CORD-19). https://www.kaggle.com/allen-institute-for-ai/CORD-19-research-challenge. Accessed on 29 Oct 2021
23. IEEE Future Directions (2021) I see you have some fever. https://cmte.ieee.org/futuredirections/2020/05/06/i-see-you-have-some-fever/. Accessed on 12 Nov 2021
24. IPFS Docs (2021) Work with blocks. https://docs.ipfs.io/how-to/work-with-blocks/. Accessed on 12 Nov 2021
25. Besançon L, Silva CFD, Ghodous P (2019) Towards blockchain interoperability: improving video games data exchange. In: 2019 IEEE international conference on blockchain and cryptocurrency (ICBC), Seoul, Korea (South), 2019, pp 81–85. https://doi.org/10.1109/BLOC.2019.8751347
26. Scheid EJ, Hegnauer T, Rodrigues B, Stiller B (2019) Bifröst: a modular blockchain interoperability API. In: 2019 IEEE 44th conference on local computer networks (LCN), Osnabrueck, Germany, 2019, pp 332–339. https://doi.org/10.1109/LCN44214.2019.8990860
27. Pang Y (2020) A new consensus protocol for blockchain interoperability architecture. IEEE Access 8:153719–153730. https://doi.org/10.1109/ACCESS.2020.3017549
28. Alizai ZA, Tareen NF, Jadoon I (2018) Improved IoT device authentication scheme using device capability and digital signatures. In: 2018 International conference on applied and engineering mathematics (ICAEM), Taxila, 2018, pp 1–5. https://doi.org/10.1109/ICAEM.2018.8536261
29. Gu T, Mohapatra P (2018) BF-IoT: securing the IoT networks via fingerprinting-based device authentication. In: 2018 IEEE 15th international conference on mobile ad hoc and sensor systems (MASS), Chengdu, 2018, pp 254–262. https://doi.org/10.1109/MASS.2018.00047
30. Johnson J (2021) Internet users in the world 2021 | Statista. [Online] Statista. https://www.statista.com/statistics/617136/digital-population-worldwide/. Accessed 09 Dec 2021

Wearable Location Tracker for Emergency Management

Rajashri Mahato⑩, S. Saadhikha Shree⑩, and S. Asha⑩

1 Introduction

In the past few years, various disasters like floods, earthquakes, tsunamis, cyclones, forest fires, etc., have rendered the world speechless due to the death toll they caused. The unexpected and mostly instantaneous nature of these occurrences leaves millions of lives and their properties in danger. This situation can be minimized if the emergency situations are examined and sufficient safety measures are taken as a precaution. Disasters can be broadly categorized by the nature of their cause as natural disasters and man-made disasters [1]. Further divisions might be presented based on the type of occurrence or cause under the primary classifications. For this particular paper, developing countries are taken into consideration and their respective emergency management systems are studied. Based on the study, an efficient tracker is developed.

Emergency management is taken care of with the help of the governments in collaboration with their citizens. The State government, in this case, has more responsibilities compared to the local governments [2]. Hence, the State acts as the substations. Emergency management has various phases including mitigation, preparedness, response, and recovery. When a solution system is designed for this problem

R. Mahato · S. Saadhikha Shree · S. Asha (✉)
Centre for Cyber Physical Systems, VIT University, Kelambakkam—Vandalur Road, Rajan Nagar, Chennai, Tamil Nadu, India
e-mail: asha.s@vit.ac.in

R. Mahato
e-mail: rajashri.mahato2019@vitstudent.ac.in

S. Saadhikha Shree
e-mail: saadhikhashree.s2019@vitstudent.ac.in

© The Author(s), under exclusive license to Springer Nature Singapore Pte Ltd. 2023   19
R. Agrawal et al. (eds.), *International Conference on IoT, Intelligent Computing and Security*, Lecture Notes in Electrical Engineering 982,
https://doi.org/10.1007/978-981-19-8136-4_2

using networking technology, it should concentrate mostly on the last two parameters. Additionally, during a disaster, it is necessary that emergency planning and response must be facilitated smoothly.

Internet of Things (IoT) is a modern, up-and-coming paradigm [3, 4] that is widely used in the field of communication to increase range and effectiveness. A typical example where such an application is a mobile phone. They allow ubiquitous communication to everyone. Mobile phones are fairly simple to use and consist of various sensors some of which help in the wireless transmission of messages. The use of IoT in networking and communication is widely appreciated. But there is a downside due to the cost associated with large-scale production and maintenance of such devices [5]. Since the introduced tracker is a small-scale device, it overcomes this disadvantage. The application of IoT is an undeniable solution in any disaster or pandemic situation, such as the COVID-19 pandemic. During this period, the critical role of IoT in the emergency management response phase was emphasized in [6]. The use of the Internet of Things can assist in managing the risk and hazardous impact of any disaster, and also in minimizing the destruction caused by the crisis.

This paper is classified as follows. Section 2 describes the background study and the review of existing literature. Section 3 outlines the methodology used to develop the proposed tracker. Sections 4 and 5 contain the results and concluding remarks achieved at the end of the process.

2 Related Work

2.1 Literature Review

Wearable devices are a great success in the areas involving the Internet of Things and Cyber-Physical Systems [7]. The idea of a 'wearable location tracker for emergency management' to help the rescue team is novel and it combines elements from several engineering domains [8]. In [9], a review of disaster management-related research papers was taken up, and 507 papers were analyzed to conclude that effective disaster management can be achieved through systems integration. The technical part of the implementation is done in [10] using global navigation satellite system data, graphic information system, and remote sensing data. It also explains the use of the above-mentioned geomatic technologies in risk and disaster management by placing more weight on the use of GPS in activities like monitoring, managing, and assessing these disaster events. Furthermore, this study emphasizes on cases where GPS was used in three disaster periods, namely pre-disaster, during the disaster, and post-disaster events to figure out the potentially vulnerable areas, detect occurrences and identify the degree of damages caused by the disaster.

A wearable system with the help of GPS is required in order to determine the location of the trackers worn by the victims of a disaster [11]. The system is supposed to point out the precise spot of the victim and display it on the receiving end, which is controlled by the disaster management unit. Adding onto the locations of the victims, the disaster management team should also receive the area of disaster by the locals for determining the above-mentioned location. Converting the victim aide system suggested in [12] to a rescuer aid system is the aim of this paper. Connecting biomedical sensors to well-trained individuals through trackers allows them to be monitored and guided exactly. Another system studied in [13] included a wrist-worn device operating as the tracker. The system reported the use of an iterative localization plan, using predetermined nodes as reference points and the distance between the unlabeled (the nodes without any defined location) nodes. This was to determine the current position of the node in focus. The location and health status of the person wearing the device were recorded regularly and transferred to be stored in a database.

Another point to note is the use of Bluetooth for communication. It is one of the most used technologies for wireless communication because it eliminates the other costly transmission means [14]. Its low energy consumption is preferred for rescue operations and also because it establishes two-way interaction. Other reasons include Bluetooth's availability since it is globally used and the frequency-hopping technique that does not allow data mix-up [15].

2.2 Difference from Existing System and Scope

The scope of this paper is to introduce novelty to the wearable tracker. A push button system is used to indicate that the user, in this case, the rescue officer's status. Additionally, it also accommodates the need for extra help. The three-button system is deployed to gain the status and statistics of the rescue team and the victims in real time from the location of the disaster. This system ensures accurate data transmission to the corresponding substation. And also proper communication between the rescue officers to avoid any loss of life or property due to negligence.

3 Proposed Methodology

3.1 System Modules

(A) Networking

With the latest development of low-power communication technology, IoT applications have been implemented using diverse communication techniques. Wireless sensor networks (WSNs) and Bluetooth modules are composed of nodes with limited computing and power resources. There are more and more applications of WSN and even Bluetooth for everyday use and also in the military, most of which need proper security and protection for the data collected by the base station from the remote

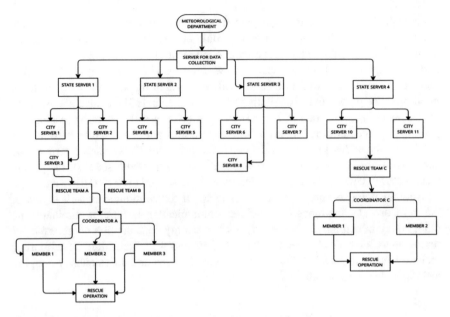

Fig. 1 Flow diagram representing the network using tree topology

sensing node. In addition, when sensor nodes transmit data to the base station, implosion occurs. Providing safety and anti-implosion measures in an environment with limited resources is a real difficulty. Reviews the aggregation methods proposed in previous literature to deal with bandwidth and security issues related to many-to-one and one-to-many transmissions on WSN and Bluetooth devices [16]. Recent contributions to desirable and lossless many-to-one communication and many-to-many communication have improved the quality of communication and provided necessary assistance in various fields (Fig. 1).

Bluetooth Low Energy (BLE) is an appealing solution that can be implemented in systems to result in reduced cost and power. Short-range wireless data transfer devices and wireless products with high adaptability have been using conventional buttons over the years [17]. Thus, to overcome the challenges faced in the realm of communication, several technologies including geospatial technologies and wireless networking technologies have been implemented through Tree topology. Therefore, hassle-free communication is possible with the integration of the WSN, BLE, and GPS under the same networking root, during emergencies.

(B) System Design

A cyber physical system (CPS) is the amalgamation of computer and tangible processes. Computers and embedded networks examine and manage substantial processes, typically with feedback loops where computations are affected by physical processes. There are significant challenges, especially as the substantial components of the cyber physical systems establish security as well as trustworthy necessities

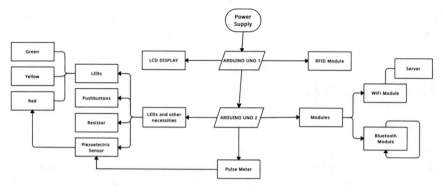

Fig. 2 Flow diagram representing the components

that are very distinct from those of the prevailing computing methods. Also, tangible components and object-oriented software components are different in nature [18]. The successful implementation of Industry 4.0 [19] can ensure that the CPS architecture design takes into account the needs of the client. The objective of this paper is to come up with a cyber physical system (CPS) design utilizing IoT to achieve a successful business system (Fig. 2).

System design analysis uses different types of information systems to hold up the processors required to perform its business purposes. Every information system has a specific purpose, and its own life cycle. This concept of 'fostering by itself' is called the development life cycle of the system [20].

Software requirements

Arduino IDE 1.8.15-Scripting
Fritzing-Circuit designing
Cisco Packet Tracer-Networking

Hardware requirements

Microcontroller-Arduino UNO R3
Sensors-Pulse Sensor
Modules used-ESP8266 (Wi-Fi module), Bluetooth (HC-05), GPS module
Display-16 × 2 LCD display
Buzzer–Piezo-electric buzzer
LEDs and push buttons
The design aims to build a portable position tracker model that combines sensors, actuators, and various modules including wireless communication and microcontrollers. The assimilation of LEDs, buttons, cables, 16 × 2 LCD screen, and Arduino UNO R3 helps to build a function that helps maintain adequate statistics for rescue operations that are carried out in an emergency situation. In addition, it helps increase emergency assistance in the event of disasters and emergencies. Other design features include frequent heart rate monitoring, location tracking, seamless communication

for rescue operations, and evacuation of rescue team members in the event of health concerns.

3.2   Components Used

Arduino Board

Arduino belongs to a family of microcontroller boards that simplifies prototyping, electronic design, and experimentation for professionals. It is designed to connect LEDs, various sensors, speakers and small motors, servo systems, etc., directly to these pins. Arduino boards can be connected to computers with the help of a USB. It can be programmed in a simple language like C, C++, in the free Arduino IDE due to the possibility of uploading the compiled code to the development board [20].

LCD

A liquid crystal display (LCD) is an electronically regulated optical device that uses the light modulation characteristics of liquid crystals in combination with polarizers. It is used to display a message on the screen.

Pulse Sensor

This sensor is used in measuring the heartbeat rate, and it can be found in any medical device that is used to measure heart rate. It can be worn on the earlobes or on the fingers.

ESP8266 (Wi-Fi module)

The ESP8266 Wi-Fi module is a self-contained SOC which has an integrated TCP/IP protocol stack. It also allows a microcontroller to access any Wi-Fi network. This module can host applications and also download all the functions of the Wi-Fi network via any other application processor.

Bluetooth Module (HC-05)

HC05 is a Bluetooth module specially intended for wireless communication. This module can explicitly be used for master station or slave station configuration. The serial Bluetooth module allows all serial compatible devices to communicate with each other via Bluetooth. Its range can reach <100 m, depending on the transmitter and receiver, the atmosphere, geographic location, and urban conditions [21].

GPS Module

The NEO-6 M GPS module with its small antenna helps in tracking the exact location by defining the latitude and longitude of that particular location where the module is present.

LEDs and Push Buttons

The light emitting diodes can be operated using push buttons and they have been implemented in the prototype for getting the desired outputs.

Piezo-Buzzer

The piezo-electric buzzer is basically a miniature speaker that can be connected directly to the Arduino. With the help of Arduino, sounds can be made using the buzzer tone.

4 System Implementation

The first segment of this section explains the efficiency of communication and handling of rescue operations over the network with the help of a portable location tracker. Tree topology has been implemented so that communication occurs at a faster and more efficient pace. The meteorological station will act as the parent node and the main server in the entire framework of networks. The State substation (client) will connect to the main server (weather station) and maintain a server-client relationship with the main server. These state substations will also be considered as servers of their respective city substations. The city substation (slave) of the corresponding state will connect to the state substation (master) and maintain a master–slave relationship with each other. The city substation will in turn be connected to the coordinator of each rescue team in that particular city. The coordinators' system will serve as the customer of the city substation, helping to communicate in an orderly and relaxed manner. The wearable location trackers that are present with each rescue team member will be connected to their respective team coordinators via Wi-Fi. These portable location trackers have built-in Bluetooth modules to support low energy and low-cost communications. This allows members of the rescue team to contact and communicate with each other at an accelerated pace. The Bluetooth module will help front-line combatants reach hand to hand in rescue operations. Thus, communication can happen at an efficient rate and also mitigate the loss caused by disasters.

The network design has been simulated in Cisco Packet Tracer. Figure 3 shows a clear outlook of the network design.

The second segment of this section focuses on building the prototype and its implementation. The assimilation of sensors, actuators, LEDs, and various other components led to the emergence of portable position trackers. The circuit is designed in Fritzing in a detailed and clearly defined way.

For convenience and clarity, the design has been divided into three parts. Figures 4, 5 and 6, represent the different parts of the circuit design, respectively.

The Bluetooth module (HC-05) and the Wi-Fi module (ESP8266) help in wireless communication at a faster rate. The 2 LEDs, namely, the yellow and the green LEDs are used for seeking assistance and for maintaining the statistics of the rescue operations being carried out, respectively, with the push buttons as the inputs.

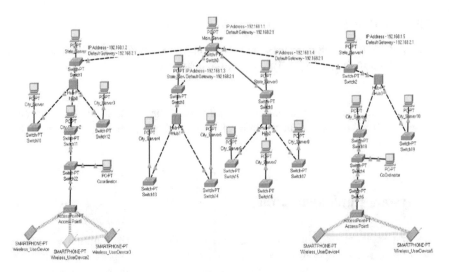

Fig. 3 Overview of the tree topology networking in Cisco Packet Tracer

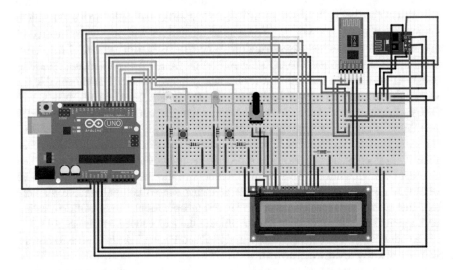

Fig. 4 Consists of the Wi-Fi and the Bluetooth modules, along with the LCD screen, the LEDs, push buttons, and Arduino UNO R3

The Pulse Sensor keeps a frequent track of the number of beats per minute (BPM) and notifies the team coordinator and the nearby team members if the pulse rate goes beyond the normal range. The Red LED starts glowing and the piezo-buzzer starts beeping if the heartbeat rate falls below 60 or goes beyond 100. Immediate replacement of the rescue warrior is made in order to ensure safety. The entire design of the process has been shown in Fig. 5.

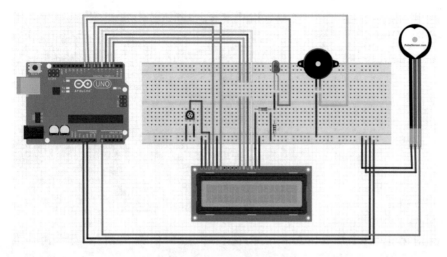

Fig. 5 Consists of the Pulse Sensor, the piezo-sensor, the LCD screen, LEDs, and Arduino UNO R3

Fig. 6 Circuit design for the GPS module

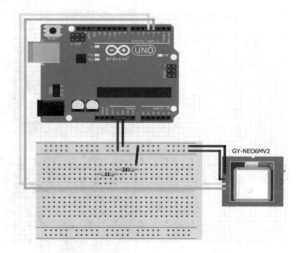

Figure 6 demonstrates the circuit design of the GPS module which helps in tracking the location of the rescue team member.

The accumulation of all three parts of the circuit leads to the development of the portable location tracker. The hardware components have been assembled which has resulted in the final outcome of the prototype (Figs. 7, 8 and 9).

Fig. 7 Comprises the piezo-buzzer, the Red LED, LCD display, Pulse Sensor, and Arduino UNO R3

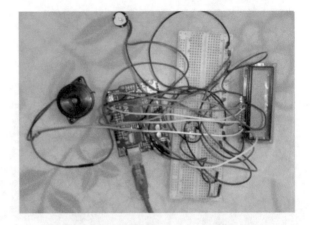

Fig. 8 Comprises the Yellow and the Green LEDs, push buttons, LCD display, Bluetooth and Wi-Fi module

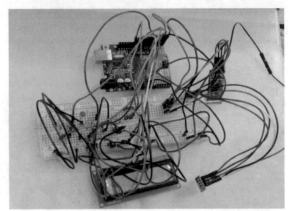

Fig. 9 Comprises of the GPS module

5 Future Work

Many different observations, tests, and analyzes have been done, which led to the results of the portable location tracker. Due to time constraints (that is, experiments using real data often take a long time, even days to complete a run), so some ideas and adaptations are left for the future. Future work involves adding an important feature to the prototype. Development of an Android app that not only helps firefighters track people but also helps to save lives at a faster rate. The app will be built using the SOS signal generator, and if there is a fire in the apartment/building and there is no way to escape, the signal generator will be released. The SOS signal will appear in the form of infrared radiations, which can be detected by the portable location tracker present with the firefighter. This will facilitate rescue efforts by shortening the time-lapse, and will also aid in improving the emergency response rate.

6 Conclusion

In this paper, a 'wearable location tracker for emergency management' has been introduced and implemented. From the implementation results, we understand the tracker's effect during the time of a disaster. Its networking facility with devices of the same kind and other devices in the substation proves the ease of data transfer at the time of emergency. Since the tracker sends the accurate and real-time status of the team member or the victims, it helps the rescue team to operate in an efficient way.

Acknowledgements The authors wish to thank the Vellore Institute of Technology, Chennai Campus, for providing the necessary means to complete the research work successfully.

References

1. Berren MR, Beigel A, Ghertner S (1980) A typology for the classification of disasters. Community Ment Health J 16(2):103–111
2. McLoughlin D (1985) A framework for integrated emergency management. Public Adm Rev 45:165–172
3. Asghari P, Rahmani AM, Javadi HHS (2019) Internet of things applications: a systematic review. Comput Netw 148:241–261
4. Kamruzzaman M, Sarkar NI, Gutierrez J, Ray SK (2017) A study of IoT-based post-disaster management. In: International conference on information networking (ICOIN). IEEE, pp 406–410
5. Stusek M, Zeman K, Masek P, Sedova J, Hosek J (2019) IoT protocols for low-power massive IoT: a communication perspective. In: 11th International congress on ultra modern telecommunications and control systems and workshops (ICUMT), 2019, pp 1–7. https://doi.org/10.1109/ICUMT48472.2019.8970868

6. Asadzadeh A, Pakkhoo S, Saeidabad MM, Khezri H, Ferdousi R (2020) Information technology in emergency management of COVID-19 outbreak. Inform Med Unlocked, p 100475
7. Lee J, Kim D, Ryoo H-Y, Shin B-S (2016) Sustainable wearables: wearable technology for enhancing the quality of human life. Sustainability 8(5):466. https://doi.org/10.3390/su8 050466
8. RMIT University Australia webpage. https://www.rmit.edu.au/news/c4de/what-are-cyber-phy sical-systems. Last accessed on 26 Aug 2021
9. Gupta S, Starr MK, Farahani RZ, Matinrad N (2016) Disaster management from a POM perspective: mapping a new domain. Prod Oper Manag 25(10):1611–1637
10. Kafi KM, Gibril MBA (2016) GPS application in disaster management: a review. Asian J Appl Sci 4.1
11. Boral SD, Das A, Khare A, Gupta A, Bhattacharyya C, Das S (2020) Wearable location tracker during disaster. Int Res J Eng Technol, pp 999–1001
12. Amutha SME, Sivagami Meera G, Priyanga T, Nisha V (2021) Location and health monitoring of human during disaster using wearable—IoT. Int Res J Eng Technol, pp 2679–2687
13. Kunnath AT, Pradeep P, Ramesh MV (2012) ER-track: a wireless device for tracking and monitoring emergency responders. Procedia Comput Sci 10:1080–1085
14. Benlghazi A, Chadli E Moussaid, D (2014) Bluetooth technologie for industrial application. In: 5th International conference on information and communication systems (ICICS). IEEE, pp 1–5
15. Gulati H, Vaishya S, Veeramachaneni S (2011) Bluetooth and Wi-Fi controlled rescue robots. In: Annual IEEE India conference. IEEE, pp 1–5
16. Kim J, Kang SK, Park J (2015) Bluetooth-based tree topology network for wireless industrial applications. In: 15th International conference on control, automation and systems (ICCAS). IEEE
17. Labib M. et al (2019) Networking solutions for connecting bluetooth low energy devices-a comparison. In: MATEC web of conferences, vol 292. EDP Sciences
18. Lee EA (2008) Cyber physical systems: design challenges. In: 11th IEEE international sympo-sium on object and component-oriented real-time distributed computing (ISORC), pp 363–369. https://doi.org/10.1109/ISORC.2008.25
19. Sony M (2020) Design of cyber physical system architecture for industry 4.0 through lean six sigma: conceptual foundations and research issues. Prod Manuf Res 8(1):158–181
20. Pavithra D, Balakrishnan R (2015) IoT based monitoring and control system for home automation. In: Global conference on communication technologies (GCCT). IEEE
21. Electronic Wings Webpage. https://www.electronicwings.com/sensors-modules/bluetooth-module-hc-05. Last accessed on 26 Aug 2021

A Review of Machine Learning Techniques (MLT) in Health Informatics

Vandana Rawat⊙, Devesh Pratap Singh⊙, Neelam Singh⊙, and Umesh Kumar Tiwari⊙

1 Introduction

1.1 Health Informatics (HI)

Health informatics is a multispectral field that associates people to revamp the technology and technology revamp people. Healthcare data is the centric part of health informatics [1]. HI (also known as Health Information Systems) which uses information technology (IT) to formulate and analyse health data records to make improvement in the output. HI is the associative study of the design, integration, advocacy and requisition of information technology-based transformation in public health and delivery of healthcare services. HI is a combination of many things like people, organizations, illness, patient care and treatment. It is a technical field that assigns storage, retrieval and sharing of biomedical info for taking decisions. Healthcare informatics is combined with Biomedical Science and with information technologies.

Healthcare informatics can be described as follows:

- It is used to describe application and research topics for biomedical informatics.
- It needs some health data sources, healthcare tools and devices, with utilization of information technology.
- It produces electronic health records for different sources like patients, doctors, hospital staff and public, by using computer science tools and techniques.

Healthcare Informatics directly deals with public health informatics (PHI). It is defined as uniform way of healthcare information, Information Technology (IT) and Computer Science in area of public health, including surveillance, prevention

V. Rawat (✉) · D. P. Singh · N. Singh · U. K. Tiwari
Department of Computer Science and Engineering, Graphic Era Deemed to be University, Dehradun, India
e-mail: vandanarawat2405@gmail.com

© The Author(s), under exclusive license to Springer Nature Singapore Pte Ltd. 2023
R. Agrawal et al. (eds.), *International Conference on IoT, Intelligent Computing and Security*, Lecture Notes in Electrical Engineering 982,
https://doi.org/10.1007/978-981-19-8136-4_3

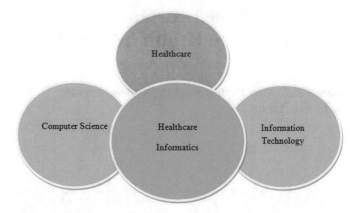

Fig. 1 Healthcare informatics

and health promotion [2]. PHI is practical based in the guidance of science and forwarded to the achievement of healthcare specific tasks. PHI needed the application of knowledge from several domains like Information Technology, Computer Science, Healthcare management, and communications with different sources. **Informatics** is generally used in association of biomedicine, health and other fields which results in terms of health informatics, biomedical informatics or legal informatics [3].

On the basis of above definitions, healthcare informatics can be divided in to three parts as follows: health care, Computer Science and Information Technology. Figure 1 represents these domains of healthcare informatics.

Health care: Health care means the good health in people via prevention, diagnosis, treatment, recovery, impairments, etc. The concept of health care uses these following 5 pillars:

(i) It improves quality, safety and efficiency by reducing health disparities.
(ii) It is engaged with patients.
(iii) It means to improve care coordination.
(iv) It means to improve public health.
(v) It means to provide adequate privacy and protection.

Healthcare is improved by health professionals with healthcare tools. Some precious parts of health care are medicine, pharmacy, public health, these all are the parts of health care. Healthcare services give best outcome if they are delivered on time. So, health care is an important aspect of improving the physical and mental health and wellness of people, worldwide.

Different contribution in healthcare industry has been done for the identification and diagnosis of diseases. Diseases are of mainly two types:

(1) **Acute Diseases**: These diseases are the short-term diseases and have very fast out start for a brief period. Acute diseases include a very rapid onset or a short course [4]. Once they are treated, patient is recovered. Acute diseases are like broken bone, common cold, respiratory infection, etc. [5].

(2) **Chronic Diseases**: These diseases are long-term diseases. In this, condition is slowly developed in human but they may progress over time, and may have any number of warning signs. Some of the chronic diseases are Alzheimer, heart disease, cancer, asthma, diabetes and many more [6]. In age group of 35–75, death rate is higher affecting from these chronic diseases [7]. Innovative solutions are required for early prediction of these diseases [8].

(3) **Computer Science and Information Technology**: It is the field of computers and computational systems dealing mostly in software and software systems. In earlier times, only doctors were the only attendant and saviours of many humans' life. Nowadays, no field is uninfluenced with computers. This is important in formation and maintenance of patient's healthcare records, for maintaining laboratory reports, health records, laboratory test, etc., for various diseases [9].

Information Technology: Information technology plays a prime role in Healthcare Industry. It helps in getting accurate, actionable and more accessible information related to patients' health to meet the needs of individuals. Health care is based increasingly on information technology (IT) to acquire, manage, analyse and disseminate healthcare information and knowledge [10].

1.2 Machine Learning (ML)

Machine Learning (ML) is a subpart of Artificial Intelligence. It is engaged with computer systems for enhancing the performance through its experiences, automatically. Machine Learning (ML) corrects the automotive learning process through training and lead towards adaptation of its algorithm [11]. It is the necessity of machines for feeding intelligence of human in it, artificially.

Machine Learning can be divided into 3 main types of learning [12]:

(a) **Supervised Machine Learning**: It uses labelled data sets to train the algorithms [11]. It is used to classify and predict outputs correctly. The two main types of supervised learning are classification and regression.

Classification of Supervised Learning Algorithms: These algorithms have several categorizations, given as follows:

i) **Linear Classifiers**: In classification, linear classifiers mean the class and group belongingness. To achieve this, we have to classify the data into labels based on a linear combination of input features. For this hyperplane issued [13].

ii) **Logistic Regression**: Logistic regression is a weighted combination of input features and it is passed through a sigmoid function which gives the output of a number between 0 and 1 [14].

iii) **Naive Bayes (NB)**: It is based upon Bayes theorem with an assumption of independence among predictors. It is used for solving classification problem. In simple words, it can be said that Naive Bayes classifier assumes that the availability of a particular characteristic in a class is unrelated to the availability of any other feature [15].

iv) **Support Vector Machines (SVMs)**: These algorithms are an alternative linear classification scheme that aims to rectify classification errors. In SVM, a classifier can be defined by the hyperplane. SVMs seek to find a hyperplane (in two dimensions, a line) that optimally separates two classes of point. The "best" classifier is the one that classifies all points correctly, such that the nearest points are as far as possible from the boundary [16].

v) **Decision Trees**: The idea behind decision tree is to split the data into "Pure" regions. In this algorithm, start with all samples at a node and partition sample input to create purest subsets. It contains root node, branches and leaf nodes. In decision tree, the input data and corresponding output data are identified with the help of training data.

vi) **Neural Networks**: In Neural Network Algorithm, each unit takes the input data from neighbours and then this data is used to compute an output signal. It is further divided into other parts, known as Test stage. Apart from this, there is the task of the adjustment of the weights that is Learning stage [17].

vii) **Unsupervised Machine Learning**: In which we have a group of unlabelled data. Dimension reduction and clustering, these two are most important types of it.

viii) **Reinforcement Machine Learning**: It is an approach which agent learns to achieve a goal in an uncertain and complex environment. In this learning process, an agent (e.g. a robot or controller) seeks to learn the optimal actions to take the outcomes of past actions. In reinforcement learning, we train ML models by using rewards and punishments.

1.3   Role of Machine Learning in Healthcare Disease

Machine Learning Algorithms improve the performance through experiences, automatically. It can be applied in various applications of our daily life scenarios like weather forecasting, image recognition, product recommendation, traffic prediction, smart car parking, etc. The intensive growth of health-related data presents unrivalled opportunities for improving patient's health. Machine Learning plays an essential role in health care as it can be applied medical image segmentation, computer-aided diagnosis of diseases, etc.

2   Literature Review and Critical Issues

Holzinger et al. [18] Health informatics is a technological field that assign with the storage, retrieval and sharing of biomedical info for taking decisions. ML algorithms learn from experiences automatically. In HI, chronic diseases can be considered.

Some of the predictive models have been presented for chronic diseases. Charleonnan et al. [19] including several ML algorithms like KNN, SVM and LR are used to predict chronic kidney disease. After applying many algorithms, it is seen that SVM classifier gives the highest accuracy in comparison to other ML algorithms. Boonchieng et al. [20] presented various supplications of Machine Learning with Software Engineering approaches, in Digital Disease Detection (DDD). The main challenge in it is to find the mark of early detection of disease because real-time monitoring and forecasting is one of the most common issues in these diseases.

Nambiar et al. [21] state that in the recent years, health is the major area of focus as many subscribers of smartphones as well as Internet users are increasing day by day so using Big Data Analytics and IOT, some advantages can be scored. In this paper, high-level concept is given about industry, opportunities, and current developments in connected health. Bhardwaj et al. [22] introduced that the perspective of applying Machine Learning Technologies in health care and in various industry initiatives. In this paper, Big Data Technology has been applied with ML Algorithms. Big Data Technology can handle structured, semi-structured and unstructured data in petabytes and more. Tafti et al. [23] stated that every day, a vast amount of data which is produced on social media platform. So to evaluate such kind of data, some techniques are to be applied that can convert the raw data into useful information using Machine Learning (ML) and Data mining techniques. In this paper, Machine Learning techniques are examined on the following most common problems like classification, clustering, regression, etc.

Nithya et al. [24] Machine Learning is the fasted evolving field in the terms of health informatics. Hence, the target is to develop new algorithms which can learn new things and can provide time to time progress to patients. L'Heureux et al., Christensen et al. and Sharma et al. [25–27] stated that various Machine Learning Techniques can also be used for prediction purpose as well it provides variety of alerting decision support systems, which are targeted to improve patient's safety and healthcare quality in many areas of healthcare applications. Mir et al. [28] Healthcare informatics is most useful area of innovation with technical advancement. In health care, Big data Analytics can be applied for better. This paper focus on implementing a classifier model using WEKA tools to predict diabetes using ML algorithms like Naive Bayes, Support Vector Machine, Random Forest, etc. Experimental results of each algorithm have been taken for evaluating the data set. It is observed that Support Vector Machine performed best in prediction of the Diabetic. Tsang et al. [29] elaborate that patients with dementia represent challenges in healthcare systems in this current century.

So by using Machine learning techniques, high quality care of patients can be given to dementia patients. Vellido et al. [30] explain the capability of ML algorithms and depicted their impact in the field of Medicine and Health care. So beyond improving the model, it requires to integrate the medical experts in the design of data analysis and interpretation strategies for improving the result in chronic diseases. Gupta et al. [31] explain that in any data set, if there is noise, then it affects the performance of result like the classification and prediction accuracy can be decreased. Therefore, there is a need to handle this problem. So for that reason, authors have investigated

79 primary studies of noise identification and noise handling techniques. Cedeno-Moreno et al. [32] took the data set of Diabetes, which is generating definite number of death annually. The result of this is in desertion in production.

Sivakani et al. [33] chronic diseases are the diseases that are long lasting so Alzheimer is one of the chronic diseases. It is one of the types of Dementia means brain disorder disease. It occurs in the age of 60–70 but nowadays, it is being attacked the young generation also. Moreb et al. [34] scrutinize the relation between Software Engineering and Machine Learning in terms of Healthcare Systems. For this one novel framework, i.e. Software Engineering for Machine Learning in health informatics (SEMLHI) has been proposed in this study. Real Healthcare data has taken from Palestine government of last three years with 7different phases. Katarya et al. [35] implemented the ML algorithms on heart disease data set for early prediction by proper monitoring of patient to get notified in an early stage using supervised Machine Learning techniques.

Alanazi et al. [36] focus is on to develop a novel Machine Learning model for providing accurate and dynamic predictions to improve the Medicine and Healthcare Field. Kumar et al. [37] applied some AI techniques on collected HER data set, sensor devices, smart devices and Internet sources to improve disease progression, prediction, clinical intervention. Govindan et al. [38]. In this paper, two criteria, age and pre-existing diseases (such as diabetes, heart problems or high blood pressure) are taken for prediction of COVID-19. For this, a practical decision support system has been proposed for COVID-19 prediction. Rayan et al. [39] taken the reviews of various smart healthcare research. A systematic pipeline of data processing is accommodated for conventional smart health using these ML models. Gnana et al. [40] taken some real-time patient data implementing Machine Learning techniques, sensors, Raspberry pi board system and camera unit. This work is aim to develop a Smart Health Monitoring System with Machine Learning techniques using Arduino. Rghioui et al. [41, 42] explain Diabetic Patient Data set from Smart phone and sensors are taken. This represents an intelligent architecture for monitoring diabetic patients by using Machine Learning algorithms with 5G technologies.

In the following Table 1, many reviews has taken from various researches in detail:

Table 1 Various key findings in Healthcare using Machine Learning Techniques (MLT)

Author's/references	Domain	What has been done		How it has been done
		Techniques	Key findings	
Alanazi et al. [13]	Medicine and Health care	Machine Learning Algorithms: • SVM • ANN • Naïve Bayes	• Emphasis is on developing a novel Machine Learning model to provide accurate and dynamic predictions	• Different disease data set has been taken for getting accuracy using MLT

(continued)

Table 1 (continued)

Author's/references	Domain	What has been done		How it has been done
		Techniques	Key findings	
Nambiar et al. [22]	Health care	• Big Data Analytics	• It combines connected smart devices with healthcare systems to provide care which will give benefit both to patients and health providers in remote areas	• Smart Devices (smart phones, personal wellness devices)personal wellnessdevices)
Nithya and Ilango [25]	Chronic diseases	• Machine Learning Algorithms: • Decision Tree Technique • Artificial Neural Network Clustering Methods	• It is showing the analysis of Machine Learning Algorithms in various domains	Chronic diseases like: • Diabetes • Cancer • Hepatitis
Ahamed et al. [29]	Personalized Health care	• IOT • Machine Learning Algorithms	• Personalized Health care (PH) applies Machine Learning Algorithms to the collect the patient health data set	• Patient data set • Electronic health records (EHR) • Sensor data set
Rayan et al. [40]	Chronic diseases	• Machine Learning Algorithms like: • ANN • SVM • Deep Learning Models	• This paper taken the reviews of various smart healthcare researches • A systematic pipeline of data processing is accommodating for conventional smart health using these Machine Learning Algorithms	Chronic diseases dataset like: • Glaucoma • Alzheimer
Sheela and Varghese [41]	Health Care	• Machine Learning • Raspberry pi board system • Camera unit	• To create as Smart Health Monitoring System with MLT using Arduino	• Real-time Patient Data
Rghioui et al. [42]	Diabetic Disease	• ESP8266-12F • Module • Machine Learning Algorithms • Sensors	• Presents an intelligent architecture for monitoring diabetic patients using 5G technologies with MLT	• Diabetic Patient Data set from smart phone and sensors

3 Area of Research and Formulation of Research Problem

In healthcare industry, various Machine Learning Techniques are being applied. In terms of diseases, many researchers have been done to achieve accuracy, early prediction of diseases and to handle unstructured and noisy data but still there are various research gaps. Some of the issues and research gaps are identified and described below:

1. Machine Learning Techniques are used for predictive modelling which gives conflicting output in terms of accuracy when same data set of chronic diseases is considered for different algorithms. So there is a need of predictive model for accurate and dynamic prediction.
2. It is difficult to deal with non-standardized and unstructured healthcare data sets in terms of Machine Learning Algorithms.
3. As biomedical data is complex in nature, thus it becomes difficult to identify testable hypothesis.
4. Complexity of biomedical data makes it challenging to build accurate model to diagnose and predict various chronic diseases.
5. In most of the cases, Machine Learning Algorithms are implemented to work on few key attributes of various chronic disease predictions. So there is a need to incorporate more key attributes in Machine Learning Algorithms to attain accuracy.
6. Machine Learning Algorithms are related to statistical analogy. In this, decision-making is done with the help of existing data. They predict diagnosis of diseases based on previous experiences. In case of diagnosis of diseases, when patient monitoring is done, then Machine Learning Algorithms analyse the situation only according to the trained data set.

4 Conclusion

In today's time, health care is one of the speedy extending areas. It needs care for people with the help of some technology revolution as the health sector becomes evident when data in hospital and technical departments (laboratory, medicine, radiology) is very large and complex to handle. So to assess and diagnose these problems, Machine Learning technologies can be applied. The government is also playing a crucial role in influencing people's health, which is not limited to the health sector. This article addressed some limitations and given some challenges faces in the healthcare field. Machine Learning is contributing to improvements for current research, with scope for considerable future contribution in each discipline.

References

1. Paul M, Nembhard HB (2016) Health data and informatics. Wiley
2. Aziz HA (2017) A review of the role of public health informatics in healthcare. J. Taibah University, 485 Medical Sciences 12, pp 78–81
3. Dalrymple PW (2011) Data, information, knowledge: the emerging field of health informatics. Bull Am Soc Inf Sci Technol 37(5):41–44
4. Gellman MD, Turner JR (2013) Encyclopedia of behavioral medicine. Springer
5. Sevick M, Trauth J, Ling B (2007) Patients with complex chronic diseases: perspectives on supporting self management. J Gen Intern Med 22(3):438–444
6. Maes S, Leventhal H, Ridder De (1996) Coping with chronic diseases. In: Zeidner M, Endler NS (eds) Handbook of coping: theory, research, applications. Wiley, pp 221–251
7. Strong K, Mathers C (2005) Preventing chronic diseases: how many lives can we save? Lancet 366:1578–1582
8. Epping-Jordan JE, Galea G (2005) Preventing chronic diseases: taking stepwise action. Lancet 366:1667–1671
9. Panth M, Acharya AS (2015) The unprecedented role of computers in improvement and transformation of public health: an emerging priority. Indian J Community Med 40(1):8
10. Stead WW, Lin HS (2009) Computational technology for effective health care: immediate steps and strategic directions. National Academies Press, Washington DC
11. Sharma K, Nandal R (2019) A literature study on machine learning fusion with IoT. In: 3rd International conference on trends in electronics and informatics (ICOEI), Tirunelveli, India, pp 1440–1445
12. Simeone O (2018) A brief introduction to machine learning for engineers. Found. Trends Signal Process. 12(3–4):200–431
13. Aldhoayan M, Leming Z (2016) An accurate and customizable text classification algorithm: two applications in healthcare. In: 6th International conference on computational advances in bio and medical sciences (ICCABS), Atlanta, GA, USA, pp 1–4. https://doi.org/10.1109/ICCABS.2016.7802778
14. Harshvardhan G, Venkateswaran N, Padmapriya N (2016) Assessment of Glaucoma with ocular thermal images using GLCM techniques and Logistic Regression classifier. In: International conference on wireless communications, signal processing and networking (WiSPNET), Chennai, India, pp 1534–1537. https://doi.org/10.1109/WiSPNET.2016.7566393
15. Ji Y, Yu S, Zhang Y (2011) A novel Naive Bayes model: packaged hidden Naive Bayes. In: 6th IEEE joint international information technology and artificial intelligence conference, Chongqing, China, pp 484–487. https://doi.org/10.1109/ITAIC.2011.6030379
16. Singh P, Singh SP, Singh DS (2019) An introduction and review on machine learning applications in medicine and healthcare. In: IEEE conference on information and communication technology, Allahabad, India, pp 1–6. https://doi.org/10.1109/CICT48419.2019.9066250
17. Brus VR, Voronova LI (2020) Neural network classification of cardiac activity based on cardiogram data for driver support system. In: Systems of signals generating and processing in the field of on board communications, Moscow, Russia, pp 1–5. https://doi.org/10.1109/IEEECONF48371.2020.9078639
18. Holzinger A (2016) Machine learning for health informatics. In: Machine learning for health informatics. Springer, pp 1–24
19. Charleonnan A, Fufaung T, Niyomwong T, Chokchueypattanakit W, Suwannawach S, Ninchawee N (2016) Predictive analytics for chronic kidney disease using machine learning techniques. In: Management and innovation technology international conference (MITicon), Bang-San, pp MIT-80–MIT-83. https://doi.org/10.1109/MITICON.2016.8025242
20. Boonchieng E, Duangchaemkarn K (2016) Digital disease detection: application of machine learning in community health informatics. In: 13th International joint conference on computer science and software engineering (JCSSE), KhonKaen, pp 1–5. https://doi.org/10.1109/JCSSE.2016.7748841

21. Nambiar AR, Reddy N, Dutta D (2016) Connected health: opportunities and challenges. In: International conference on big data (Big Data), Boston, MA, pp 1658–1662. IEEE. https://doi.org/10.1109/BigData.2017.8258102

22. Bhardwaj R, Nambiar AR, Dutta D (2017) A study of machine learning in healthcare. In: 41st Annual computer software and applications conference (COMPSAC), Turin. IEEE, pp 236–241. https://doi.org/10.1109/COMPSAC.2017.164

23. Tafti AP, Larose E, Badger JC, Kleiman R, Peissig P (2017) Machine learning- as-a-service and its application to medical informatics. In: Machine learning and data mining in pattern recognition, pp 206–219. https://doi.org/10.1007/978-3-319-62416-715

24. Nithya B, Ilango V (2017) Predictive analytics in health care using machine learning tools and techniques. In: International conference on intelligent computing and control systems (ICICCS), Madurai, pp 492–499. https://doi.org/10.1109/ICCONS.2017.8250771

25. L'Heureux A, Grolinger K, Elyamany HF, Capretz MAM (2017) Machine learning with big data: challenges and approaches. IEEE Access 5:7776–7797. https://doi.org/10.1109/ACCESS.2017.2696365

26. Christensen T, Frandsen A, Glazier S, Humpherys J, Kartchner D (2017) Machine learning methods for disease prediction with claims data. In: International conference on healthcare informatics (ICHI), New York, NY. IEEE, pp 467–4674. https://doi.org/10.1109/ICHI.2018.00108

27. Sharma H et al (2018) Portable phenotyping system: a portable machine-learning approach to i2b2 obesity challenge. In: International conference on healthcare informatics workshop (ICHI-W), New York, NY. IEEE, pp 86–87. https://doi.org/10.1109/ICHI-W.2018.00032

28. Mir A, Dhage SN (2019) Diabetes disease prediction using machine learning on big data of healthcare. In: 4th International conference on computing communication control and automation (ICCUBEA), Pune, India, pp 1–6. https://doi.org/10.1109/ICCUBEA.2018.8697439

29. Tsang G, Xie X, Zhou SM (2020) Harnessing the power of machine learning in dementia informatics research: issues, opportunities, and challenges. IEEE Rev Biomed Eng 13: 113–129. https://doi.org/10.1109/RBME.2019.2904488

30. Vellido A (2019) The importance of interpretability and visualization in machine learning for applications in medicine and health care. Neural Comput Appl. Last accessed 4 Feb 2019

31. Gupta S, Gupta A (2019) Dealing with noise problem in machine learning data-sets: a systematic review. Procedia Comput Sci 161:466–474

32. Cedeno-Moreno D, Vargas-Lombardo M (2019) Application of machine learning with supervised classification algorithms: in the context of health. In: 7th International engineering, sciences and technology conference (IESTEC), Panama, Panama, pp 613–618. https://doi.org/10.1109/IESTEC46403.2019.00115

33. Sivakani R, Ansari GA (2010) Machine learning framework for implementing Alzheimer's disease. In: International conference on communication and signal processing (ICCSP), Chennai, India, pp 0588–0592. https://doi.org/10.1109/ICCSP48568.2020.9182220

34. Moreb M, Mohammed A, Bayat O (2020) A Novel software engineering approach toward using machine learning for improving the efficiency of health systems. IEEE Access 8:23169–23178. https://doi.org/10.1109/ACCESS.2020.2970178

35. Katarya R, Srinivas P (2020) Predicting heart disease at early stages using machine learning: a survey. In: International conference on electronics and sustainable communication systems (ICESC), Coimbatore, India, pp 302–305. https://doi.org/10.1109/ICESC48915.2020.9155586

36. Alanazi HO, Abdullah AH, Qureshi KN (2017) A critical review for developing accurate and dynamic predictive models using machine learning methods in medicine and health care. J Med Syst 41(4):69

37. Kumar S, Gaur MS (2019) Handoff prioritization to manage call admission control in mobile multimedia networks for healthcare. In: 2019 10th International conference on computing, communication and networking technologies (ICCCNT), Kanpur, India, pp 1–7. https://doi.org/10.1109/ICCCNT45670.2019.8944618

38. Govindan K, Mina H, Alavi B (2020) A decision support system for demand management in healthcare supply chains considering the epidemic outbreaks: a case study of coronavirus disease 2019 (COVID- 19). Transp Res Part E: Logistics Transp Rev
39. Rayan Z, Alfonse M, Salem A (2019) Machine learning approaches in smart health. Procedia Comput Sci 154:361–368
40. Sheela KG, Varghese AR (2020) Machine learning based health monitoring system. Mater Today: Proc 24:1788–1794
41. Rghioui A, Lloret J, Sendra S, Oumnad A (2020) A smart architecture for diabetic patient monitoring using machine learning algorithms. Healthcare 8(3):348
42. Rawat V (2019) A classification system for diabetic patients with machine learning techniques. Int J Math Eng Manage Sci 4:729–744
43. Ahamed F, Farid F (2018) Applying internet of things and machine-learning for personalized healthcare: issues and challenges. In: International conference on machine learning and data engineering (iCMLDE), Sydney, Australia, pp 19–21. https://doi.org/10.1109/iCMLDE.2018.00014

A Task Scheduling Algorithm for Optimizing Quality of Service in Smart Healthcare System

Prabhdeep Singh, Vikas Tripathi, Kiran Deep Singh, M. S. Guru Prasad, and H. Aditya Pai

1 Introduction

Many aspects of our everyday lives have been changed by new technology. Traditional health care has evolved into smart health care as a result of the benefits of using information and communication technology (ICT) in today's healthcare system. When it comes to health care, smart technology means developing better diagnostic tools, better treatments for patients, and products that enhance the quality of life for everyone. Consumers are becoming more involved and aware of their health as healthcare technology advance [1–3]. The need for remote care is only going to grow in popularity. There are no technologies in existing healthcare ecosystems that can enhance patient care by providing them with real-time patient information and enabling them to adopt proactive treatment steps. Cloud computing is computing performed on the Internet, whereas other resources that are configurable and shared are provided as services with computers and other devices. Examples of configurable shared resources are infrastructure, platform, and software. Services that are based on cloud computing over the Internet, commit to consumer computation, software, and data. The consumers, who wish to utilize the cloud services, can get those services over the network [4–6]. The cloud service users can buy the services or use the services which are available free of cost in the cloud. Many types of services are available in the cloud such as infrastructure as a service, platform as a service, and

P. Singh (✉) · V. Tripathi · M. S. Guru Prasad · H. Aditya Pai
Department of Computer Science & Engineering, Deemed to be University, Graphic Era, Dehradun, India
e-mail: ssingh.prabhdeep@gmail.com

K. D. Singh
Chitkara University Institute of Engineering and Technology, Chitkara University, Rajpura, Punjab, India

R. Agrawal et al. (eds.), *International Conference on IoT, Intelligent Computing and Security*, Lecture Notes in Electrical Engineering 982,
https://doi.org/10.1007/978-981-19-8136-4_4

software as a service. These cloud services are effectively worked in the healthcare sector [7–10].

In cloud computing, the level of user level is not equal, and they do not give a similar level of services, based on the location of the user's in the institution, the significant users generate the tasks. The significance of the request relies upon the position level of who commands that request [11–14]. There is a substantial gap between the level of users and task scheduling algorithms in cloud computing.

One of the main challenges in distributed systems like the grid, peer-to-peer, and cloud environments is task scheduling. The task scheduling method provides adequate services in cloud computing. It is executed by mapping job requirements for the possible resources to the users and decreasing total response time [15–19]. Many jobs need the processor to remain idle for a length of time while they are being processed. If idle time slots can be fully used, job processing times will be reduced, and overall work processing efficiency will increase.

A more complex task scheduling technique is needed since a simple job scheduling approach may not be able to fully use idle resources. As part of our investigation, we look at how heterogeneous cloud computing system's schedules work. The scheduling method that performed the execution of a job in the least execution time is termed as Shortest Job First (SJF) Scheduling Algorithm or Min-Min Algorithm, while the Round-Robin (RR) scheduling algorithm scheduling implements all jobs in the least time unit running within a circular queue [20–23].

In the rest of the paper, Sect. 2 consists of a brief review of the existing health system based on the cloud. The proposed cloud-based health system and ant colony optimization algorithm for task scheduling, the concept of green computing, and the detailed description of the power consumption of the proposed system are given in Sect. 3. The experiments are simulated and evaluate the performance of the proposed framework in Sect. 4. The conclusion is given in Sect. 5.

2 Literature Review

The Energy-Aware Distributed Hybrid Flow Shop Scheduling Problem with Multiprocessor Tasks (EADHFSPMT) are addressed by concurrently addressing two goals, namely makespan and total energy usage [24]. It is divided into three subproblems: work assignment between factories, job sequencing inside factories, and machine assignment for each job. We offer a mixed interlinear programming model and a Novel Multi-Objective Evolutionary Algorithm (NMOEA/D) based on decomposition.

The authors [25] design and develop a resource-aware dynamic task scheduling method. The simulation tests were conducted using the cloudsim simulation program with three well-known datasets: HCSP, GoCJ, and Synthetic workload. The suggested approach's findings are then compared to those obtained from RALBA, Dynamic MaxMin, DLBA, and PSSELB scheduling methods in terms of average resource utilization (ARUR), makespan, throughput, and average reaction time (ART). The DRALBA method resulted in substantial gains in ARUR, throughput, and makespan.

The authors [26] present a DNN task scheduling algorithm that maximizes job parallelism by minimizing communication delay variation caused by the HMC connection characteristics. Task partitioning is used to maximize parallelism while maintaining inter-HMC traffic within the link's reasonable bandwidth. To conceal the average communication delay of intermediate DNN processing outcomes, task-to-HMC mapping is done. To speed DNN inference while optimizing resource usage, a work plan is created using retiming. The suggested method's efficacy was shown via simulations utilizing a variety of actual DNN applications on a ZSim x86-64 simulator.

The authors [27] propose a task duplication-based scheduling method, dubbed TDSA, for optimizing makespan in cloud-based workflows with limited budgets. Two new mechanisms are used in TDSA: (1) a dynamic sub-budget allocation mechanism, which is responsible for recovering unused budget for scheduled workflow tasks and redistributing remaining budget, thereby enabling the use of more expensive/powerful cloud resources to accelerate the completion time of unscheduled tasks; and (2) a duplication-based task scheduling mechanism, which aims to exploit idle slots on resources to selectively duplicate tasks' predecessors, thereby accelerating the completion time of these tasks while remaining within budget.

The authors [28] present a partitioning-based method, P-EDF-omp, that automatically guarantees the tied constraint as long as an OpenMP task system can be partitioned effectively to a multiprocessor platform. Additionally, we undertake extensive tests using both synthetic workloads and well-known OpenMP benchmarks to demonstrate that our method regularly outperforms the work in even without changing the TSCs.

The authors [29] present a cloud workflow scheduling method that combines particle swarm optimization (PSO) with idle time slot-aware rules in order to optimize the execution cost of a workflow application that is subject to a deadline restriction. A novel particle encoding is developed to describe the virtual machine (VM) type needed by each job and the task scheduling sequence. To decode a particle into a scheduling solution, an idle time slot-aware decoding method is suggested. To deal with jobs that have incorrect priority as a result of PSO's unpredictability, a repair technique is utilized to restore those priorities and generate proper task scheduling sequences.

The authors [30] introduce the MDVMA metaheuristic framework for dynamic virtual machine allocation and job scheduling optimization in a cloud computing context. The MDVMA is focused on developing a multi-objective scheduling method based on the non-dominated sorting genetic algorithm (NSGA)-II algorithm for task scheduling optimization with the goal of minimizing energy consumption, time, and cost while providing a trade-off to cloud service providers based on their requirements.

The authors [31] propose an asynchronous-based dynamic and elastic task scheduling scheme that avoids device underutilization, load imbalance between devices, and frequent kernel launches, inter-device data transfers, and inter-device synchronizations by dynamically adjusting the chunk size in response to runtime performance changes.

The authors [32] define task assignment and power allocation together in order to minimize overall task execution delay by taking into consideration user mobility, dispersed resources, task characteristics, and the user device's energy limitation. We begin by proposing an evolutionary method based on genetic algorithms (GAs) for solving our stated mixed-integer nonlinear programming (MINLP) issue. Then, they propose a heuristic called mobility-aware task scheduling (MATS) for efficiently assigning tasks.

The authors [33] suggest a real-time dynamic scheduling (RTDS) system for ensuring the transmission of emergency jobs and optimizing scheduling efficiency via scheduling sequence modification and task management. To begin, relative SDRN system models are presented, followed by a dynamic scheduling method that accounts for disturbance variables throughout the scheduling process. The RTDS system is then executed in three phases based on the dynamic scheduling mechanism: initial scheduling, dynamic scheduling for emergency jobs, and final scheduling.

3 Proposed Work

Many metaheuristic methods have been developed recently by cloud computing researchers to address the issues that arise in task scheduling activities. The scheduling of work in a cloud computing environment is a significant source of worry. Otherwise, the system will be much less efficient. As a result, a novel work scheduling method is required to improve cloud performance. The proposed cloud-based health system is shown in Fig. 1. Various types of cloud repositories and cloud servers are placed in the cloud. The data storage and processing of the data are the primary responsibility of these servers. A task scheduler is situated near all cloud data servers which is responsible for distributing the task and providing the scheduling. Task scheduler is one of the major components of the proposed framework that cannot schedule the job using single criteria. It schedules the task based on several criteria and rules. The service provider and the user agreed on rules and criteria. There is a substantial gap in the designing of the task scheduling algorithm and the level of service provided to users. The proposed framework makes a priority of the user that has the highest level to receive the best services in separate cloud computing and also utilizes similar network speed, storage capacity, and processor speed.

(A)   Task Scheduling of Proposed Cloud-based health framework

The following framework makes use of an ant colony optimization method.

Step 1:   To begin, set the task list, VM list, and minimum value all to 999
Step 2:   Initialize the pheromone trail and visibility for each job and VM edge.
Step 3:   Go to Step 4 if the current iteration is more than the max iteration, or Step 8 if it is not.

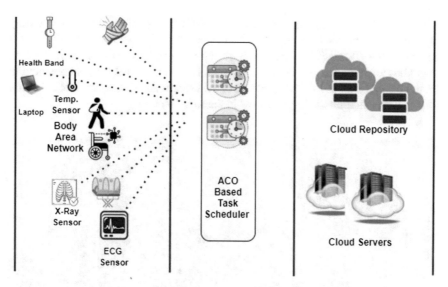

Fig. 1 Proposed framework of the cloud-based health system

Step 4: Step 5 is what I do for every job.
Step 5: Execute them by binding them together using Eq. (2) to calculate each task's VM.
Step 6: Calculate the duration of the trip by adding up the time it took to complete job I on each VM.
Step 7: If all jobs have been assigned, go to step 9, otherwise continue to step 4.
Step 8: If the schedule's execution time is less than one minute, then the current iteration will be one minute longer than the previous iteration.
Step 9: Make the schedule's duration min
Step 10: End

4 Experimental Results

The ant colony optimization algorithm is implemented in Java-based open source tool CloudSim. In the experiment two metrics, Average Response Time (ART) and Average Waiting Time (AWT) is used to estimate the algorithm performance in the cloud scenario. In the cloud-based health system, 40 virtual machines and 90 tasks are created.

The virtual machines are configured with 3000–5200 MIPS. Those machines utilized 1–5 processing elements with 1000–3000 bandwidths. The length of the task is ranging from 10,000 to 50,000, and the input file size ranged from 100 to 800. Initial pheromone deposition is 0.8, Q is taken to be 100, the value is 0.3, and is 2. However, it was changed to see the effect on the makespan, and these values were found best to use for experimentation. The performances of proposed and existing algorithms for task scheduling have been recorded. The results in Figs. 2 and 3 show that the Average Response Time of the proposed algorithm is always better than traditional algorithms. Average Waiting Time is recorded less as compared to traditional algorithms. In the second fold of the experiment, the approach of reduction of power consumption is implemented in the same scenario.

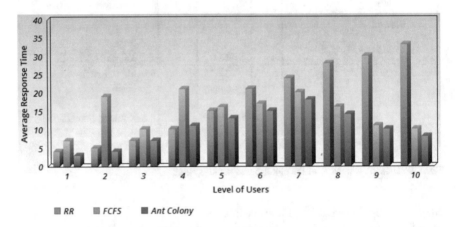

Fig. 2 Comparison of Average Response Time

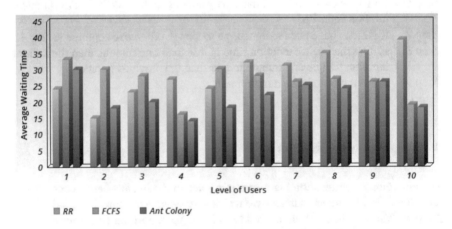

Fig. 3 Comparison of Average Waiting Time

5 Conclusion and Future Work

Additionally, diverse resources and scheduling methods from other areas were analyzed to get a deeper grasp of the idea. It goes into more detail on CloudSim and its architecture, which provides an environment that can be used directly to simulate cloud computing infrastructures. This paper demonstrates how it may be used to mimic how clouds operate by building data centers, virtual machines, and brokers. It describes the scheduling methods used on CloudSim to compare the interactions of different entities. The suggested method in this paper is based on the ant colony optimization algorithm, which is a heuristic-based approach. The main goal of adopting the proposed method was to shorten schedules' makespan. When the simulation results of the proposed algorithm are compared to those of other conventional algorithms, the proposed method has a shorter makespan time, demonstrating that it is more superior and efficient than others.

References

1. Zhang F, Ge J, Li Z, Li C, Wong C, Kong L, Luo B, Chang V (2018) A load-aware resource allocation and task scheduling for the emerging cloudlet system. Futur Gener Comput Syst 87:438–456
2. Saleh H, Nashaat H, Saber W, Harb H (2018) IPSO task scheduling algorithm for large scale data in cloud computing environment. IEEE Access 7:1–1
3. Safari M, Khorsand R (2018) Energy-aware scheduling algorithm for time-constrained workflow tasks in DVFS-enabled cloud environment. Simul Model Pract Theory 87:311–326
4. Makaratzis T, Filelis-Papadopoulos CK, Giannoutakis KM, Gravvanis GA, Tzovaras D (2017) A comparative study of CPU power consumption models for cloud simulation frameworks. In: Proceedings of the 21st Pan-Hellenic conference on informatics
5. Liu L, Fan Q, Buyya R (2018) A deadline-constrained multi-objective task scheduling algorithm in mobile cloud environments. IEEE Access 6:52982–52996
6. Choe S, Li B, Ri I, Paek C, Rim J, Yun S (2018) Improved hybrid symbiotic organism search task-scheduling algorithm for cloud computing. KSII Trans Internet Inf Syst (TIIS) 12:3516–3541
7. Rezaei-Mayahi M, Rezazad M, Sarbazi-Azad H (2019) Temperature-aware power consumption modeling in Hyperscale cloud data centers. Future Gener Comput Syst 94: 130–139
8. Chen H, Liu G, Yin S, Liu X, Qiu D (2018) Erect: energy-efficient reactive scheduling for real-time tasks in heterogeneous virtualized clouds. J Comput Sci 28:416–425
9. Arunarani R, Manjula D, Sugumaran V (2019) Task scheduling techniques in cloud computing: a literature survey. Futur Gener Comput Syst 91:407–415
10. Alworafi MA, Mallappa S (2018) An enhanced task scheduling in cloud computing based on deadline-aware model. Int J Grid High Perform Comput (IJGHPC) 10:31–53
11. Kaur T, Chana I (2018) GreenSched: an intelligent energy-aware scheduling for deadline-and-budget constrained cloud tasks. Simul Model Pract Theory 82:55–83
12. Alla HB, Alla SB, Touhafi A, Ezzati A (2018) A novel task scheduling approach based on dynamic queues and hybrid meta-heuristic algorithms for cloud computing environment. Clust Comput 21:1797–1820
13. Abd SK, Al-Haddad SAR, Hashim F, Abdullah ABHJ, Yussof S (2017) An effective approach for managing power consumption in cloud computing infrastructure. J Comput Sci 21: 349–360

14. Kaur R, Singh PD, Kaur R, Singh KD (2021) A delay-sensitive cyber-physical system framework for smart health applications. In: Advances in clean energy technologies. Springer, Singapore, pp 475–486
15. Tang C, Hao M, Wei X, Chen W (2018) Energy-aware task scheduling in mobile cloud computing. Distrib Parallel Databases, 36: 529–553
16. Kaur S, Singh KD, Singh P, Kaur R (2021) Ensemble model to predict credit card fraud detection using random forest and generative adversarial networks. In: Emerging technologies in data mining and information security. Springer, Singapore, pp 87–97
17. Sood SK, Singh KD (2021) Identification of a malicious optical edge device in the SDN-based optical fog/cloud computing network. J Opt Commun 42(1):91–102
18. Angurala M, Bala M, Bamber SS, Kaur R, Singh P (2020) An internet of things assisted drone based approach to reduce rapid spread of COVID-19. J Saf Sci Resilience 1(1):31–35
19. Sood SK, Singh KD (2019) Hmm-based secure framework for optical fog devices in the optical fog/cloud network. J Opt Commun
20. Seth J, Nand P, Singh P, Kaur R (2020) Particle swarm optimization assisted support vector machine based diagnostic system for lung cancer prediction at the early stage. PalArch's J Archaeol Egypt/Egyptol 17(9):6202–6212
21. Sood SK, Singh KD (2019) Optical fog-assisted smart learning framework to enhance students' employability in engineering education. Comput Appl Eng Educ 27(5):1030–1042
22. Singh PD, Kaur R, Singh KD, Dhiman G, Soni M (2021) Fog-centric IoT based smart healthcare support service for monitoring and controlling an epidemic of Swine Flu virus. Inform Med Unlocked 26:100636
23. Singh KD, Sood SK (2020) Optical fog-assisted cyber-physical system for intelligent surveillance in the education system. Comput Appl Eng Educ 28(3):692–704
24. Jiang E, Wang L, Wang J (2021) Decomposition-based multi-objective optimization for energy-aware distributed hybrid flow shop scheduling with multiprocessor tasks. Tsinghua Sci Technol 26(5):646–663
25. Nabi S, Ibrahim M, Jimenez JM (2021) DRALBA: dynamic and resource aware load balanced scheduling approach for cloud computing. IEEE Access 9:61283–61297
26. Lee YS, Han TH (2021) Task parallelism-aware deep neural network scheduling on multiple hybrid memory cube-based processing-in-memory. IEEE Access 9:68561–68572
27. Yao F, Pu C, Zhang Z (2021) Task duplication-based scheduling algorithm for budget-constrained workflows in cloud computing. IEEE Access 9:37262–37272
28. Wang Y, Jiang X, Guan N, Guo Z, Liu X, Yi W (2020) Partitioning-based scheduling of OpenMP task systems with tied tasks. IEEE Trans Parallel Distrib Syst 32(6):1322–1339
29. Wang Y, Zuo X (2021) An effective cloud workflow scheduling approach combining PSO and idle time slot-aware rules. IEEE/CAA J Automatica Sinica 8(5):1079–1094
30. Alsadie D (2021) A metaheuristic framework for dynamic virtual machine allocation with optimized task scheduling in cloud data centers. IEEE Access 9:74218–74233
31. Wan L, Zheng W, Yuan X (2021) Efficient inter-device task scheduling schemes for multi-device co-processing of data-parallel kernels on heterogeneous systems. IEEE Access 9:59968–59978
32. Zhang J, Zhou X, Ge T, Wang X, Hwang T (2021) Joint task scheduling and containerizing for efficient edge computing. IEEE Trans Parallel Distrib Syst 32(8):2086–2100
33. Dai CQ, Li C, Fu S, Zhao J, Chen Q (2020) Dynamic scheduling for emergency tasks in space data relay network. IEEE Trans Veh Technol 70(1):795–807

Comparative Study of Machine Learning Models for Early Detection of Parkinson's

Mohammad Abdullah Tahir and Zamam Farhat

1 Introduction

Parkinson's disease is the second most common neurodegenerative disorder and the most common movement disorder after Alzheimer's [1], essential tremor (ET) [2], and dementia [3, 4]. We ignore the tremors, the shaking, the breaking of voice, punching, or flailing in sleep that happened [5, 6]. Like any other disease symptom, symptoms of PD don't happen overnight, which means it gets worse and worse over time. You barely notice the tremors or voice shaking in the beginning [7]. PD is a neurodegenerative disorder means the neuron of central nervous system decay with time. And due to loss of these neurons, the production of dopamine present in the brain also decreases about 70–80%. Dopamine is a neurotransmitter which is responsible for the movement and the motor functions of our body [8]. As the amount or density of dopamine-producing cells started to diminish, PD increases. Scientists are still trying to figure out what causes these cells that produce dopamine to die [9]. Due to the loss of neurons, people also lose the nerve endings that produce norepinephrine, the main chemical messenger of the sympathetic nervous system. Norepinephrine increases heart rate, pumping of blood from the heart, and blood pressure. Some non-movement effects of PD such as fatigue, difficulty in speaking, voice shaking, lower pitch while speaking, depression are due to loss of norepinephrine [10–12] There are more than 1 million in India that are affected by PD and 10 million people worldwide who suffer from this disease. It is estimated that around 4% of people with PD are diagnosed before 50 [7]. About 1900 patients per 10,000 are over 80 years. The symptoms of this disease evolve moderately and get nasty over time. Most of the time symptoms of PD appear at the age of 60 or after. People with PD often show symptoms related to Parkinsonism, i.e. hyperkinesia (i.e. lack of movement), bradykinesia (i.e. slowness

M. A. Tahir (✉) · Z. Farhat
Aligarh Muslim University, Aligarh, India
e-mail: abdullahtahir919@gmail.com

R. Agrawal et al. (eds.), *International Conference on IoT, Intelligent Computing and Security*, Lecture Notes in Electrical Engineering 982,
https://doi.org/10.1007/978-981-19-8136-4_5

51

of movement), rigidity (arms and neck), and rest tremors (lack of dopamine) [13, 14] These symptoms start with small tremors on one side of the body that won't affect the day-to-day activities of the patient and gradually progresses to both sides [5, 8]. That results in repeated tremors and continuous shaking. Usually of a limb, often their hand or fingers. As their movement started getting steady, the speed of doing things will increase like chewing off food or swapping clothes.

They'll still move and walk but very leisurely and sometimes even fall [15]. At this point, they'll need help. And as time go by, symptoms intensify [8]. The person with PD will need someone to be with them 24×7, to help them with daily activities and needs. They may not be able to stand because of the inflexibility of muscles in the leg or might even sit for that matter. Many of the patients are bedbound [14]. They are needed to be taken care of all the time. Parkinson's just does not only affect your body physically but internally as well. As a result of PD constipation, weaken the sense of smell, soft voice, and dream enactment can take place years before motor symptoms of Parkinson's [6, 11, 12, 16]. Regardless the sex, everyone's affected by this neurodegenerative disorder, but men have more tendency to get affected rather than women. The symptoms of PD have a different rate of progression among different individuals [7]. The cause of PD is still unknown, i.e. why the dopamine-producing neurons die [17]. But several factors appear to play a role namely:

- Environmental factors and exposures [18, 19].
- Genetics [20, 21].

As of today, there is no cure, treatment options vary and include medications and surgery. Doctors usually start with medications that would increase several amounts of dopamine but it might certainly have side effects, along with the prescriptions to control your tremors as well [22]. Levodopa, a naturally occurring chemical developed 30 years ago is said to be the most vigorous drug for Parkinson's therapy and controlling its symptoms for a while. It infuses within the blood vessels and travels to the brain and then converts into dopamine [23]. Earlier in the clinical trials, patients faced side effects like nausea and vomiting. Since most of the levodopa were absorbed inside the arteries before it reaches the brain and very few after crossing the blood barrier, a drug called Carbidopa was developed. It helps in passing Levodopa progressively through the blood barrier and into the brain and also reduces the side effects caused by Levodopa. Levodopa only helps in compressing down the symptoms such as Tremors, muscle rigidity, and slowness of movement however it does not stops the progression of the disease.

And in some cases, if Levodopa doesn't work or the patient is irresponsive to it, there is an option of Deep Brain Stimulation. In DBS, electrodes are embedded typically to only one side of the brain, but if the condition is serious, then on both side of it, which then connects to a generator (very much like a pacemaker) placed under the skin near to the abdomen, which sends an electrical impulses from time to time to control the symptoms. It is also not a cure neither does it slows the progression of PD. Speech problems like voice shaking also don't improve by this. And like other brain surgeries, DBS also carries a little bit of a risk of infection or seizures [24, 25].

Therapies help too [26, 27]. Here are some other medications used for the treatment of PD:

- Carbidopa-Levodopa Infusion [28]
- Dopamine-Agonists [29]
- Selegiline (MAO-B inhibitors) [30]
- Catechol-O-methyltransferase(COMT inhibitors) [31]
- Anticholinergics [32]
- Amantadine [33]

All these medications can work more efficiently if doctors can diagnose The PD in early stages.

2 Related Work

The early diagnosis of any disease is always better. And for the disease like Parkinson's, it is much better to diagnose in the early stage [13] so that the patient can get proper medication. In a country like India, where resources are insubstantial, it is good to have some methods to diagnose Parkinson's in the early stages efficiently. Researchers are using different machine learning techniques in the diagnosis of Parkinson's disease, and in the past decade, they get some groundbreaking results. In a study [4], researchers used several features like cerebrospinal fluid data, rapid eye movement. Then proposed a deep learning model and 12 machine learning models to diagnose Parkinson's in the early stages. Also, in a study [8], researchers achieved 93.84% accuracy using supervised machine learning. Gao, C., Sun, H., Wang, T.et, al present an article about which type of machine learning models can be better in the diagnoses of Parkinson's. In 2014, Yahia A. et al used Naive based and KNN classification algorithms using speech data sets for the diagnoses of Parkinson's. The Naive Bayes algorithm performed better than KNN with an accuracy of 93.3%. In 2013, Farhad S. proposed multi-layer perceptron for the early diagnoses of Parkinson's. This model performed with 93.22% of accuracy. In this paper, we have done a comparative analysis of logistics regression, support vector machine, random forest algorithm and artificial neural networks.

3 Methodology

3.1 Data Set and Their Analysis

The data set used for this study was acquired from an online data repository of the University of California, Irvine. This data set was created by the researcher of the University of Oxford Max Little with the collaboration of the Centre of National

Voice and Speech, Denver, Colorado [34]. The data set contains the reading of 23 different biomedical voice measurements of 31 persons, of which 23 persons have Parkinson's disease, and the remaining persons are healthy. Around one person has six recordings, and the total data set has 195 different readings. The purpose of this data set is to classify whether a person has Parkinson's disease or not based on the column name Status. Data set has 24 columns in which the first column contains the name and recording number of the person in ASCII. The column name "Status" has the health status of a person. The "Status" column is our target column. Data set has zero null values for each column. The features used for the training and testing for the machine learning models are the remaining 22 columns present in the data set [34] (Figs. 1, 2 and 3).

Fig. 1 Distribution of the entire data set in terms of status column

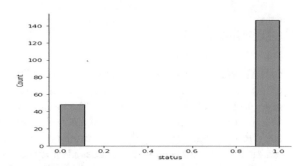

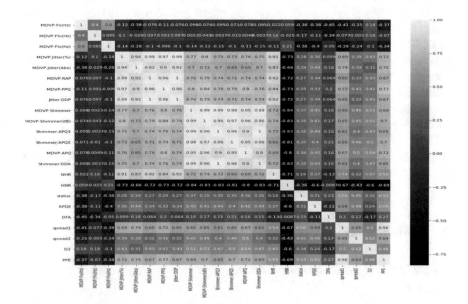

Fig. 2 Correlation between features used for the training and testing of machine learning models

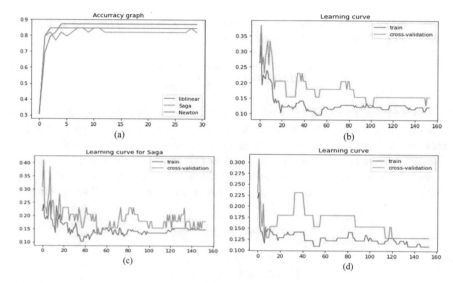

Fig. 3 **a** Accuracy graph of all logistic regression models **b** Loss versus number of examples curve for logistic regression with liblinear as a solver **c** Loss versus number of examples curve for logistic regression with saga as a solver **d** Loss versus number of examples curve for logistic regression Newton as a solver

Based on the "STATUS" column, as we move down the heatmap, if it is positively correlated, the value above it will be in an increasing method showing the features from negatively correlated to positively correlated. As we move down after passing that positively correlated feature, these traits become further negatively correlated to other features of the column that don't have any characteristics as same as the feature we are looking at. Likewise, if we look from left to right, we see that the features are showing the association from being negatively correlated to positively correlated but the values don't reduce much after a feature shows maximum correlation with itself. These values are still showing being positively correlated to other features of the STATUS column. Data pre-processing and splitting of data into a test and train data set have been done using Python script and Scikit-learn. All the features in the data set were normalized using the Standard Scaler method of Scikit-learn, which scales the distribution by subtracting the mean from all values then divide their difference by Standard Deviation. The entire data set was divided into train and test set using the test_train_split method of Scikit-learn. The train set consists of 156 rows of data, and the test set consists of 39 rows of data.

3.2 Machine Learning Models and Metrics for Evaluation

For this study, we consider 7 different machine learning algorithms. For the evaluation of these machine learning models, we used Accuracy, Precision, Recall, and F1-score as the evaluation metrics. Accuracy is the division of correct prediction upon total number of prediction made by the classification model.

$$\text{Accuracy} = \frac{\text{Correct Prediction}}{\text{Total Prediction}} \tag{1}$$

But accuracy is not a correct evaluation metrics to use when the given data is very imbalanced. It is always better to use Specificity and Sensitivity as the performance measure for imbalanced classification tasks. Specificity is the number of true negative (TN) divided by the sum of true negative (TN) and false positives (FP), while Sensitivity is the number of true positives divided by the sum of true positives (TP) and false negatives (FN). And F1-score is the harmonic mean of recall and precision.

$$\text{Sensitivity} = \frac{\text{TP}}{\text{TP} + \text{FN}} \tag{2}$$

$$\text{Specificity} = \frac{\text{TN}}{\text{TN} + \text{FP}} \tag{3}$$

3.2.1 Logistic Regression

Logistic regression is a supervised machine learning algorithm used for classification problems. It is one of the most used and simple machine learning algorithms [35]. We use 3 different versions of logistic regression. For the first version, we used liblinear as the solver, for the second version, we used saga and for the last one, we used newton-CG. Then, we plotted a graph for the number of iteration versus accuracy for all the versions of the model. After 11 iterations, the accuracy of all versions became constant.

3.2.2 Support Vector Machine

Support vector machine is the learning algorithm proposed by Vladimir Vapnik [36] in 1997. It is one of the powerful machine learning algorithm. It can also be use in nonlinear classification problems with different kernels. In addition to this, support vector machine can be also used in regression problems. The version used in regression problems is called support vector clustering. This version was proposed by Hava Siegel Mann and Vladimir Vapnik.

For this study, we used three different version of SVM. First is linearSVC, then with Linear kernel [37], and for the last version, we used RBF as a kernel. We ran all the SVM model for 100 iteration and plotted accuracy versus number of iteration graph (Fig. 4a). Then, after 40 iterations accuracy of SVM with RBF kernel

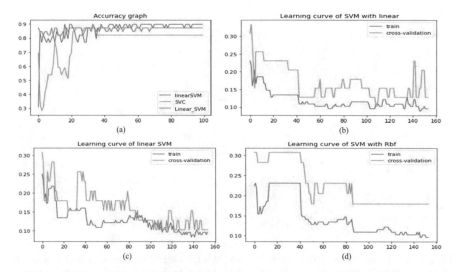

Fig. 4 **a** Accuracy graph of all SVM models **b** Loss versus number of examples curve of SVM with linear kernel **c** Loss versus number of examples curve of linear SVM **d** Loss versus number of examples curve of SVM with Rbf as the kernel

became constant. For both linearSVC and SVM with linear kernel, it takes around 70 iterations.

3.2.3 Decision Tree Classification Algorithm

Decision tree classifier is a supervised learning approach. It can be used for both classification and regression problems. It is preferred to use it for classification problems [38]. We use three different decision tree classifiers. First classifier we uses $\log_2$ number of estimators, the second classifier uses square root number of times of number of estimators given to the model and the last uses the same number of estimators given to the model. Then, we plotted accuracy versus max_depth and max_leaf_node of all versions (Fig. 5).

3.2.4 Random Forest Classifier

Random forest classifier is learning technique that is based on multiple decision tree. As the name suggests, random forest classifier is a set of many individual decision trees. Deep decision tree can be prone to overfitting that is the reason we prefer random forest because it control overfitting. For this study, we use three different versions of random forest classifier. The first random forest tree classifier uses $\log_2$ number of estimators, the second classifier uses square root number of times of

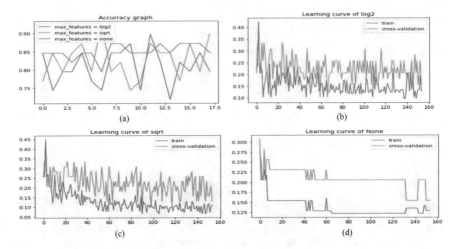

Fig. 5 **a** Accuracy graph of all decision tree models **b** Loss versus number of examples curve of decision tree model with maximum features equal to log2 of original features **c** Loss versus number of examples curve of decision tree model with maximum features equal to square root of original features and **d** Loss versus number of examples curve of decision tree model with maximum features equal to original features

number of estimators given to the model, and the last uses the same number of estimators given to the model (Fig. 6).

3.2.5 Artificial Neural Network

Usually artificial neural network consists of input layer, hidden layers, and output layer. The neural network implementation in this study has been done using keras [33]. The detail of every layer used in the neural network is in Table 1. We use dense layer with 256 features followed by batch normalization with 256 features. Next dense layer has 128 features followed by batch normalization layer with 128 features. Next up we have dense layer with 64 features followed by drop out by 0.5. Next dense layer has only feature with sigmoid as an activation function. Other than last, all previous layers have ReLU as activation function. And we use drop out to reduce overfitting (Figs. 7 and 8).

4 Result and Discussion

The main objective of this paper was to compare several machine learning techniques. Specifically, logistic regression, support vector machine, decision tree classifier, random forest, and artificial neural networks for the early diagnosis of Parkinson's disease.

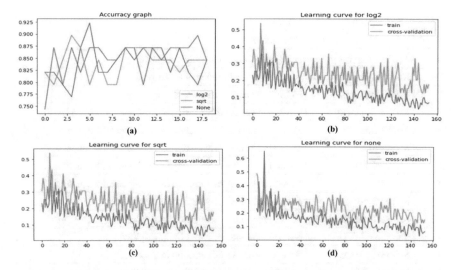

Fig. 6 a Accuracy graph of all random forest models **b** Loss versus number of examples curve of random forest model with log2 number of estimators given to the model. **c** Loss versus number of examples curve of random forest model with square root number of estimators given to the model. **d** Loss versus number of examples curve of random forest model with same number of estimators given to the model

4.1 Logistic Regression

Confusion metric is the metric where all the prediction made by machine learning models are classified in term of their correctness. y-axis represents the true labels of the data and x-axis represents the predicted labels of the data (Figs. 9, 10, 11 and 12).

Logistic regression with Newton-cg outperformed all of the versions on both test and train data with 87% and 89% of accuracy, respectively.

4.2 Support Vector Machine

We have 3 version of SVM and the confusion metrics of first version of SVM are given below (Figs. 13, 14 and 15)

LinearSVC outperformed all of the versions of SVM on both test and train data with 90.3% and around 89.73% of accuracy, respectively.

Table 1 Description of all the features

S. no	Name of the feature	Description
1	MDVP:Fo (Hz)	Average vocal fundamental frequency
2	MDVP:Fhi (Hz)	Maximum vocal fundamental frequency
3	MDVP:Flo (Hz)	Minimum vocal fundamental frequency
4	MDVP:Jitter (%)	Several measures of variation in fundamental frequency
5	MDVP:Jitter (Abs)	Several measures of variation in fundamental frequency
6	MDVP:RAP	Several measures of variation in fundamental frequency
7	MDVP:PPQ	Several measures of variation in fundamental frequency
8	Jitter:DDP	Several measures of variation in fundamental frequency
9	MDVP:Shimmer	Several measures of variation in amplitude
10	MDVP:Shimmer (dB)	Several measures of variation in amplitude
11	Shimmer:APQ3	Several measures of variation in amplitude
12	Shimmer:APQ5	Several measures of variation in amplitude
13	MDVP:APQ	Several measures of variation in amplitude
14	Shimmer:DDA	Several measures of variation in amplitude
15	NHR	Two measures of ratio of noise to tonal components in the voice
16	HNR	Two measures of ratio of noise to tonal components in the voice
17	RPDE	Two nonlinear dynamical complexity measures
18	D2	Two nonlinear dynamical complexity measures
19	DFA	Signal fractal scaling exponent
20	Spread1	Three nonlinear measures of fundamental frequency variation
21	Spread2	Three nonlinear measures of fundamental frequency variation
22	PPE	Three nonlinear measures of fundamental frequency variation

```
Layer (type)                    Output Shape        Param #
=================================================================
dense_12 (Dense)                (None, 256)         5888
_____
batch_normalization_6 (Batch    (None, 256)         1024
_____
dense_13 (Dense)                (None, 128)         32896
_____
batch_normalization_7 (Batch    (None, 128)         512
_____
dense_14 (Dense)                (None, 64)          8256
_____
dropout_3 (Dropout)             (None, 64)          0
_____
dense_15 (Dense)                (None, 1)           65
=================================================================
Total params: 48,641
Trainable params: 47,873
Non-trainable params: 768
```

Fig. 7 Detail summary of proposed artificial neural networks

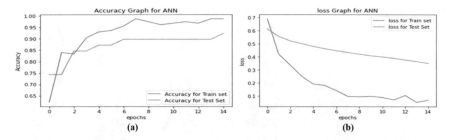

Fig. 8 **a** Accuracy graph of ANN **b** Loss graph of ANN

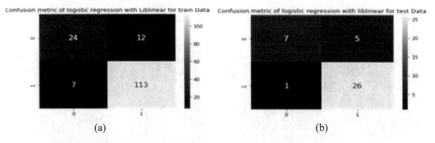

Fig. 9 Confusion matrix with liblinear as a solver for **a** train data **b** test data

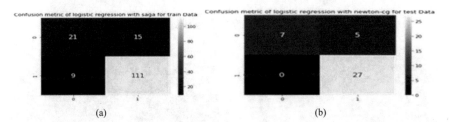

Fig. 10 Confusion matrix with Saga as a solver for **a** train data **b** test data

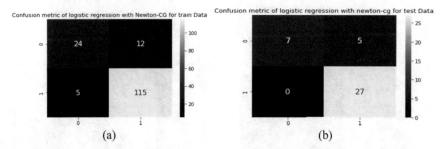

Fig. 11 Confusion matrix with newton-CG as a solver for **a** train data **b** test data

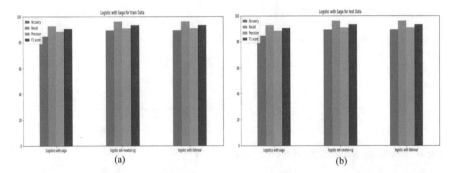

Fig. 12 Result of logistic regression models on **a** train data **b** test data

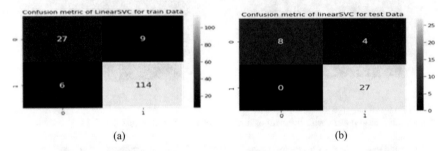

Fig. 13 Confusion matrix of SVM by LinearSVC for **a** train data **b** test data

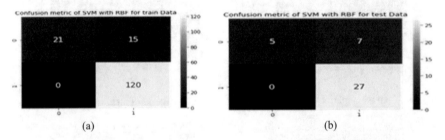

Fig. 14 Confusion matrix of SVM by RBF kernel on **a** train data **b** test data

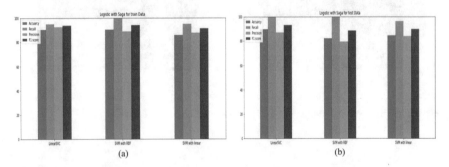

Fig. 15 Results of SVM models on **a** train data **b** test data

4.3 Decision Tree

We have 3 version of decision tree classifier with different depth, number of leaf nodes, and maximum features (Figs. 16, 17, 18, 19 and 20).

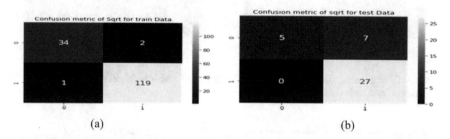

Fig. 16 Confusion matrix with depth and number of leaf nodes of 6 and square root of parameters for **a** train data **b** test data

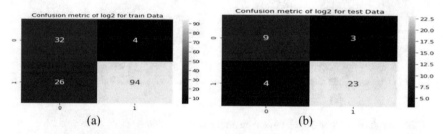

Fig. 17 Confusion matrix with depth and number of leaf nodes of 3 and log of total parameters for **a** train data **b** test data

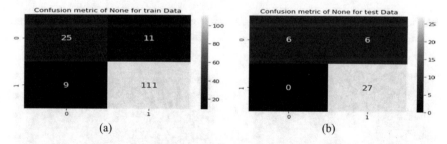

Fig. 18 Confusion matrix with depth and number of leaf nodes of 2 and maximum features of parameters for **a** train data **b** test data

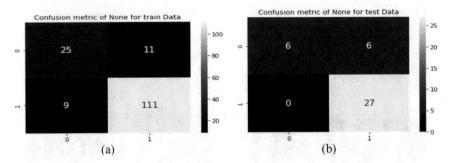

Fig. 19 Confusion matrix with depth and number of leaf nodes of 2 and maximum features of parameters for **a** train data **b** test data

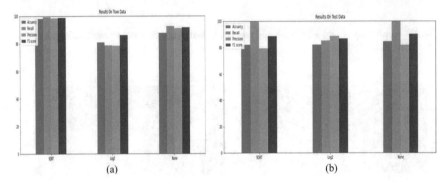

Fig. 20 Result of decision tree classifiers on **a** train data **b** test data

4.4 Random Forest

We have 3 version of random forest classifier with different number of estimators and maximum features (Figs. 21, 22, 23, 24, 25 and 26).

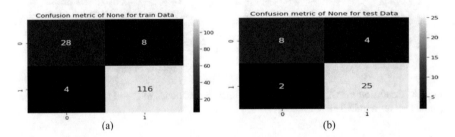

Fig. 21 First version confusion matrix with number of estimators of 1 max features of square root of total parameters for **a** train data **b** test data

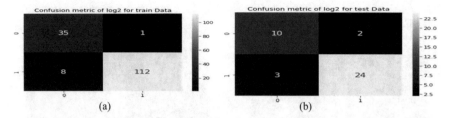

Fig. 22 Second version confusion matrix with number of estimators of 2 and max features are log of total parameters for **a** train data **b** test data

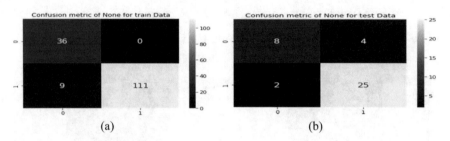

Fig. 23 Second version confusion matrix with number of estimators of 2 and max features equals to total parameters for **a** train data **b** test data

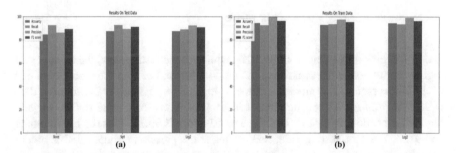

Fig. 24 Results of random forest classifiers on **a** train data **b** test data

4.5 Artificial Neural Network

From Table 2, it can be inferred that the decision tree has the least accuracy while classifying both the training and testing data and that of 87.17 and 84.61%. The training data in the decision tree has a precision of 90.09%, while in the testing data, it is only 81.81%. Moreover, the recall value of the decision tree for the training set is 92.5% and the recall value of the testing data for all the algorithms beside random forest is 100%. F1-score of the decision tree has more value in training data than in testing data, i.e. 91.73% in training data and 90% in testing data. Nevertheless, the accuracy of logistic regression for the training data is moderately better than

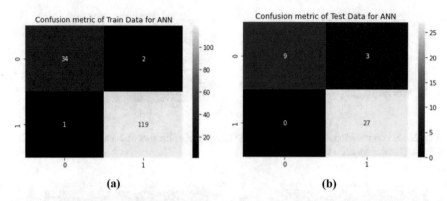

Fig. 25 Confusion metrics for ANN for **a** train data **b** test data

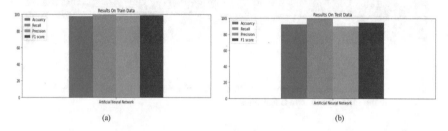

Fig. 26 Result of ANN on **a** train data **b** test data

that of a decision tree that is, of 89.10%, while for the testing data, the accuracy of logistic regression is the same as of the random forest, i.e. both the model obtained 87.17% accuracies. The precision value of logistic regression for the training data is 90.58%, and for the testing data, it is 84.37%. The recall value for the logistic regression of training data is 95.83%. The F1-score is 93.11% for the training data and 91.52% for the testing data, making logistic regression a more reliable model than the decision tree. The accuracy of support vector machine (or SVM solely) is 90.38% for the training data and 89.74% for the testing data which is a little more than the previously introduced models. SVM is more precise than both logistic regression and decision tree producing precision value for the training data equals 92.68% and testing data equals 87.09%. The recall value of training data for SVM is 95%. The F1-Score for the training and testing data of the SVM is 93.82% and 93.10%, respectively. Random forest's accuracy for the training data as well as the testing is 94.23 and 87.17%, sequentially which is more than the previously defined models for the training data but for testing data, the accuracy of SVM is better than the random forest. The precision value of the random forest for training data is 96.13%, and for testing data, it equals 92.30%. The recall value is 93.33% for the training data and 88.88% for the testing data. Random forest having the F1-Score of 96.13 and 90.56% for the training and testing data, respectively, which is the best F1-Score yet specified in Table 2. However, ANN was able to attain more accuracy

Table 2 Results from the models

Model	Training data				Testing data			
	Accuracy	Precision	Recall	F1-Score	Accuracy	Precision	Recall	F1-Score
Logistic regression	89.10	90.58	95.83	93.11	87.17	84.37	100	91.52
SVM	90.38	92.68	95	93.82	89.74	87.09	100	93.10
Decision tree	87.17	90.09	92.5	91.73	84.61	81.81	100	90
Random forest	94.23	96.13	93.33	96.13	87.17	92.30	88.88	90.56
ANN	98.07	98.34	99.16	98.75	92.30	90	100	94.73

for the training as well as testing data than any previously considered models which are 98.07% and 92.30%, respectively. The precision value of ANN is significantly more than any model in training data and that of 98.34% but the precision value for testing data of ANN is 90% which is much better than logistic regression, SVM, and decision tree but the random forest has more precision in testing data than ANN. Additionally, the recall value of the ANN model was also noticeably higher in both training data and testing data than the preceding values, i.e. the recall value for both is approximately 100%. Finally, the F1-Score of ANN for the training data is 98.75%, and for the testing, data is 94.73% which is much more for both training and testing data in the earlier considered models. This shows that in the end, ANN results in more reliable performance than logistic regression, SVM, decision tree, and random forest on Parkinson's disease.

5 Conclusion

The natural development of Parkinson's disease is vary, although it is frequently faster in late-onset individuals. Because there are no definitive diagnostic tests for it, practitioners must have a good understanding of the clinical manifestations of PD in order to distinguish it from other conditions. Poor assessment of symptoms has an impact on the cost of care for individuals with the condition in society; developing integrated techniques to evaluate symptoms will aid in future identification and implementation of innovative therapeutic approaches.

In this analysis, we used several supervised machine learning approaches and we found that that artificial neural network outperformed every one of them. This can be extremely beneficial in areas where there are a limitation of medical institutes and specific specialists. Each classification algorithm was built and tested using a training set sample from the data set. Thereafter, on test data, all of the algorithms perform well, with accuracy ranging from 84 to 93%. However, artificial neural networks outperformed all other algorithms in the study, with test data accuracy of 92.31%, precision of 90%, recall of 100%, and F1-score of 94.73%, and training data accuracy of 98.07%, precision of 98.34%, recall of 99.16%, and F1-score of 98.75%.

In our opinion, early detection of any illness, not only Parkinson's, can help doctors treat patients more effectively. In the case of Parkinson's disease, this might be extremely valuable to both physicians and patients.

We only performed on these algorithms, many more complicated neural networks and other advanced machine learning methods will be available in future, and it is indeed expected that these additional algorithms will be preferred for constructing a more exact model for the early detection of Parkinson's disease.

References

1. Fahn S (2003) Description of Parkinson's disease as a clinical syndrome. Ann-NY Acad Sci 991:1–14
2. Lesage S, Brice A (2009) Parkinson's disease: from monogenic forms to genetic susceptibility factors. Hum Mol Genet 18(R1):R48–R59
3. Lau D, Lonneke ML, Breteler MMB (2006) Epidemiology of Parkinson's disease. Lancet Neurol 5(6): 525–535
4. Schapira AH (1999) Parkinson's disease. BMJ 318(7179):311–314
5. Sveinbjornsdottir S (2016) The clinical symptoms of Parkinson's disease. J Neurochem 139:318–324
6. Peng CYJ, Lee KL, Ingersoll GM (2002) An introduction to logistic regression analysis and reporting. J Educ Res 96(1):3–14
7. Hawkes CH, Del Tredici K, Braak H (2010) A timeline for Parkinson's disease. Parkinsonism Relat Disord 16(2):79–84
8. Fahn S (2003) Description of Parkinson's disease as a clinical syndrome. Ann-NY Acad Sci 991:1–14
9. Ben-Shlomo Y (1996) How far are we in understanding the cause of Parkinson's disease. J Neurol Neurosurg Psychiatry 61(1):4
10. Aminoff MJ (1994) Treatment of Parkinson's disease. West J Med 161(3):303
11. Rao SS, Hofmann LA, Shakil A (2006) Parkinson's disease: diagnosis and treatment. Am Fam Physician 74(12):2046–2054
12. Ricciardi C, Amboni M, De Santis C, Ricciardelli G, Improta G, Iuppariello L, D'Addio G, Barone P, Cesarelli M (2019) Classifying different stages of Parkinson's disease through random forests. In: Mediterranean conference on medical and biological engineering and computing. Springer, Cham, pp 1155–1162
13. Bind S et al (2015) A survey of machine learning based approaches for Parkinson disease prediction. Int J Comput Sci Inf Technol (IJCSIT) 6(2):1648–1655
14. Gao X, Fan L, Xu H (2018) Multiple rank multi-linear kernel support vector machine for matrix data classification. Int J Mach Learn Cybern 9(2):251–261
15. Chaudhuri KR, Naidu Y (2008) Early Parkinson's disease and non-motor issues. J Neurol 255(5):33–38
16. Jankovic J (2008) Parkinson's disease: clinical features and diagnosis. J Neurol Neurosurg Psychiatry 79(4):368–376
17. Choudhary S, Vaya S, Parvesh PS (2016) Understanding essential tremors: a review. International J Res Sci Innov 3(5):21–22
18. Noyce AJ, Bestwick JP, Silveira-Moriyama L, Hawkes CH, Giovannoni G, Lees AJ, Schrag A (2012) Meta-analysis of early nonmotor features and risk factors for Parkinson disease. Ann Neurol 72(6): 893–901

19. Espay AJ, LeWitt PA, Kaufmann H (2014) Norepinephrine deficiency in Parkinson's disease: the case for noradrenergic enhancement. Mov Disord 29(14):1710–1719
20. Vazey E, Aston-Jones G (2012) The emerging role of norepinephrine in cognitive dysfunctions of Parkinson's disease. Front Behav Neurosci 6:48
21. Cortes C, Vapnik V (1995) Support-vector networks. Mach Learn 20(3):273–297
22. Connolly BS, Lang AE (2014) Pharmacological treatment of Parkinson disease: a review. JAMA 311(16):1670–1683
23. Deng H, Wang P, Jankovic J (2018) The genetics of Parkinson disease. Ageing Res Rev 42:72–85
24. Surmeier DJ, Guzmán JN, Sánchez-Padilla J, Goldberg JA (2010) What causes the death of dopaminergic neurons in Parkinson's disease? Prog Brain Res 183:59–77
25. Volkmann J (2004) Deep brain stimulation for the treatment of Parkinson's disease. J Clin Neurophysiol 21(1):6–17
26. Lang AE, Lozano AM (1998) Parkinson's disease. N Engl J Med 339(16):1130–1143
27. Waters C (2000) Catechol-O-Methyltransferase (COMT) inhibitors in Parkinson's disease. J Am Geriatr Soc 48(6):692–698
28. Katzenschlager R, Sampaio C, Costa J, Lees A (2002) Anticholinergics for symptomatic management of Parkinson s disease. Cochrane Database Syst Rev
29. Schwab RS, England AC, Poskanzer DC, Young RR (1969) Amantadine in the treatment of Parkinson's disease. JAMA 208(7):1168–1170
30. Bucher ML, Barrett CW, Moon CJ, Mortimer AD, Burton EA, Greenamyre JT, Hastings TG (2020) Acquired dysregulation of dopamine homeostasis reproduces features of Parkinson's disease. NPJ Parkinson's Dis 6(1):1–13
31. Samii A, Nutt JG, Ransom BR (2004) Parkinson's disease. Lancet 363(9423):1783–1793. https://doi.org/10.1016/S0140-6736(04)16305-8.PMID15172778.S2CID35364322
32. Francois Chollet K (2015) http://keras.io. 2017
33. Dezsi L, Vecsei L (2017) Monoamine oxidase B inhibitors in Parkinson's disease. CNS Neurol Disord-Drug Targets (Formerly Curr Drug Targets-CNS Neurol Disord), 16(4):425–439
34. Davie CA (2008) A review of Parkinson's disease. Br Med Bull 86(1):109–127
35. Olanow CW, Kieburtz K, Odin P, Espay AJ, Standaert DG, Fernandez HH, Vanagunas A, Othman AA, Widnell KL, Robieson WZ, Pritchett Y (2014) Continuous intrajejunal infusion of levodopa-carbidopa intestinal gel for patients with advanced Parkinson's disease: a randomised, controlled, double-blind, double-dummy study. Lancet Neurol 13(2):141–149
36. Little M, McSharry P, Hunter E, Spielman J, Ramig L (2008) Suitability of dysphonia measurements for telemonitoring of Parkinson's disease. Nat Proc, pp 1–1
37. Luquin MR, Scipioni O, Vaamonde J, Gershanik O, Obeso JA (1992) Levodopa-induced dyskinesias in Parkinson's disease: clinical and pharmacological classification. Mov Disord Official J Mov Disord Soc 7(2):117–124
38. Chaudhuri KR, Healy DG, Schapira AH (2006) Non-motor symptoms of Parkinson's disease: diagnosis and management. Lancet Neurol 5(3):235–245
39. Seppi K, Weintraub D, Coelho M, Perez-Lloret S, Fox SH, Katzenschlager R, Hametner EM, Poewe W, Rascol O, Goetz CG, Sampaio C (2011) The Movement Disorder Society evidence-based medicine review update: treatments for the non-motor symptoms of Parkinson's disease. Mov Disord 26(S3):S42–S80

Mining Repository for Module Reuse: A Machine Learning-Based Approach

Preeti Malik and Kamika Chaudhary

1 Introduction

The research area of Mining Software Repositories (MSR) concerned with analysing the data stored in repositories. The purpose of this analysis is to find out interesting facts about the data, software system, and projects. Component repositories have been playing a major role in various research areas ranging from software distribution to development of applications for more than two decades [1]. All real free and open source software distributions are sorted out around huge repositories of software components. In this paper, the terms component and module are utilized conversely. In spite of various phrasings, and a wide assortment of solid organizations, every one of these archives utilizes metadata that permits recognizing modules, with their forms and interdependencies. These modules can be reused for the undertakings in future. Our paper focuses on the issue of reusing appropriate modules.

Software Repositories (SR) built during software evolution have an abundance of important data with respect to the developmental history of a product venture and can be utilized by designers to deal with their undertaking [2–4]. Many researchers have shown the usefulness of software repositories during software development. It can be used to identify hidden code dependencies suggested by Gall et al. [5], can assist management in structuring reliable software systems by forecasting bugs and effort suggested by Graves et al. [6] and Port et al. [7], and can assist developers in understanding large systems suggested by Chen et al. [8].

Development in various domains has been growing rapidly since the inception of computers. However, there is no automated repository available with any of the

P. Malik
Department of Computer Science and Engineering, Graphic Era University, Dehradun, India

K. Chaudhary (✉)
Department of Computer Science, M.B. Government P.G. College, Haldwani, Nainital, India
e-mail: kamikaduhoon@gmail.com

© The Author(s), under exclusive license to Springer Nature Singapore Pte Ltd. 2023
R. Agrawal et al. (eds.), *International Conference on IoT, Intelligent Computing and Security*, Lecture Notes in Electrical Engineering 982,
https://doi.org/10.1007/978-981-19-8136-4_6

biggest software development companies which can ease their software developers work. Every time one has to rely on digging the brain, hunting through the files on the computer or hunting through a pool of reports submitted to the end users. A very tedious task indeed engineers normally design modules which can be reused in the applications which are developed in a company but they too are not properly documented if the team breaks. In that case, none of software developer or software architect would know what the modules were designed for. The authors propose a machine learning-based approach to facilitate software developers to ease their work. The proposed model first reads the folders of the developed applications and records the created files. Same steps will be followed on all the machines of the developers and data will be collected on the cloud. The raw data will be classified on the basis of applications and modules which are marked present and absent.

Rest of the paper is organized as follows; Sect. 2 discusses types of software repositories. Section 3 includes existing research done in this field. Section 4 consists of the proposed approach for creating and mining repository. Section 5 describes a mathematical model of the proposed approach; the next section shows an experimental set-up then Sect. 7 analyses various algorithms and their accuracy. Section 8 concludes the work.

2 Software Repositories

The SRs are a capacity area kept up on the web or disconnected by a few programming improvement associations where programming bundles, source code, bugs, and numerous other data identified with programming and their improvement procedure are kept up. Because of open source, the quantity of these archives and its uses is expanding at a fast rate. Anybody can separate numerous kinds of information from here, contemplate them, and can roll out improvements as indicated by their need.

2.1 Historical Repositories

These repositories consist of a heterogeneous and enormous measure of programming building information created between significant lots of time. These are called Source Control Repositories (SCR). SCR documents the advancement history of an undertaking. These stores record every one of the progressions done in the source code and its metadata about each change amid upkeep, e.g. engineer name that rolled out the improvement, the moment at which modification was done and a small message depicting the expectation of progress, and the progressions performed. The most regularly accessible, available, and utilized archives for programming ventures are source control vaults. Some generally utilized source control stores are perforce, clear case, concurrent versions system, subversion, and so on.

2.2 Run Time Repositories

The storehouses consist of identified data with the execution and the use of an application at a solitary or numerous diverse sending site, for example, arrangement logs, deployment logs (DL). Programming arrangement is entire arrangements of exercises that make a product item accessible for utilizing [9]. These exercises can happen at the maker or buyer or the two sides. The data identified with the execution of a specific arrangement of a product venture or heterogeneous organizations of similar activities is recorded in this kind of stores [10].

2.3 Code Repositories

CR storehouses are kept up by the gathering of the source code of countless tasks. Some of the examples of CR are Sourceforge.net, GitHub, and Google code.

3 Related Work

Designers can use previously mentioned archives by applying mining strategies to them to computerize and enhance the extraction of data to pick up learning to a deal with their activities. Mining programming archives are another rising field and spotlight on extricating of both basic and important data with respect to programming qualities from heterogeneous surviving programming storehouses [11, 12]. These kinds of archives are mined to separate the concealed actualities by various benefactors with respect to achieving the diverse targets. The utilization of information mining is increasing constant fame in programming designing situations because of agreeable outcomes since a decade ago [13, 14] and its application incorporate zone as bugs forecast [15], Co-advancement of generation and test code [16], affect investigation [17], exertion expectation [18, 19], similitude examination [20], forecast of programming structural change [21], programming knowledge [10], likewise used to diminish complexities in Global Software Development [22], and some more.

Poncin [23] first utilized process mining to find process delineates consolidating occasions from various stores. Sunindyo [24] dissected bug revealing arrangement of Red Hat Enterprise constrained and performed conformance checking for process change. Gupta and Surekha [25] mined the bug report history of open source venture, for example, Firefox and Mozilla to consider process wasteful aspects. Gupta et al. [26] investigated different storehouses, for example, ITS, peer code survey subsystem and variant control subsystem to enhance programming imperfection determination process. So also, an examination was led by Zimmermann et al. [27] on Windows working framework to arrange likely explanations of revived bugs.

Ganesha et al. [28] proposed an approach for source code mining. The key thought is to investigate the mining undertaking to distinguish and expel the unimportant segments of the source code, preceding running the mining errand. Creators accomplish all things considered a 40% decrease in the assignment calculation time.

Corbellini et al. [29] examine a way to deal with find programming web benefits that work on diagrams with client benefit connections and utilize lightweight topological measurements to survey benefit similitude. At that point, "socially" comparable administrations, which are resolved abusing express connections and mining understood connections in the chart, are grouped by means of model-based bunching to at last guide revelation.

Tiwari et al. [30] proposed a stage called Canodia for building and sharing MSR devices. Candoia stage gives information extraction devices to curating custom data sets for client undertakings, and information deliberations for empowering uniform access to MSR curios from dissimilar sources, which makes applications versatile and adoptable crosswise over assorted programming venture settings of MSR analysts and specialists.

4 Methodology

Some product designing examinations target hundreds or thousands of activities in their assessments [31-33]. For these examinations, projects can be chosen exclusively based on meta-information and straightforward measurements, for example, the programming dialect utilized. A variety of tools exist to assist scientists with choosing projects in this way. The proposed approach makes use of prediction modelling algorithm to identify the modules that already exist and the developer can reuse them. This process saves the time and cost of the development. The proposed approach consists of three phases; creation of repository, train and test the system and last phase is to predict module using prediction algorithms (Fig. 1).

Repository Creation: Proposed approach starts by importing the existing projects and their meta-data. In this step, a log file is created for each existing project. This log file contains project structure and various metrics. Mining algorithms then scan the log file and identify the modules from project meta-data (Fig. 2).

The output of the first phase is a repository file. From this repository file, we have created a module matrix. Suppose we have n projects in a repository. Each project has a structure and log in repository. Each column in module matrix is corresponding to a project and each row is corresponding to a module if ith module is present in jth project then place 1 in ijth cell otherwise a 0 value is placed in that cell. In this way, we have created a module matrix of 14*21 dimensions.

Train and Test: The data we get from first phase is used for training and testing. 70% is used for training and 30% is used for testing the algorithm.

Prediction Phase: This phase starts with an input of a new requirement or new project meta-data. It then reads this project meta-data and performs prediction on the received output and suggests the module required for the new project and the old

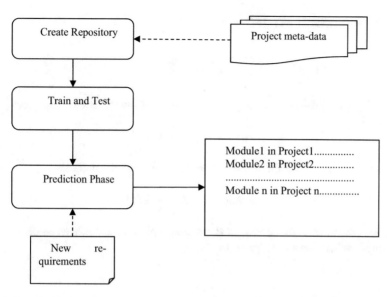

Fig. 1 Proposed approach

Steps in Repository Creation
Step 1: Start.
Step 2: Import meta-data for the existing projects and create log file.
Step 3: Identify the modules in each project using mining algorithm.
Step 4: Create a module matrix of dimension m*n
Step 4.1: if i^{th} module is present in j^{th} project then place 1 in ij^{th} cell
Step 4.2: else Place a 0 value in ij^{th} cell
Step 5: Stop.

Fig. 2 Repository creation phase

project where it is located in the repository. The proposed process is depicted below in the form algorithm (Fig. 3).

Steps in Prediction
Step 1: Start.
Step 2: Get a new requirement.
Step 3: Get the proposed meta-data and assign 1 to the modules required.
Step 4: Train and test using prediction algorithms on the model.
Step 5: Perform prediction on the input using the received output.
Step 6: Stop.

Fig. 3 Prediction phase

5 Mathematical Model

$$\forall P_n = \sum_{i=1}^{n} \text{module}_i \tag{1}$$

Here, $\forall P_n$ is the project repository module$_i$ which is the modules in a project

$$\{\text{module}_j\} = m \in P \tag{2}$$

Here, m is the module in the Project P module$_j$ which is the modules

$$\{\text{module}_p\} = ML(m, \forall P_n) \tag{3}$$

Here, m is the module in the Project P module$_p$ which is the modules predicted after applying machine learning algorithm.

6 Experimental Set-Up

In our experiments, the forecast calculation is to recognize wrong and right programming modules. Keeping in mind the end goal to do as such, we have made a 70% or 30% stratified split up of the first data sets into preparing and test data sets. Since the data sets utilized are generally substantial, early ceasing can be connected. Approval data sets are then required, and accordingly, one third of the preparation data sets is separate for validation purposes. To compare the results, state-of-the-art classification techniques K-nearest neighbour (KNN), logistic regression (LR), linear discriminant analysis (LDA), decision tree classifier (DTC), support vector machine (SVM), Naive Bayes (NB), and random forest (RF) are implemented in Python language. Following Fig. 4 shows mean, standard deviation, and min parameters.

Figure 5 shows the output received after applying predictive modelling to the data set. It has been observed that the modules required for a project, if present in another module are displayed. As known to all, **admin** is the most common module in many projects. In the proposed algorithm, if a module is found in more than one projects, then the most suitable module is predicted base on the feature match.

The training set contains the modules with their project names and testing set is the set of modules required for new projects. It is observed that if a required module is present in any of the project already done by the company, it is predicted by the algorithm. Health monitoring has modules like online query, FAQ, online Form which are required for Online_Medical_Assitant and Job_Assistant. Similarly, Online_Course and Online_Booking have modules Admin, Reports, FAQ, Query, Online Application which are required for Speedway and Job_Assitant projects taken up by the company.

Fig. 4 Mean, standard deviation, min and max

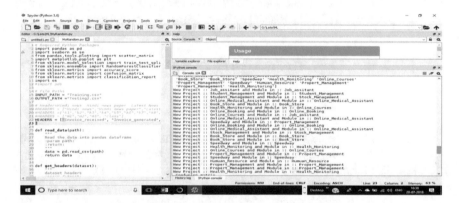

Fig. 5 Output of prediction algorithm

7 Performance Analysis

The performance of the proposed approach is evaluated by using the following statistical measures. The True Positive (TP) denotes the correctly identified module; P indicates the total identified modules. The positive (P) can also be calculated as the sum of True Positive (TP) and False Negative (FN). False Negative can be defined as an incorrectly identified module. Using the above-mentioned parameters, the Sensitivity can be evaluated as mentioned in Eq. (4). True Positive Rate (TPR) or Sensitivity is described as the ratio of the identified True Positive legal words and can be evaluated as:

$$\text{TPR} = \text{TP}/\text{P} = \text{TP}/(\text{TP} + \text{FN}) \tag{4}$$

True Negative Rate (TNR) or Specificity is described as the ratio of the true negatives modules and is described in Eq. (5):

$$TNR = TN/N = TN/(TN + FP) \tag{5}$$

where TN denotes True Negatives means correctly rejected modules, N indicates total rejected negative modules and can be calculated as the sum of TN and FP, FP represents False Positive that is incorrectly rejected modules. Using the above-calculated parameters, accuracy can be evaluated using the formula in Eq. (6):

$$Acc = (TP + TN)/(TP + FP + TN + FN) \tag{6}$$

To additionally assess the execution of the proposed approach, the review and exactness is likewise ascertained. Precision is the level of the recovered records which are really pertinent to the query, recall is the level of the important reports that are actually recovered [34-35]. Recall and precision are calculated using Eqs. (7) and (8):

$$Precision = TP/(TP + FP) \tag{7}$$

$$Recall = TP/(TP + FN) \tag{8}$$

The F1-score can be deciphered as a weighted normal of the precision and recall, where a F1-score achieves its best an incentive even from a pessimistic standpoint score at 0. The formula for the F1-score is given in Eq. (9) (Table 1).

$$F1 = 2^*(precision^*recall)/(precision + recall) \tag{9}$$

The process discussed above is applied by using the above-mentioned prediction algorithms. Following Fig. 6 presents the accuracy of each algorithm. RF algorithm gives most accurate results.

8   Conclusion

Reusability is a very good concept while developing software from the scratch. Software repositories can ease the work of development by reusability. Every time one has not to rely on digging the brain, hunting through the files on the computer or hunting through a pool of reports submitted to the end users. A very tedious task indeed, engineers normally design modules which can be reused in the applications which are developed in a company but they too are not properly documented if the team breaks no one would know what the modules were designed for. This paper proposed an approach to provide a solution to this problem. The approach moves

Table 1 Precision, recall, F1-score, and support for random forest algorithm

	Precision	Recall	F1-score	Support
Employee_Managment	1.00	1.00	1.00	1
Health_Management	0.50	0.50	0.50	2
Hotel_Booking	1.00	1.00	1.00	2
Inventory	1.00	1.00	1.00	1
Library_Management	1.00	1.00	1.00	4
Online_Toll	1.00	1.00	1.00	3
Online_Training	0.00	0.00	0.00	2
Placement_assitant	0.50	0.33	0.40	3
Real_Estate_Management	1.00	1.00	1.00	3
Student_Management	1.00	1.00	1.00	3
avg / total	0.81	0.79	0.80	24

Fig. 6 Accuracy of different prediction algorithm

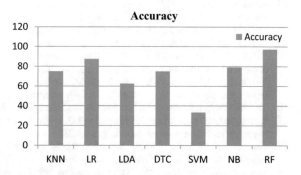

in three phases; repository creation phase, training and testing phase, and prediction phase. This approach gives you the existing project in which one of the modules of a new project can be found. Hence, developer doesn't need re-develop that module. The work has been tested on various machine learning techniques and it has been found that random forest worked better than SVM, and Naïve Bayes algorithms.

References

1. Pietro A, Roberto Di C, Louis G, Fabrice F, Ralf T, Stefano Z (2015) Mining component repositories for installability issues. In: 12th working conference on mining software repositories, pp 24–33
2. Rituraj JCK (2011) Bhensdadia, "study of importance of data mining on software repositories." Int J Adv Res Comput Sci 2(3):406–408
3. Prasanth A, Mladen V (2009) On mining data across software repositories. In: MSR, 2009 6th IEEE international working conference on mining software repositories, pp 171–174

4. Hassan AE (2008) The road ahead for mining software repositories. In: Proceedings ICSM, FoSM track, pp 48–57
5. Harald G, Karin H, Mehdi J (1998) Detection of logical coupling based on product release history. In: ICSM, 14th IEEE international conference on software maintenance (ICSM'98), pp 190
6. Graves TL, Karr AF, Marron Js, Siy H (2000) Predicting fault incidence using software change history. IEEE Trans Softw Eng 26(7):653–661.https://doi.org/10.1109/32.859553
7. Port D, Nikora A, Hayes JH, Huang LG (2011) Text mining support for software requirements: traceability assurance. In: Proceedings of 43rd IEEE Hawaii international conference on system sciences (HICSS 2011), pp 1–11
8. Chen A, Chou E, Wong J, Yao AY, Zhang Q, Zhang S, Michail A (2001) CVSSearch: searching through source code using CVS comments. In: ICSM, 17th IEEE international conference on software maintenance (ICSM'01), pp 364
9. Software deployment. https://en.wikipedia.org/wiki/Software_deployment
10. Hassan AE, Xie T (2010) Software intelligence: the future of mining software engineering data. In: Proceedings of the FSE/SDP workshop on future of software engineering research, ACM, pp 161–166
11. Keivanloo I (2012) A linked data platform for mining software repositories. In: 9th IEEE working conference on mining software repositories
12. Arora R, Garg A (2018) Analysis of software repositories using process mining. Springer Nature Singapore Pte Ltd. In: Satapathy SC et al. (eds) Smart computing and informatics, smart innovation, systems and technologies, vol 78. pp 637–643
13. Halkidi M (2011) Data mining in software engineering. Intell Data Anal 15(3):413–441
14. Xie T, Pei J, Hassan AE (2007) Mining software engineering data. In: 29th international conference on software engineering companion, ICSE
15. Vandecruys O, Martens D, Baesens B, Mues C, De Backer M, Haesen R (2008) Mining software repositories for comprehensible software fault prediction models. J Syst Softw 81:823–839
16. Zaidman A, Van Rompaey B, Demeyer S, van Deursen A (2008) Mining software repositories to study co-evolution of production and test code. In: 1st international conference on software testing, verification, and validation, pp 220–229
17. Canfora G, Cerulo L (2005) Impact analysis by mining software and change request repositories. In: 11th IEEE international symposium on software metrics
18. Moser R (2008) A model to identify refactoring effort during maintenance by mining source code repositories. In: Product focused software process improvement, Springer, Berlin, pp 360–370
19. Weiss C (2007) How long will it take to fix this bug?" In: Fourth international workshop on mining software repositories, ICSE Workshops MSR '07
20. Sager T (2006) Detecting similar Java classes using tree algorithms. In: Proceedings of the 2006 international workshop on mining software repositories, ACM, pp 65–71
21. Ratzinger J (2007) Mining software evolution to predict refactoring. In: First international symposium on empirical software engineering and measurement
22. Kandjani H, Tavana M, Bernus P, Wen L, Mohtarami A (2015) Using extended axiomatic design theory to reduce complexities in global software development projects. Comput Ind 67:86–96
23. Poncin W, Serebrenik A, Brand MVD (2011) Process mining software repositories. In: 15th European conference on software maintenance and reengineering, IEEE, Germany, pp 5–14
24. Sunindyo W, Moser T, Winkler D, Dhungana D (2012) Improving open source software process quality based on defect data mining. In: Biffl S, Winkler D, Bergsmann J (eds) LNBIP, vol 94. Springer, Berlin, Heidelberg, pp 84–102
25. Gupta M, Surekha A (2014) Nirikshan "mining bug history for discovering process maps, inconsistencies and inefficiencies". In: International conference on software engineering, ACM, Chennai, pp 1
26. Gupta M, Surekha A, Padmanabhuni S (2014) Process mining multiple repositories for software defect resolution from control and organizational perspective. In: Proceedings of the 11th working conference on mining software repositories, ACM, Hyderabad, pp 122–131

27. Zimmermann T, Nagappan N, Guo PJ, Murphy B (2012) Characterizing and predicting which bugs get reopened. In: Proceedings of the 34th international conference on software engineering, IEEE, Zurcih, pp 1074–1083
28. Upadhyaya G, Rajan H (2018) On accelerating source code analysis at massive scale. Accepted in IEEE Trans Softw Eng
29. Corbellini A, Godoy D, Mateos C, Zunino A, Lizarralde I (2017) Mining social web service repositories for social relationships to aid service discovery. IEEE/ACM 14th international conference on mining software repositories (MSR), pp 75–79
30. Tiwari NM, Upadhyaya G, Nguyen HA, Rajan H (2017) Candoia: a platform for building and sharing mining software repositories tools as apps. In: IEEE/ACM 14th international conference on mining software repositories (MSR), pp 53–63
31. Alexander T, Trautsch F, Herbold S, Ledel B, Grabowski J (2020) The smartshark ecosystem for software repository mining. In: Proceedings of the ACM/IEEE 42nd international conference on software engineering: companion proceedings, pp 25–28
32. Ray B, Posnett D, Filkov V, Devanbu P (2014) A large scale study of programming languages and code quality in github. In: Proceedings of the 22Nd ACM SIGSOFT international symposium on foundations of software engineering, FSE 2014, New York, NY, USA, ACM
33. Fan Y, Xia X, Lo D, Hassan AE, Li S (2021) What makes a popular academic AI repository? Empir Softw Eng 26(1):1–35
34. Huang H, Badar JS (2009) Precision and recall estimates for two hybrid screens. J Bioinform 25(3):372–378
35. Siblini W, Fréry J, He-Guelton L, Oblé F, Wang YQ (2020) Master your metrics with calibration. In: Berthold M, Feelders A, Krempl G (eds) Advances in intelligent data analysis XVIII, IDA, pp 457–469.

An Approach to Mine Low-Frequency Item-Sets

Reshu Agarwal⬤, Arti Gautam⬤, Amrita Rai⬤,
Shylaja VinayKumar Karatangi⬤, and Eesha Verma⬤

1 Introduction

In today's world, data is growing day by day with the advent of information technology. To manage and analyse this data, there is need of data mining techniques. Data mining is a technique to detect patterns or rules from the huge volume of data, which is helpful in decision making in various businesses. There are various types of data mining techniques like association rule mining, sequence mining, temporal association rule mining, utility mining, etc. Association rule mining is used to retrieve association rules based on support and confidence, but considering this technique, we cannot retrieve rules based on the utility of items. To cope up, with this problem, high utility mining comes into focus. High utility mining is a technique to retrieve the item-sets with high profits. Two-phase algorithm was the first algorithm that applies transaction weighted utilization for mining processes. But this algorithm was suitable for static database and is not able to handle dynamic database. Later on, incremental high utility pattern mining algorithm is proposed for handling dynamic databases. However, this method requires a lot of database scans and operations. Later on, some additional high utility mining algorithms were proposed to scan the complete database only in a single scan, but it also consumes high computational time. Many algorithms were proposed like efficient high utility item-set mining (EFIM), UP-Growth, and HUI-Miner to mine high utility item-sets. However, all these algorithms did not considers that low-frequency items can also be profitable. Moreover, above algorithms does not considers factors like discounts also. For example, in a restaurant, food items like pizza or pasta might earn high profit, but also sells other

R. Agarwal (✉) · E. Verma
Amity Institute of Information Technology, Amity University, Noida, India
e-mail: agarwal.reshu3@gmail.com

A. Gautam · A. Rai · S. V. Karatangi
G L Bajaj Institute of Technology and Management, Greater Noida, India

items like beverages (Coke or Pepsi), which have comparatively less profit. So if we merge items like pasta or pizza having high utility (less frequency) with less profit items like Pepsi or Coke (high frequency), we can increase the utility of the transaction by giving some discount offers on Pepsi or Coke. For effective decision making, there is a need of algorithm that can consider both high-frequency and low-frequency items considering discount also as a factor along with less computational time.

In this paper, we suggested an algorithm that is a combination of low frequent item-sets and high frequent item-sets considering factors like discount and frequency patterns. Moreover, a numerical example is used to clarify the given method. Experiments on real world data sets were conducted to explain the utility of the proposed work.

2 Related Research

When transaction database was updated dynamically, it is difficult to work with static algorithms. To overcome this problem, an algorithm was proposed to mine association rules from dynamic databases [1]. Most of the algorithms focus on mining frequent item-set, but in real world, there are items which are not frequent, also add to the profit for the business. So, an algorithm was proposed to find infrequent items that play a significance role in business [2]. To add feature of privacy and efficiency to mine, association rules were proposed [3]. The main limitation of above proposed algorithms is that they did not take into concern about quantitative attributes while mining association rules. An algorithm based on quantitative approach to mine association rules was proposed which have a great significance. Further, an algorithm was proposed to distinguish between useless rules and losing useful rules including negative items [4]. Further, to mine meta-association rules considering dynamic transaction databases have been proposed using AR-Markov model. Several researchers work to mine association based on level-wise hierarchy. This mining helps to mine items according to brand, category, and items. Supports are used, i.e. different support at each level. Further, distributed data mining is very interesting technique in which rules are discovered over items that are located across the network. An algorithm was proposed for finding rare association rules in distributed environment. It used utilizing measurement percentile to create different minimum supports to mine uncommon association rules. This method finds association rules at very low cost. High utility item mining is an emerging technique to find patterns having a high importance in databases [5]. There are certain extensions of HUIM algorithms. In this way, the uncommon item issue happens, which is that couples of patterns are discovered containing less frequent or productive items with other more frequent or beneficial items. By letting the user set an alternate threshold for every item, this issue can be remove. Moreover, a quantitative approach was applied to mine frequent high utility item-sets [6, 6]. Many algorithms have been proposed to see HUIs considering situations when transaction database is updated [8, 9]. Still researchers are working to improve efficiency of these algorithms.

3 Proposed Work

The algorithm consists of four steps:

Step 1: Derive candidate 1-item-sets from transaction database. Also, calculate their frequency and utility values.

Step 2: Generate k-item-sets by using FP tree method. Now, classify these item-sets as high-frequency or low-frequency item-sets dependent on the frequency value min_supp of k –item-sets.

Step 3: Now classify item-sets got from step 2 as (i) high-frequency high utility item-sets (HFHU) (ii) high-frequency low utility item-sets (HFLU) (iii) low-frequency high utility item-sets (LFHU) (iv) low-frequency low utility item-sets (LFLU).

Step 4: Lastly, derive the association rules for the different combinations of above item-sets using the Confidence min_conf measure.

Definition: If the frequency of an item-set is greater than or equal to minimum support, it is considered as high-frequency item-set and if the frequency of an item-set is less than minimum support, it is considered as low-frequency item-set.

Definition: Utility of each item-set can be calculated using formula:

$$\text{Utility}(C_K) = \sum \text{Frequency } i_j \times \text{Utility } i_j$$

Definition: HFHU item-set is the item-set whose utility of a high-frequency item-set is greater than or equal to minimum utility.

Definition: HFLU item-set is the item-set whose utility of a high-frequency item-set is less than minimum utility.

Definition: LFHU item-set is the item-set whose utility of a low-frequency item-set is greater than or equal to minimum utility.

Definition: LFLU item-set is the item-set whose utility of a low-frequency item-set is less than minimum utility.

4 Numerical Example

Table1 gives sample transaction database containing items with their respective profit values.

Table 1 Transaction database with profit values

Transaction	Items	Utility
T10	P, Q, R	3,3,2
T20	P, R, S	4,3,8
T30	P,Q, R,T	6,2,4,10

Table 2 Frequent item-sets (min_support > = 0.50)

Item-set	Support
(P)	1.0
(Q)	0.66
(R)	1.0
(P, Q)	0.66
(P, R)	1.0
(Q, R)	0.66
(P, Q, R)	0.66

Suppose minimum support (min_support) is 0.50, frequent item-sets have been found using FP tree following the condition support value >= 0.50 as given in Table 2. The item-sets which are having support value less than 0.50 is given in Table 3.

The proposed algorithm gives four different types of item-sets. Using frequent and less frequent item-sets, utility values are calculated. Suppose minimum utility (min_utility) is 18. The item-sets having utility value >= min_utility are high utility item-sets and remaining item-sets are low utility item-sets. Using Tables 2 and 3 and high utility and low utility item-sets, four different types of item-sets are generated which are high-frequency high utility (HFHU) item-sets, high-frequency low utility (HFLU) item-sets, low-frequency high utility (LFHU) item-sets, low-frequency low utility (LFLU) item-sets as given in Tables 4, 5, 6 and 7, respectively.

Table 3 Less frequent item-sets (min_support < 0.50)

Item-set	Support
(S)	0.33
(T)	0.33
(P, S)	0.33
(P, T)	0.33
(Q, T)	0.33
(R, S)	0.33
(R, T)	0.33
(P, R, S)	0.33
(P, Q, T)	0.33
(Q, R,T)	0.33
(P,Q,R,T)	0.33

Table 4 HFHU item-sets

Item-set	Utility
(P, R)	22
(P, Q, R)	20

Table 5 HFLU item-sets

Item-set	Utility
P	13
Q	5
R	9
(P, Q)	14
(Q, R)	11

Table 6 LFHU item-sets

Item-set	Utility
(P, Q, T)	18
(P,Q,R, T)	22

Table 7 LFLU item-sets

Item-set	Utility
S	8
T	10
(P, S)	12
(P, T)	16
(Q, T)	12
(R, S)	11
(R, T)	14
(P, R, S)	15
(Q, R, T)	16

Hence, first phase of algorithm is completed. Now in second phase, four types of association rules are generated using item-sets from phase 1 as shown below. Assume that confidence is 30%.

(1) LFLU ➔ HFHU: According to this rule, profit and selling count of LFLU item-sets can be increased if such item-sets will be combined with HFHU item-sets as shown in Table 8. More offers can be given to increase sell of LFLU items like discounts and buy one get one offer on HFHU item-sets.

(2) LFHU ➔ HFHU: According to this rule, selling count of LFHU item-sets can be increased if such item-sets will be combined with HFHU item-sets as given in Table 9. More offers can be given to increase sell of LFHU items like discounts and buy one get one offer on HFHU item-sets.

Table 8 Association rules for LFLU➔HFHU

Association rule	Confidence
(P,R)➔ (P,R,S)	33.33

Table 9 Association rules for LFHU➜HFHU

Association rule	Confidence
(P,R)➜ (P,Q,R,T)	33.33
(P,Q,R)➜ (P,Q,R,T)	50.00

Table 10 Association rules for LFHU➜HFLU

Association rule	Confidence
P➜ (P, Q, T)	33.33
P➜ (P,Q,R,T)	33.33
Q➜(P, Q, T)	50
Q➜(P,Q,R,T)	50
R➜(P,Q,R,T)	33.33
(P,Q)➜(P,Q,T)	50
(P,Q)➜(P,Q,R,T)	50
(Q,R)➜(P,Q,R,T)	50

Table 11 Association rules for LFLU➜HFLU

Association rule	Confidence
P➜ (P, S)	33.33
P➜(P, T)	33.33
P➜(P, R, S)	33.33
Q➜(Q, T)	50
Q➜(Q, R, T)	50
R➜(R, S)	33.33
R➜(R, T)	33.33
R➜ (P, R, S)	33.33
R➜(Q, R, T)	33.33
(Q,R)➜ (Q,R,T)	50

(3) LFHU ➜ HFLU: According to this rule, selling count of LFHU item-sets and profit of HFLU item-sets can be increased if LFHU item-sets will be combined with HFLU item-sets as given in Table 10. More offers can be given to increase sell of LFHU items and profit of HFLU items like discounts and buy one get one offer on HFLU item-sets.

(4) LFLU ➜ HFLU: According to this rule, profit of both item-sets and selling count of LFLU item-sets can be increased if LFLU item-sets will be combined with HFLU item-sets as given in Table 11. More offers can be given to increase utility of LFLU and HFLU items like discounts and buy one get one offer on HFLU item-sets (Table 12).

Table 12 Number of rules for assumed database	Database	Number of rules
	LFLU➔HFHU	1
	LFHU➔HFHU	2
	LFHU➔HFLU	8
	LFLU➔HFLU	10

5 Experimental Analysis with Large Data Sets

There are four types of item-set, viz. HFHU item-set, HFLU item-set, LFHU item-set, and LFLU item-set. We perform various investigations to discover the association rules for the distinctive combinations of item-sets created utilizing the proposed technique. We perform experiment on a large data set of a retail store. There is no pruning criterion in whole process of generation of low-frequency and high-frequency item-sets, as we use FP growth algorithm. Hence, here low support and low utility threshold value is used to prune certain item-sets based on their less frequency of occurrence. Here, we use 1000, 2000 and 5000 transactions with 10 items per transaction. We compare both methods of Apriori algorithm and FP growth implementation of proposed approach. We find that our method is more efficient than Apriori implementation. Moreover, our method provides more useful association rules considering both low-frequency and high-frequency item-sets. The proposed method also takes less time to find desired association rules as compared to other methods. Figure 1 shows the comparison of our proposed method implementation and Apriori implementation.

Fig. 1 Experimental result on retail store data set

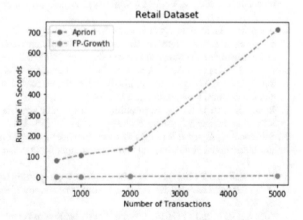

6 Conclusion and Future Scope

HUIM is a data mining technique that returns high profit when sold together or alone that encounters a user-specified minimum utility threshold. There are certain limitation of frequent pattern mining like quantity of items purchased, importance of items, discounts on items and frequency of occurrence of items, not taken into consideration while designing business strategies. To remove this problem, this paper suggests a method for mining association rules based on combination of low-frequency item-sets with high-frequency item-sets considering different factors like discounts and frequency patterns. Experimental experiments on real world data sets have been done to prove that different types of association rules derived considering above factors are important for decision making. Moreover, a numerical example is devised to explain the proposed model. In future, advanced algorithms can be used to calculate the utility item-sets with less time and without candidate item-set generation.

References

1. Ouyang W, Huang Q (2009) discovery algorithm for mining both direct and indirect weighted association rules. In: International conference on artificial intelligence and computational intelligence, Shanghai, India, pp 322–326
2. Wang P, Shi L, Bai J, Zhao Y (2009) Mining association rules based on Apriori algorithm and application. In: International forum on computer science-technology and applications, Chongqing, China, pp 141–143
3. Jingjing F, Qingfei Z, Zhonglin Z (2010) A method of mining the meta-association rules for dynamic association rule based on the model of AR-Markov. In: Second international conference on networks security, wireless communications and trusted computing, Wuhan, Hubei, China, pp. 210–214
4. Agarwal R, Gautam A, Dixit P, Rana A (2020) An approach to mine frequent item sets considering negative item values. In: 8th international conference on reliability, Infocom technologies and optimization (Trends and Future Directions), Noida, India, pp 208–211
5. Agarwal R, Gautam A, Saksena AK, Rai A, Karatangi SV (2020) Method for mining frequent item sets considering average utility. In: International conference on emerging smart computing and informatics, Pune, India, pp 275–278
6. Ryang H, Yun U (2016) High utility pattern mining over data streams with sliding window technique. In: Expert systems with applications 57:214–231
7. Saksena AK, Agarwal R (2021) Methods for classification of items for inventory management. In: International conference on computer communication and informatics, Coimbatore, India, pp 1–4
8. Junrui Y, Jingyi Y (2021) Frequent item sets mining algorithm for uncertain data streams based on triangular matrix. In: IEEE international conference on power electronics, computer applications, pp 327–330
9. Agarwal R, Mittal M (2019) Inventory classification using multi-level association rule mining. Int J Decision Support Syst Technol 11(2):1−12

Forecasting Floods in the River Basins of Odisha Using Machine Learning

Vikas Mittal, T. V. Vijay Kumar, and Aayush Goel

1 Introduction

During past five decades, significant increase in the number of natural hazards has been observed. Among these natural hazards, floods occurred more frequently and affected large number of population in the world [39]. India, with more than twelve percent of its land area susceptible to floods, is the second largest flood affected country in the world. More than one lakh people have died, and economic losses of 347 thousand crore rupees have been reported by Central Water Commission (*CWC*) from 1953 to 2016 due to floods [6]. Further, malnutrition and stunted growth among children have also been observed as an after effect of floods [25, 38, 39]. Therefore, efficient flood mitigation strategies are required to extenuate the losses due to floods. The objective of flood mitigation strategies is to take structural and non-structural measures [24] wherein construction of dikes and dams on the river can be carried out as a part of structural measures, and designing of early warning systems (*EWSs*) and defining land use policies can be carried out as non-structural measures. Primary objective of the *EWS* is to forecast floods and issue early warnings to the vulnerable community. Early flood forecasting and related warnings to the various stakeholders can give them sufficient time to execute the preparedness plans. Machine learning (*ML*), a branch of artificial intelligence (*AI*), can be used to design such flood fore-casting systems. Supervised machine learning techniques can be used to train learning

V. Mittal (✉) · T. V. V. Kumar
School of Computer and Systems Sciences, Jawaharlal Nehru University, New Delhi, India
e-mail: vikas.mittal.10@gmail.com

T. V. V. Kumar
e-mail: tvvijaykumar@hotmail.com

A. Goel
Bharati Vidyapeeth's College of Engineering, New Delhi, India
e-mail: aayushgoel1996@gmail.com

models considering past flood's data. In this paper, historical meteorological data of ten flood affected districts of Odisha, India for the period (1991–2002) obtained from the Climatic Research Unit, University of East Anglia [1, 19] have been used to design such *ML*-based flood forecasting models. Classification-based *ML* techniques, viz., artificial neural network (*ANN*) [11], *k*-nearest neighbor (*KNN*) [12, 16], logistic regression (*LR*) [5, 28], Naive Bayes (*NB*) [29], support vector machine (*SVM*) [9, 40], and random forest (*RF*) [3, 4] classifiers have been applied on this flood data for designing flood forecasting models.

This paper is organized as follows: An overview of the flood forecasting models and related work are presented in Sect. 2. *ML*-based flood forecasting models are discussed in Sect. 3. Section 4 is experimental results discussing the comparison of the *ML*-based flood forecasting models on various performance metrics. Section 5 is the conclusion.

2   Related Work

Flood forecasting (*FF*) models are classified into three major categories: physical models, conceptual models, and data-driven models [8, 14, 34]. Physical and conceptual models are traditionally used for forecasting floods. Physical models use hydrological equations that characterize physical properties of the catchment area and require parameters related to river geometry, channel roughness, etc., to predict water levels of the river [26]. Conceptual models use calibration of model parameters that requires large amount of hydrological and meteorological data for forecasting floods. However, calibration is a complex process, and its interpretation requires domain expertise [26]. In order to forecast floods, data-driven flood forecasting models apply *ML* techniques on historical data related to floods. *ML* techniques establish a relationship between parameters and identify the patterns in the data. Therefore, *ML*-based data-driven models are less complex and are widely used for forecasting floods [10, 18, 20, 22, 26, 31, 36, 37].

In [10, 26, 36], a three layer *ANN* model is proposed for forecasting floods in three different regions in India. In [10], a backpropagation *ANN* model predicts hourly runoff for Govindpur basin on Brahmani River in Odisha, India. In [26], feed forward neural network (*FFANN*) model forecasts floods in Kushabhadra branch of the Mahanadi delta in Odisha, India, with a lead time of up to 5 h. In [36], *FFANN* model forecast floods in the Bhasta River, Maharashtra, India, with a lead time of up to 3 h. In [18], modified Takagi Sugeno (*T-S*) fuzzy inference system (*FIS*) [35] forecasts rare and frequent flood conditions in upper Narmada basin with lead times of up to 6 h. In [22], an adaptive neurofuzzy inference system (*ANFIS*) forecasts floods in the Kolar basin, Madhya Pradesh, India. The proposed *ANFIS* model combined the features of fuzzy inference system (*FIS*) and the neural networks. The combined model was shown to perform comparatively better for larger lead time [21]. In [31, 37], wavelet transform was used to decompose the time series data to form subcomponents. Further, in [31], *ANN* model, optimized using genetic algorithm, was

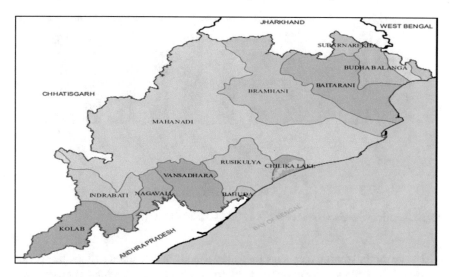

Fig. 1 River Basins in Odisha (*Source* https://gisodisha.nic.in)[1]

used to forecast floods in Kosi and Gandak rivers of Bihar, India, with a lead time of up to 24 h. In [37], sub-components obtained using wavelet transform were re-sampled using bootstrapping. Further, *ANN* model was trained using re-sampled sub-components to forecast floods in Mahanadi river basin, Maharashtra, India with a lead time of up to 10 h. In [20], *SVM* model used meteorological parameters for forecasting floods in urban areas with a lead time of up to 48 h.

Lead time of flood forecasting models discussed above ranges between 1 and 48 h, which is inadequate for effective and efficient mitigation strategies. This paper attempts to address this by focusing on designing *ML*-based flood forecasting models, based on monthly mean of meteorological parameters, having larger lead time.

3 *ML*-Based Flood Forecasting Model

Odisha is a state located in the east coast of India. The state has 482 km-long coastline which makes the state vulnerable to frequent floods, cyclones, and tsunamis [7]. The state is also drained by eleven major rivers, namely Mahanadi, Brahamani, and Baitarani, etc. River basins of these rivers are shown in Fig. 1.

---

[1] Disclaimer: The presentation of material and details in maps used in this chapter does not imply the expression of any opinion whatsoever on the part of the Publisher or Author concerning the legal status of any country, area or territory or of its authorities, or concerning the delimitation of its borders. The depiction and use of boundaries, geographic names and related data shown on maps and included in lists, tables, documents, and databases in this chapter are not warranted to be error free nor do they necessarily imply official endorsement or acceptance by the Publisher or Author.

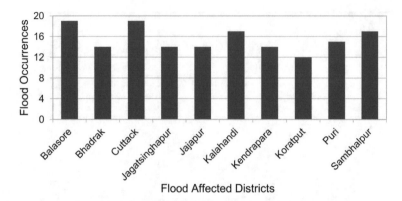

Fig. 2 District-wise distribution of floods in Odisha (1991–2002) [7]

Heavy rainfall during the monsoon period usually causes flood or flood like conditions in almost all the river basins and delta areas in the state. Typhoons and cyclones in the coastal areas of Odisha are also reasons for floods in the state. High tides during the monsoon period also poses a threat of floods in flat coastal areas with poor drainage system. Three major rivers, viz., Mahanadi, Brahamani, and Baitarani rivers constitute a catchment area of more than one lakh square kilometers, which is more than 65% of the total area of the Odisha state, and flood in these river basins affects 72% of the total population of the state. These rivers also form a common delta where flood water interlace and brings about disruption. The ten most flood affected districts in Odisha are Balasore, Bhadrak, Cuttack, Jagatsinghapur, Jajpur, Kendrapara, Puri, Koraput, Kalahandi, and Sambalpur [7]. The flood occurrences in these districts are shown in Fig. 2 [7].

The flood data of these districts have been considered for designing *ML*-based flood forecasting models. Historical flood data comprise of seven meteorological parameters, viz., precipitation, temperature, evapotranspiration, crop evapotranspiration, cloud cover, wet day frequency, and vapor pressure of these flood affected districts have been obtained from Climatic Research Unit, University of East Anglia [1, 19] and labeled using flood information available at state's disaster management portal (https://dowrodisha.gov.in). Meteorological data for the period 1991–2002 have been obtained for designing flood forecasting models. There are total 1440 data instances in the labeled dataset out of which 155 instances are flood instances. Meteorological parameters in the dataset have values in varying ranges which might affect the classification performance of the *ML* model [32]. In the proposed methodology, the values of the meteorological parameters are normalized using min–max scaler [5], as defined below:

$$X' = \frac{X - X_{min}}{X_{max} - X_{min}}$$

where X_{max} and X_{min} denote the maximum and minimum values, respectively, of the parameter (X). The value of X' varies between 0 and 1.

Classification-based *ML* techniques, viz., artificial neural network (*ANN*) [11], k-nearest neighbor (*KNN*) [12, 16], logistic regression (*LR*) [5, 28], Naive Bayes (*NB*) [29], support vector machine (*SVM*) [9, 40], and random forest (*RF*) [3, 4] classifiers have been used to design flood forecasting models that use the normalized flood forecasting dataset. These models are briefly discussed below:

Logistic Regression (*LR*): Logistic regression uses a logistic function to classify data points into different classes [5, 28]. The logistic function $L_\theta(X)$ is a sigmoid function which is defined as: $L_\theta(X) = sigmoid(Z)$, where Z is a linear function of input features X and is defined as: $Z = \beta_0 + \beta_1 X$, where β_0 is the intercept and β_1 is the weight.

Support Vector Machine (*SVM*): *SVM* creates a hyperplane to classify data points into different classes [9, 40]. The objective of the *SVM* classifier is to choose a hyperplane that has maximum distance from the data points in a N-dimensional space in order to distinctly classify the data points.

k-Nearest Neighbors (*KNN*): *KNN* classification technique uses plurality of vote from k-nearest neighbors of a data point in order to decide its class [12, 16]. Different distance measures like Euclidean distance, Manhattan distance, etc., are used to find the nearest neighbors.

Naive Bayes (*NB*): Assuming that attributes or parameters in the dataset are independent of each other, Naive Bayes (*NB*) classifier uses Bayes theorem for classification of data points into discrete classes [29].

Random Forest (*RF*): Random forest is an ensemble classifier which uses multiple decision trees to classify a data point. Each decision tree in the forest classifies and votes for the output class for the given data point [3, 4].

Artificial Neural Network (*ANN*): *ANN* is a classification technique, which comprises a network of computational nodes called neurons [11]. Each node in the network is connected with the other node in the network through synapses. Each connection/synapse in the network has a weight, which is optimized using the backpropagation algorithm.

The performance of the abovementioned *ML* models is compared on various performance metrics such as accuracy, precision, recall, F-measure, and *AUC-ROC*. These comparisons are discussed as part of experimental results.

4 Experimental Results

The abovementioned *ML* classification techniques: *ANN, KNN, LR, NB, SVM,* and *RF,* were applied on the normalized flood forecasting (*FF*) dataset discussed in Sect. 3. The details of the experimental setup and simulations are given in Table 1.

For each *ML* model, fivefold cross-validation [2] has been used to obtain the experimental results. True positives (*TP*), false positives (*FP*), false negatives (*FN*), and true negatives (*TN*) [17] have been used to compute the value of performance metrics

Table 1 Experimental setup

Operating system	Windows 10
Processor	Intel i7@2.80 GHz
RAM	16 GB
Tool	Python 3.7.7
Features	Monthly average of precipitation, Vapor pressure, Temperature, Wet day frequency, Evapotranspiration, Crop evapotranspiration and Cloud cover
Number of folds	5
Learning rate in *ANN*	0.05
Number of decision trees in *RF*	20
Value of *k* in *k-NN*	5

such as accuracy, precision, recall, *F*-measure, and *AUC-ROC* values, across all folds. The comparison of *ML* models for flood forecasting (*FF*): *ANN_FF*, *KNN_FF*, *LR_FF*, *NB_FF*, *RF_FF* and *SVM_FF*, based on afore-mentioned performance metrics, is discussed below:

Accuracy: Accuracy is computed as [17]:

$$\text{Accuracy} = \frac{TP + TN}{TP + TN + FP + FN}$$

Mean accuracy and standard deviation across all folds of the abovementioned *ML* models is given in Table 2. It can be noted from Table 2 that mean accuracy of all the models ranges between 0.794 and 0.905. The *RF_FF* model has the highest mean accuracy while the *NB_FF* model has the least mean accuracy. Mean accuracy of *RF_FF* and *KNN_FF* models is very close, but standard deviation of *RF_FF* model is less. Further, *SD* value was zero in the case of *ANN_FF* and *SVM_FF* models, i.e., no variation in the accuracy among all folds was observed for *ANN_FF* and *SVM_FF* models. Thus, *RF_FF* model can be considered for forecasting floods in case when accuracy is the key performance metric.

Precision: Precision is computed as [17]:

Table 2 Accuracy of all *ML*-based *FF* models

Model	Accuracy	
	Mean	SD
ANN_FF	0.892	0
KNN_FF	0.903	0.012
LR_FF	0.885	0.003
NB_FF	0.794	0.031
RF_FF	0.905	0.004
SVM_FF	0.892	0

Table 3 Precision of all *ML*-based *FF* models

Model	Precision	
	Mean	SD
ANN_FF	0	0
KNN_FF	0.571	0.075
LR_FF	0.263	0.14
NB_FF	0.319	0.038
RF_FF	0.65	0.063
SVM_FF	0	0

$$\text{Precision} = \frac{TP}{TP + FP}$$

Mean precision and standard deviation across all folds of the abovementioned *ML* models is given in Table 3. It can be noted from Table 3 that the mean precision value of the *RF_FF* model is highest among all *ML* models. However, mean precision value of the *ANN_FF* and *SVM_FF* models is lowest (zero) among all *ML* models. *SD* value for *NB_FF* model is the lowest among *KNN_FF*, *LR_FF*, *NB_FF*, and *RF_FF* models. *SD* value of *RF_FF* is not high. Thus, *RF_FF* model can be considered for forecasting floods in case when precision is the key performance metric.

Recall: Recall is computed as [17]:

$$\text{Recall} = \frac{TP}{TP + FN}$$

Mean recall and standard deviation across all folds of the abovementioned *ML* models is given in Table 4. It can be noted from Table 4 that the mean recall value of the *NB_FF* model is the highest among all the *ML* models, and mean recall value of *ANN_FF* and *SVM_FF* model is observed the lowest (zero) among all the *ML* models. Further, the standard deviation of the *NB_FF* models is not high. Thus, *NB_FF* model can be considered for forecasting floods in case when recall is the key performance metric.

Table 4 Recall of all *ML*-based *FF* models

Model	Recall	
	Mean	SD
ANN_FF	0	0
KNN_FF	0.452	0.046
LR_FF	0.058	0.043
NB_FF	0.768	0.055
RF_FF	0.29	0.114
SVM_FF	0	0

Table 5 *F*-Measure of all *ML*-based *FF* models

Model	F-Measure	
	Mean	SD
ANN_FF	0	0
KNN_FF	0.502	0.049
LR_FF	0.092	0.062
NB_FF	0.448	0.033
RF_FF	0.381	0.115
SVM_FF	0	0

As observed above, the mean accuracy of all the models is comparable except *NB_FF* model which has lowest mean accuracy value. *RF_FF* model has obtained highest mean precision value, but its mean recall value is low. *NB_FF* model has obtained highest mean recall value, but its mean precision value is low. Therefore, *F*-measure and *AUC-ROC* have been used to evaluate and compare the performance of the *ML* models.

F-measure: *F*-measure is computed as [17]:

$$F - \text{Measure} = \frac{2 \times \text{Precision} \times \text{Recall}}{\text{Precision} + \text{Recall}}$$

Mean *F*-measure values of all models are given in Table 5. It can be noted from Table 5 that the mean *F*-measure value of *KNN_FF* model is the highest among all *ML* models. Both *SVM_FF* and *ANN_FF* models have obtained lowest mean *F*-measure values because of lowest mean precision and mean recall values. Among *KNN_FF*, *LR_FF*, *NB_FF*, and *RF_FF* models, mean *F*-measure of *LR_FF* models is lowest. Further, *KNN_FF* model has better *SD* when compared with that of *LR_FF* and *RF_FF* models. Thus, *KNN_FF* model can be considered for forecasting floods in case when *F*-measure is the key performance metric.

AUC-ROC: *AUC-ROC* is the area under the *ROC* curve plotted between true-positive rate *(TPR)* and false-positive rate *(FPR)*, where *TPR* and *FPR* are defined as [17]:

$$TPR = \frac{TP}{TP + FN}$$

$$FPR = \frac{FP}{FP + TN}$$

To assess the overall performance of the *ML* models across various thresholds, the *AUC-ROC* has been computed as shown in Table 6. It can be noted from Table 6 that *RF_FF* model has the highest mean *AUC-ROC* value while *KNN_FF* model has the least mean *AUC-ROC* value among all *ML* models. Further, standard deviation of *RF_FF* models is better than standard deviation of *KNN_FF* model. Thus, *RF_FF*

Table 6 AUC-ROC of all *ML*-based *FF* models

Model	AUC-ROC	
	Mean	SD
ANN_FF	0.895	0.016
KNN_FF	0.882	0.029
LR_FF	0.893	0.013
NB_FF	0.886	0.012
RF_FF	0.914	0.019
SVM_FF	0.897	0.014

model can be considered for forecasting floods in case when *AUC-ROC* is the key performance metric.

It can be noted from the above that the performance of *RF_FF* model is comparatively better than other *ML* models in terms of accuracy, precision, and *AUC-ROC* values across all folds whereas *NB_FF* and *KNN_FF* models perform significantly better than all *ML* models in terms of recall and *F*-measure values, respectively.

5 Conclusion

This paper emphasized on designing *ML*-based flood forecasting models with larger lead times for the flood affected districts of Odisha, India. For forecasting floods, *ML* techniques like *NB, LR, SVM, KNN, RF,* and *ANN* were applied on the meteorological data characterized by seven features, viz., precipitation, vapor pressure, temperature, evapotranspiration, crop evapotranspiration, wet day frequency, and cloud cover of ten most flood affected districts of Odisha, India, during the period 1991–2002. The normalized labeled dataset was re-sampled using stratified fivefold cross-validation, and six *ML* models were trained and tested. Five performance metrics: accuracy, precision, recall, f-measure, and *AUC-ROC,* were used to analyze the performance of these models. Among these *FF* models, *RF_FF* model performed comparatively better in terms of accuracy, precision, and *AUC-ROC* values whereas *NB_FF* and *KNN_FF* models performed comparatively better in terms of recall and *F*-measure values, respectively. Furthermore, while comparing the overall performance of all the *ML* models, *RF_FF* model performs comparatively better and thus can be preferred over others for forecasting floods.

References

1. Allen RG, Pereira LS, Raes D, Smith M (1998) Crop evapotranspiration: guidelines for computing crop water requirements, FAO Irrigation and drainage paper 56. Italy, Rome

2. Berrar D (2018) Cross-validation. Encyclopedia of Bioinform Computational Biol: ABC of Bioinform 1–3:542–545
3. Biau G (2012) Analysis of random forests model. J Mach Learn Res 13:1063–1095
4. Breiman L (2001) Random forests. Mach Learn 45:5–32
5. Cabrera AF (1994) Logistic regression analysis in higher education: an applied perspective. In: Smart JC (ed) Higher education handbook of theory and research, vol 10. Pp 225–256
6. Central Water Commission (2018) Flood damage statistics (Statewise and for the Country as a whole) for the period 1953 to 2016; Central Water Commission (CWC), Flood Forecast Monitoring Directorate, Government of India. 3:37. Retrieved from http://www.indiaenvironmentportal.org.in/content/456110/flood-damage-statistics-statewise-and-for-the-country-as-a-whole-for-the-period-1953-to-2016/
7. Department of Water Resources, Odisha. Retrieved from https://www.dowrodisha.gov.in/
8. Devia GK, Ganasri BP, Dwarakish GS (2015) A review on hydrological models. Aquatic Proc 4(ICWRCOE):1001–1007
9. Evgeniou T, Pontil M (2001) Support vector machines: theory and applications. In: Paliouras G, Karkaletsis V, Spyropoulos CD (eds) Machine learning and its applications. ACAI 1999. Lecture notes in computer science, vol 2049. pp 249–257
10. Ghose DK (2018) Measuring discharge using back-propagation neural network: a case study on Brahmani River Basin. Intell Eng Inform 591–598
11. Gurney K (1997) In: An introduction to neural networks. CRC Press
12. Guo G, Wang H, Bell D, Bi Y, Greer K (2003) KNN model-based approach in classification. In: Meersman R, Tari Z, Schmidt DC (eds) On the move to meaningful internet systems 2003: CoopIS, DOA, and ODBASE. OTM 2003. Lecture notes in computer science, vol 2888. pp 986–996
13. Jain SK, Agarwal PK, Singh VP (2007) Physical environment of India. In: Jain SK, Agarwal PK, Singh VP (eds) Hydrology and water resources of India, vol 57.
14. Jain SK, Mani P, Jain SK, Prakash P, Singh VP, Tullos D, Dimri AP (2018) A Brief review of flood forecasting techniques and their applications. Int J of River Basin Managem 16(3):329–344
15. Jayalakshmi T, Santhakumaran A (2011) Statistical normalization and back propagation for classification. Int J Comput Theory Eng 3(1):89–93
16. Laaksonen J, Oja E (1996) Classification with learning k-nearest neighbors. In: IEEE international conference on neural networks—conference proceedings, vol 3. pp 1480–1483
17. Liu Y, Zhou Y, Wen S, Tang C (2014) A Strategy on Selecting Performance Metrics for Classifier Evaluation, International Journal of Mobile Computing and Multimedia. Communications 6(4):20–35
18. Lohani AK, Goel NK, Bhatia KKS (2014) Improving real time flood forecasting using fuzzy inference system. J Hydrol 509:25–41
19. Mitchell TD, Jones PD (2005) An improved method of constructing a database of monthly climate observations and associated high-resolution grids. Int J Climatol 25(6):693–712
20. Nayak MA, Ghosh S (2013) Prediction of extreme rainfall event using weather pattern recognition and support vector machine classifier. Theoret Appl Climatol 114(3–4):583–603
21. Nayak PC, Sudheer KP, Ramasastri KS (2005) Fuzzy computing based rainfall-runoff model for real time flood forecasting. Hydrol Process 19(4):955–968
22. Nayak PC, Sudheer KP, Rangan DM, Ramasastri KS (2005) Short-term flood forecasting with a neurofuzzy model. Water Resour Res 41(4):1–16
23. NDMA (2017) National Disaster Management Authority (NDMA). In Ndma. Retrieved from http://ndma.gov.in
24. NIDM ((2002)) Disaster management-terminology 8:1–4. Retrieved from https://nidm.gov.in/PDF/Disaster_terminology.pdf
25. NIDM (2018) Safety tips for floods, cyclones and Tsunamis. Retrieved from https://nidm.gov.in/safety_flood.asp
26. Panda RK, Pramanik N, Bala B (2010) Simulation of river stage using artificial neural network and MIKE 11 hydrodynamic model. Comput Geosci 36(6):735–745

27. Parida Y, Saini S, Chowdhury JR (2021) Economic growth in the aftermath of floods in Indian states. Environ Dev Sustain 23(1):535–561
28. Peng CJ, Lee KUKL, Ingersoll GM (2002) An introduction to logistic regression analysis and reporting. J Educ Res 96(1):3–14
29. Rish I (2001) An empirical study of the naive Bayes classifier. 3(22):4863–4869
30. Rosasco L, De Vito ED, Caponnetto A, Piana M, Verri A (2004) Are loss functions all the same? Neural Comput 16(5):1063–1076
31. Sahay R, Srivastava A (2014) Predicting monsoon floods in rivers embedding wavelet transform, genetic algorithm and neural network. Water Resour Manage 28(2):301–317
32. Singh D, Singh B (2020) Investigating the impact of data normalization on classification performance. Appl Soft Comput 97(B)
33. Singh SK (2013) Flood management information system. In: Government of Bihar. Retrieved from http://fmis.bih.nic.in
34. Sitterson J, Knightes C, Parmar R, Wolfe K, Muche M, Avant B (2017) An overview of rainfall-runoff model types an overview of rainfall-runoff model types
35. Takagi T, Sugeno M (1985) Fuzzy identification of systems and its applications to fault diagnosis systems. IEEE Trans Syst Man Cybernet 15(1):116–132
36. Thirumalaiah K, Deo MC (1998) Real-time flood forecasting using neural networks. Comput-Aided Civil Infrastruct Eng 13(2):101–111
37. Tiwari MK, Chatterjee C (2010) Development of an accurate and reliable hourly flood forecasting model using wavelet–bootstrap–ANN (WBANN) hybrid approach. J Hydrol 394(3–4):458–470
38. UNDRR (2020) Human cost of disasters. In: Human cost of disasters
39. Wallemacq P, Below R, McLean D (2018) UNISDR, CRED.: economic losses, poverty and disasters (1998–2017), CRED, 2018, available at https://www.cred.be/unisdr-and-cred-report-economic-losses-poverty-disasters-1998-2017
40. Zhang Y (2012) Support vector machine classification algorithm and its application. Commun Comput Inform Sci 308CCIS(PART 2):179–186

Emo-Spots: Detection and Analysis of Emotional Attributes Through Bio-Inspired Facial Landmarks

V. S. Bakkialakshmi⊙, T. Sudalaimuthu⊙, and B. Umamaheswari⊙

1 Introduction

Facial landmark extraction is the fundamental component for many face-related analyses such as face recognition, pose variation, face analysis, and emotion recognition. Face landmarks act as an independent and single variable for partial face analysis. Face images are important for most multimedia applications: for age estimation, feedback systems, real emotion detection, etc. Automated prediction of facial attributes is important to analyze emotions accurately. The face is the index of the mind. What is felt inside the mind and heart is clearly expressed via face. Face landmarks like showing eyebrows, compressed smiles, lowered chins, expanded noses, and related small actions in these areas convey unique hints of facial expressions.

Facial images are extracted from the videos. Facial landmark spotting is separated by passing a picture or video outline through a course of pre-prepared regression trees. After examining different arrangements of facial components in face direction location procedures and testing the consequences of each, a bunch of six facial focuses nose tip, jaw, corner points of the eyes, and corner points of the mouth are viewed as enough for the calculation to have the option to identify the direction of the face in a wide scope of view with lesser calculations. Facial feature extraction

V. S. Bakkialakshmi (✉) · T. Sudalaimuthu
Department of Computer Science and Engineering, Hindustan Institute of Technology and Science, Chennai, India
e-mail: bakkyam30@gmail.com

T. Sudalaimuthu
e-mail: sudalaimuthut@gmail.com

B. Umamaheswari
Department of Computer Science and Engineering, Amrita Vishwa Vidyapeetham, Amrita School of Engineering, Bengaluru, India
e-mail: b_uma@blr.amrita.edu

© The Author(s), under exclusive license to Springer Nature Singapore Pte Ltd. 2023   103
R. Agrawal et al. (eds.), *International Conference on IoT, Intelligent Computing and Security*, Lecture Notes in Electrical Engineering 982,
https://doi.org/10.1007/978-981-19-8136-4_9

using deep neural networks and cascaded regression trees were found effective in recent developments. Depending on the head pose facial features are extracted [1].

Expressions of the face are related to human perception. The systematic analysis gives correlated results and uncorrelated outcomes that enable the analysis group to get highlighted hints on real emotions. The challenges in the perception are the effective gap. To address the affective gap discussed in previous works, the system considered facial landmarks as an emotional attribute. Face landmarks are detected by extracting the face iconic parts such as eyes, eyebrows, nose, mouth, jawline, etc., the so-called steps are previously discussed with the Haar-Cascade model that acts as the standard detection model for face detection. Facial landmark detection is the mission of detecting key benchmarks on the face and chasing them (being strong to rigid and non-rigid facial buckles due to head travels and facial terminologies) [2].

In existing work, multiple modality-based convolutional networks (MTCNN*) to extract the multi-perceptional face detection and face landmark confinement in complex conditions are discussed. The detector can work out well at high accuracy on the face database, a unique benchmark for face detection data, and (AFLW) Annotated Facial Landmarks in the Wild benchmark dataset for facial landmark detection [2]. Algorithms that extract better like Gaussian-guided landmark maps from a large set of tree mappings and predict outstanding facial landmarks through a regression algorithm are discussed. Gaussian landmark feature maps act as prior information models in which the regression network effectively formulates the direction of the refined landmarks on the 300 W test dataset [3].

1.1 Bio-Inspired Learning

Bio-inspired computing is a unique method of learning and organizing the solutions of certain computerized functions using biological signals in the form of facial expressions, physical structures, etc. The procedure includes the following steps. Extraction of biological signals, merging of extracted biological hints, framing the correlated solutions, problem search, etc. Bio-inspired computing is a continuous process that extracts the bio-inspired data, learns the key points, adjusts the bias, and pretends the solution according to dynamic inputs [4].

1.2 Face Landmarks

Face detection is used to detect the frontal human face and works with the commonly used algorithm named viola jones algorithm and Haar cascade classifier. The face frontal part has a unique pattern of key points called landmarks that is extracted from the expansion of eyebrows, mouth length, eyeball and eyelid length, forehead area, jawlines, and their angle formation with nose position many more. These landmark points are measured to make the unique pattern of the face. Face landmarks are highly helpful in finding facial expressions. Face landmarks are detected using automated models and manual models [5].

2   Literature Survey

2.1   Landmark Detection

Created a task-oriented landmark extraction technique with deep convolutional neural networks that modify the requirement of landmark spots depending on the complexity of the facial features [6]. The intention is to create a lightweight architecture and reduce the number of tasks needed to extract facial landmarks. The system focused on reducing the error rate and regardless of head pose the landmark extraction is developed. Presented a generalized adverse network for facial landmark detection and focused on developing an encoder to reduce the occlusion in detecting the facial landmarks [7]. The system focuses on reducing the error rate with less processing time. The presented model is compared with existing deep recurrent neural networks with an error rate of 5.82.

The author addressed landmark detection and enhanced prediction through the EAR aspect ratio and wink rate of an eye. The learned pattern of occurrence is processed and trained to provide real-time controls [8]. The presented system utilized a thermal image dataset collected from UND for facial expression detection through landmarks [9]. A multi-task learning architecture is developed and focused with detailed landmark points of 68 points. U-Net architecture is developed for analysis and obtained an error rate of 6.84 comparatively.

EEG signals are primary physiological data that express the unique peaks concerning changes in emotions [10]. The author presented a system in which deep learning neural network is combined with a principal component analysis framework to get the covariate points from the given signal sequence. EEG is a noisy signal that needs to filter before processing. The author presented a real-time evaluation framework in which a smartphone is used to capture the facial images and expression features are grabbed with smartphone APIs [11]. Expression depends on the mental stimulus and physical changes that resembled much in the face. The study helps us understand the integrity between physiological signals and facial expressions.

3   System Design

The system model is developed using an image processing toolbox from MATLAB 2017 software [12]. The dataset considered here is the JAFFE [13] contents of numerous female facial expressions. The need for automated and robust prediction of landmarks and analysis of landmarks with emotional recognition is considered as

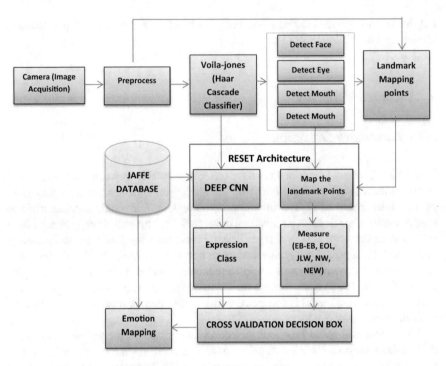

Fig. 1 The system architecture of the proposed RESET architecture

the problem statement. Figure 1 shows the proposed model detects the human face using the Viola-Jones algorithm with the Haar Cascade classifier. MATLAB image processing toolbox has a dedicated command set for training the input images. Landmark detection is planned using a point mapping model which is initially trained with manual points and finally extracted using automated localizations.

3.1 Problem Finding

Facial landmark detection techniques use various formulations, in which the prime motive is to reduce the error rate of prediction comparing the traditional approaches. Each face has unique variations in the landmarks and the localization of points gets deviated through head pose variations. Extracting unique landmark points with less computation time and reduced mapping points is the primary requirement of the presented system.

3.2 Proposed Method

The proposed system is focused on deriving a robust Emo-spot extraction technique called RESET, in which landmark points of 13 points are considered for evaluation. The detailed spotting of important landmarks that are responsible for expressions is discussed here. The proposed work considers a JAFFE dataset with 213 images of various facial expressions of 10 females. Six basic facial expressions are marked as targets for our proposed RESET algorithm.

4 Methodology

4.1 Preprocessing Block

The preprocessing module consists of image acquisition using a Camera. Real-time images are captured using the webcam. The images are resized to the dimension that matches the database image. The steps of checking the correlation of facial image points such as eyebrows, nose, and jaw lines make it an automated detection process, initial few images are tested manually. The preprocessed image was restored to the same dimension as the database image.

4.2 Face Detection Block

Viola-Jones algorithm (VJA) is one of the standard models that calculate the integral image, Haar-like features, and finally a classifier to correlate it. VJA works on the principle of extracting the face from the non-face from the images. It compares the practical difficulties with face detection and performs a robust operation. The Haar cascade feature extracts the pixels related to the adjacent one within the rectangular block. The primitive blocks hold the relevant pixels that capture the eye portion, nose portion, and mouth separately.

4.3 Landmark Mapping Block

The process of landmark mapping is initiated after the detection of the face, eye, mouth, and nose separately. The landmark mapping is initially localized using the user routine that stacks the pixel locations near eyebrows, mouth, nose, and jaw lines approximately. The routine we call the point mapping model (PMM). From the point mapping model, the estimation needs to be derived by measuring the distance between the mapped points example eyebrows to eyeball (EB-EB), End of Lips

(EOL), Jawline widths (JLW), Nose width (NW), Nose ear width (NEW), etc. The same set of steps repeated for the JAFFE dataset images to make it a reference model for comparing the landmarks during expressions.

4.4   Classification Block

The classification block consists of a deep learning convolution neural network in which the JAFFE database image is trained with a fusion of a few real-time images taken from the web camera. For creating the classification model, the database is split into 70% for training, 20% for testing, and 10% for validation. Based on the accuracy of the training stage, the testing of real-time data is performed. The iterative looping of checking the landmark threshold happens with the randomly looped KNN model [14]. KNN is normally used as the supervised and unsupervised classification depending on the data input. Here we adopted randomly looped KNN since the landmark weights are not exactly labeled with the dataset, whereas the threshold keeps the predicted value with a 5% tolerance of the original value. Hence for this place, KNN is the best-suited one and we applied novelty to the existing KNN by tuning the looped model, iterative checking the feedback of the obtained weight of the threshold, and correlating the threshold to form a new threshold with 5% tolerance.

4.5   Validation Block

The quantitative measures are estimated using the confusion matrix with a true positive value, true negative value, false-positive value, false-negative value, etc. the classification process also consists of a cross-validation block that checks the correlation between the obtained result based on the [15] JAFFE database and real-time data. The higher the correlation the higher the positive result would be.

4.6   Testing Summary

The test procedure is planned to take the image of the individual captured from the webcam and test the expression-based emotional attribute. Instead of simply classifying it as Happy and sad, our proposed system detects resilient expressions such as disgust, moody, excited, contempt, fear, sad, and happy these detection thresholds are iteratively tested with the database images and finally update the weight to the real-time images. He correlated a result that pretends false answers are occurring in between as performance loss [16].

5 Summary of RESET Algorithm

5.1 CNN Framing

Convolutional neural networks are one of the robust techniques used to classify the input images according to the database images. The normal CNN architecture classifies the database images concerning the test images. Despite creating a novel architecture, the landmark feature-based mapping technique called point mapping technique PM in which dominates the classification process of CNN. A dual-stacked CNN architecture is used to detect the face structure that correlates with the database as well as the mapped 13 Landmark points that correlate with the detected fee structure.

Detecting the facial skeleton is the first step, and mapping the extracted points in the structure is the second step. The fusion of face structure with landmark points is the third step. Now the secondary CNN detects the maximum correlation with the database structure. In case of higher correlation, the suggested she and output are further compared with the point adjusting technique called randomly loop KNN. Here the randomly looped KNN (RLKNN) algorithm is used for two processes, namely, classifying the facial image data with the appropriate images in the database as well as adjusting the landmark points concerning maximum and minimum ranges of benchmark values [17].

5.2 Layers of CNN

5.2.1 Convolutional Layer

Convolutional layers act as the foremost layer in the CNN architecture in which the input features are mapped as a feature vector into the N*N matrix fetched into the CNN. In the convolutional layer, the simple operation called convolution or dot operation will happen for the given raw data. The convolution layer is followed by the filter that rolls over the input matrix and matches the correlated patches concerning the database images. Zero padding is the process of evaluating the borders present in the test image.

5.2.2 Max-Pooling Layer

Based on the hidden neutrons present in the convolution the number of computations is concerned. The Max pooling layer is used to reduce the computations or connections between the input features with the preferred neurons. Complex input images need more connectivity that is handled by the max-pooling layers.

5.2.3   Fully Connected Layers

The fully collected layer is used to extract the required bias weights and connect them to the output layer to make the final suggestion. It acts as a bridge between the max-pooling layer and the output layer. Some of the values are flattened based on the maximum metric point and fetch to the output layer at this stage.

5.3   Proposed Layer Configurations

The number of layers in a 'Deep CNN' is indicated by the word 'deep'. In a typical CNN, there will be normally 5–10 or perhaps more feature learning layers. Networks of more than 50–100 layers are common in modern architectures used in cutting-edge applications. Table 1 shows the configuration of the layers utilized in the proposed model.

Pseudocode—Steps of RESET Algorithm

Start Process
Load DB(JAFFE);
Extract EB-EB, EOL, JLW, NW, NEW;
Map landmark points (PPM)
Save New_Db
Save New_Db_test=New Image
Configure CNN_1(Jaffe_train,Jaffe_test);
Configure CNN_2(New_Db, New_Db_test);
Start RLKNN_loop
Compare (New_Db, New_Db_test)
If bias (New_Db(i)= New_Db_test) @5% tol
Det_image= New_Db_test;
Else
Update_bias(New_Db_test);
End
Repeat
Det_image= New_Db_test;
End process

Table 1 Layers configurations utilized in the proposed model

Layers configured	Dimensions
Input layer	$100 \times 100 \times 11 \times 62$
Stride	10×10
Fully connected layer	384×62
Fully connected layer	384×42
Fully connected layer	384×1
Output layer	7

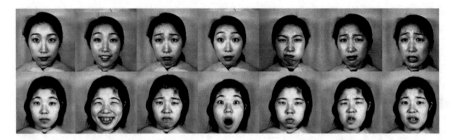

Fig. 2 Training images from the JAFFE dataset

6 Results and Discussions

6.1 Training Images

Figure 2 shows the input training images of [15] JAFFE dataset consists of 216 images of ten different female faces with unique expressions. The trained data contains six reliable target expressions.

6.2 Detection of Emo-Spots and Mappings

Figure 3 shows the results of Emo-Spots mappings using the PPM technique with 13 unique landmark points that are responsible for basic expressions [18]. Emotions such as happy, excitement, sad, contempt, disgust, and fear are focused. These emotions merely need the utilization of important landmark components called eyes, nose, mouth, and jaws. Variation in emotional changes produces obvious changes with the mentioned landmark components. Despite reducing the landmark points for fewer computation works, 13 important landmark points are considered [19].

Fig. 3 Detection of Emo-Spots and mappings

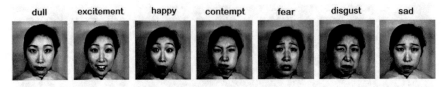

| dull | excitement | happy | contempt | fear | disgust | sad |

Fig. 4 Predicted Emotions using Emo-Spots

6.3   Predicted Results

Figure 4 shows the predicted emotions using the RESET algorithm with detailed 13 Emo-Spots identification. The proposed RESET algorithm consumes less computation time and found a reduced error rate of 2.42.

6.4   Weight Updates from Randomly Looped KNN (RLKNN)

Weights of the images are updated randomly with the loop from the algorithm K-nearest neighbor with support of 62 images of JAFFE dataset images after the PPM technique using the RLKNN-Randomly looped K-nearest neighbor algorithm. The weights of the images are updated randomly with the loop from the algorithm K-nearest neighbor. The weights are continuously modified by iteratively going through the provided database and randomly plotting 195 test photos. The accuracy of the proposed model is compared with the labels that were actually collected. The proposed RESET algorithm is also compared with existing systems discussed [6, 7, 9] of TSDCN task-specific deep convolutional neural network, DRN-GAN deep recurrent network with generalized adversarial network, and MTL-U-Net multi-task learning-based U-Net architecture, respectively. The performance measures on the various emotions are tabulated in Table 2.

Table 2 Summary of detected emotions with landmark points

JAFFE Dataset

Tested emotions	Landmark points	EB-EB	LOL	JLW	NW	NEW	Correlated results with accuracy
Sad	13	10	22	95	10	35	0.95
Disgust	13	11	23	92	10	36	0.92
Contempt	13	10	24	97	11	34	0.9
Fear	13	12	21	96	10	39	0.95
Moody	13	12	25	92	12	35	0.95
Excited	13	17	29	92	11	36	0.95
Happy	13	15	27	91	14	35	0.94

7 Challenges

The proposed model is planned to implement as a lightweight facial expression detection system through Emo-Spots or Emotional landmarks. The primary challenge faced with the proposed system is that the face structure of the real-time images varies apparently. Keeping the standard threshold is not applicable all the time. Initial stages of implementations are started with static real-time images captured and stored with the local server. Further real-time image face structure that matches with the real-time trained image structure is smoothly correlated. The unstructured face example overweight peoples and skinny peoples are very difficult to pretend to. In such cases, we further recommend improving the size of the training dataset, also considering more unique parameters in the face as emo-spots. More unique details at the eyes, mouth, and ears to map more features helpful in making the accurate classification.

8 Conclusion

Impacted emotional spot detection using facial landmarks to recognize the real emotion of humans is discussed in the study. The proposed system focused on real-time challenges, automated detection, lightweight design solutions, etc. while choosing the architecture. The proposed structure uses Voila-jones and Haar cascade classifiers for face extraction. A deep learning convolution network for static image correlation and Novel RESET algorithm are created with 13 detailed landmark points using Deep CNN and Emo-Spot extractor with lightweight points mapping routine. The novel algorithm is created by tuning the bias weights of the landmark points that vary from one face to another. Here the static dataset is tuned for testing the novel algorithm. The proposed structure works with a static dataset namely JAFFE. Expressions such as happy, excitement, sad, contempt, disgust, and fear are focused. The static images are trained and tested with a Deep convolutional neural network and achieve an average accuracy of 95% with a reduced computation error rate of 2.42. The dynamic tuning of images with point mapped using PMM, correlated and Tested with RLKNN to make a high recognition rate [20]. The proposed system performs with less error rate of 2.42 comparatively with the state-of-art approach discussed in Table 3. The usage of RLKNN improves the performance rate by customizing the processing delay. Further, the system should be improved by adding more real-time images and training them with more focused Emo-Spots on the eye portion, and Lips portion alone to converge more landmark points. Further, the Emo-Spot identification is improved with a fusion of real-time and static data for analyzing spatial factors.

Table 3 Comparative analysis of proposed RESET architecture with existing system in terms of error rate

S. No.	Reference	Methodology	Landmark points	Dataset	Error rate
1	[6]	TSDCN	29 points	AFLW	8.2
2	[7]	DRN-GAN	56 points	COW	2.94
3	[9]	MTL-U-Net	68 points	UND-TD	6.84
4	Proposed	RESET-(DCNN + RKNN)	13 points	JAFFE	2.42

References

1. Shah HM, Dinesh A, Sharmila TS (2019) Analysis of facial landmark feature to determine the best subset for finding face orientation. In: 2019 international conference on computational intelligence in data science (ICCIDS), 2019, pp 1–4. https://doi.org/10.1109/ICCIDS.2019.8862093
2. Ma M, Wang J (2018) Multi-view face detection and landmark localization based on MTCNN. Chinese Automation Congress (CAC) 2018:4200–4205. https://doi.org/10.1109/CAC.2018.8623535
3. Lee Y, Kim T, Jeon T, Bae H, Lee S (2019) Facial landmark detection using gaussian guided regression network. In: 2019 34th international technical conference on circuits/systems, computers and communications (ITC-CSCC), 2019, pp 1–4. https://doi.org/10.1109/ITC-CSCC.2019.8793317
4. Wu D, Zhang J, Zhao Q (2020) Multimodal fused emotion recognition about expression-EEG interaction and collaboration using deep learning. IEEE Access 8:133180–133189. https://doi.org/10.1109/ACCESS.2020.3010311
5. Medium (2021) Viola-Jones algorithm and haar cascade classifier. [online] Available at <https://towardsdatascience.com/viola-jones-algorithm-and-haar-cascade-classifier-ee3bfb19f7d8>. Accessed 28 October 2021
6. Zhang Z, Luo P, Loy CC, Tang X (2014) Facial landmark detection by deep multi-task learning. In: Fleet D, Pajdla T, Schiele B, Tuytelaars T (eds) Computer vision—ECCV 2014. ECCV 2014. Lecture Notes in Computer Science, vol 8694. Springer, Cham. https://doi.org/10.1007/978-3-319-10599-4_7
7. Liu H, Zheng W, Xu C, Liu T, Zuo M (2020) Facial landmark detection using generative adversarial network combined with autoencoder for occlusion. Math Probl Eng 2020:1–8
8. Rakshita R (2018) Communication through real-time video oculography using face landmark detection. In: 2018 second international conference on inventive communication and computational technologies (ICICCT), 2018, pp 1094–1098.https://doi.org/10.1109/ICICCT.2018.8473269
9. Chu W, Liu Y (2019) Thermal facial landmark detection by deep multi-task learning. In: 2019 IEEE 21st international workshop on multimedia signal processing (MMSP), 2019, pp 1–6. https://doi.org/10.1109/MMSP.2019.8901710
10. Jirayucharoensak S, Pan-Ngum S, Israsena P (2014) EEG-based emotion recognition using deep learning network with principal component-based covariate shift adaptation. The Scientific World J pp 1–10
11. Cuimei L, Zhiliang Q, Nan J, Jianhua W (2017) Human face detection algorithm via Haar cascade classifier combined with three additional classifiers, In: 2017 13th IEEE international conference on electronic measurement and instruments (ICEMI), 2017, pp 483–487. https://doi.org/10.1109/ICEMI.2017.8265863
12. Zeng G, Zhou J, Jia X, Xie W Shen L (2018) Hand-crafted feature guided deep learning for facial expression recognition. In: 2018 13th IEEE international conference on automatic face and gesture recognition (FG 2018), 2018, pp 423–430. https://doi.org/10.1109/FG.2018.00068

13. Michael L, Kamachi M, Global J (2017) Japanese female facial expression (JAFFE) database
14. Taunk K, De S, Verma S, Swetapadma (2019) A brief review of nearest neighbor algorithm for learning and classification. In: 2019 international conference on intelligent computing and control systems (ICCS), 2019, pp 1255–1260. https://doi.org/10.1109/ICCS45141.2019.906 5747
15. Ullah A, Wang J, Anwar MS, Ahmad U, Wang J, Saeed U (2018) Nonlinear manifold feature extraction based on spectral supervised canonical correlation analysis for facial expression recognition with RRNN. In: 2018 11th international congress on image and signal processing, biomedical engineering and informatics (CISP-BMEI), 2018, pp 1–6. https://doi.org/10.1109/CISP-BMEI.2018.8633244
16. Bakkialakshmi VS, Sudalaimuthu T (2022) Dynamic cat-boost enabled keystroke analysis for user stress level detection. In: 2022 international conference on computational intelligence and sustainable engineering solutions (CISES), 2022. https://doi.org/10.1109/cises54857.2022.984 4331
17. Suk M, Prabhakaran B (2015) Real-time facial expression recognition on smartphones. In: 2015 IEEE winter conference on applications of computer vision, 2015, pp 1054–1059.https://doi.org/10.1109/WACV.2015.145
18. Muttu Y, Virani HG (2015) Effective face detection, feature extraction and neural network-based approaches for facial expression recognition. In: 2015 International conference on information processing (ICIP), 2015, pp 102–107. https://doi.org/10.1109/INFOP.2015.7489359
19. Sadu C, Das PK (2020) Swapping face images based on augmented facial landmarks and its detection. In: 2020 IEEE region 10 conference (TENCON), 2020, pp 456–461. https://doi.org/10.1109/TENCON50793.2020.9293884
20. Bakkialakshmi VS, Sudalaimuthu T (2022) Emo-net artificial neural network: a robust affective computing prediction system for emotional psychology using amigos. Indian J Comput Sci Eng 13(4):1040–1055. https://doi.org/10.21817/indjcse/2022/v13i4/221304034

Rapid Face Mask Detection and Person Identification Model Based on Deep Neural Networks

Abdullah Ahmad Khan, Mohd. Belal, and Ghufran Ullah

1 Introduction

As the coronavirus case are increasing day by day although the vaccination drive has been going on at the full speed but still only 29% of total world population has been fully vaccinated yet. In India, only 12% of people have been vaccinated and it is assumed that next wave of newly mutated coronavirus is coming in mid-October, so getting vaccinated and wearing a face mask is still a requirement. As we have also seen the rise of pollution in different cities so wearing a face mask is also beneficial for one's health. By implementing our proposed project, we can avoid spread of COVID-19 which is necessary and beneficial for large institutions and businesses so that they can maintain their productivity while avoid getting infected to COVID-19 and also encourage people to wear mask while commuting so that they don't breathe polluted air and be safe. As the quality of air is getting worst day and in some cities the air is so bad that the government has suggested people to wear a face mask so that people can breathe safely and does not inhale air which is mixed with toxins. But since 2020 the world has been dealing up with the pandemic i.e., COVID-19. The earlier symptoms of it were having high fever and difficulty in breathing etc. and the people who had no previous medical records are also getting infected with the virus.

It is advised by every major Health Organizations that wearing face mask and getting vaccinated and washing hands with alcohol-based sanitizers. The first two parts we need to be governed by ourselves, the proposed model will make aware people about the safety of wearing a face mask not only for avoiding polluted air but also for others safety. In this paper, we have tried to minimize the time taken by detection model to detect and predict the person and his/her identity. The InsightFace

A. A. Khan (✉) · Mohd. Belal · G. Ullah
Department of Computer Science, Aligarh Muslim University, Aligarh, Uttar Pradesh, India
e-mail: aakhan3@myamu.ac.in

© The Author(s), under exclusive license to Springer Nature Singapore Pte Ltd. 2023   117
R. Agrawal et al. (eds.), *International Conference on IoT, Intelligent Computing and Security*, Lecture Notes in Electrical Engineering 982,
https://doi.org/10.1007/978-981-19-8136-4_10

module has been incorporated in place of previous module to reduce effective time taken during the computation. InsightFace is an incorporated Python library for 2D and 3D face analysis. InsightFace effectively executes a rich variety of best in-class algorithms of face recognition, face identification, and face alignment, which upgraded for both training and deployment. The contribution of this research work is as follows:

- A SoftMax loss-based module is incorporated, i.e., InsigtFace for Detection and Prediction.
- The simulation is performed on wider face benchmark to check the efficiency of the proposed module.

The images we have used in this project for visual object recognition are gathered from ImageNet. ImageNet has a mega database of different ordinary images classification from having hundreds of images of "strawberry" to "inflatable. It is database which mainly focuses on visual object recognition. It contains about 20,000 classifications and about 14 million pictures. The presented project is measured on ImageNet. For getting prediction on masked and non-masked face and person identification we have used deep learning network. Deep Learning is part of Machine Learning which is neural network that has three more layers. These neural networks in the deep learning model help us to simulate the behavior of human brain like learning huge amounts of data and matching the ability to the real human brain. In previous generations, we have seen AI and machine learning working on some amazing tasks like self-driving car, feasible working of websites, etc. These things have become so common today we see them a lot in our real life but we ignore them.

The rest of the paper is divided into seven sections given as Sect. 2 discusses related work on different modules of face detection and recognition. In Sect. 3, we presented the problem formation focusing in the existing system. Section 4 presents the proposed work, including the overview of ArcFace algorithm based on Soft Max loss. Section 5 gives Simulation study for benchmark on the proposed module. Section 6 presents the analysis and results. Finally, Sect. 7 talks about the conclusion and future work part of the paper.

2   Related Work

Yadav and Javad [1] this paper titled "Deep learning based safe social distancing and facemask detection in public areas for COVID-19 safety guidelines adherence" in this research paper author has built a real-time system integrated with security cameras in the public places which detects whether the person is wearing a face mask or not and also check if heaa/she is maintaining proper social distance. The system reports directly to the authorities and it was built using raspberry pi and OpenCV.

Loey et al. [2] they have designed a hybrid deep-learning model using two components. Resnet50 being the first component is used for feature extraction and for classification they have used decision tree and Support Vector Machine (SVM) algorithms. For simulation study, they have used three datasets to benchmarking their model Simulated Mask Face Dataset (SMFD) is the first dataset Labeled Faces in the Wild (LFW) is the second dataset and Real-World Masked Faced Dataset (RMFD) is the third one. The results achieved LFW in the SVM algorithm was 100% for the RMFD it achieved 9.64% and for SMFD it gets 99.49% accuracy.

Jiang et al. [3] in this paper, a high accuracy and efficient detector called as "Retina Face" was developed on stage which consists of feature pyramid network to combine several feature maps. Transfer learning is applied to reduce the shortage of datasets. The approach of this model was to eliminate the projections of poor confidence and high Union Intersection. The accuracy result for the retina face was high developed on the dataset of face mask. It achieved 2.3% which was 1.5% higher than standard results and achieved detection precision of 1.0% which was 5.9% higher than previous standard results.

The paper proposed by the author Toshanlal Meenapal [4] Put together couple of face classifier which identifies any face in any conditions. This paper proposed a novel technique uses that uses convolutional neural network model VGG-16 predefined weights etc. Convolution Networks are used to fragment the faces from the image/picture. For utilizing misfortune work Binomial Cross Entropy is used and for preparing they have used Angle Descent. FCN is used to eliminate undesirable noise and to avoid false expectations. Results of the model got the accuracy it got for the portion where the face mask is used to cover the face was 93.844%.

Leung et al. [5] in this paper, previous built state-of-the-art deep learning model named InceptionV3 is fine tuned. For training the dataset the author has used SFMD, i.e., Simulated Face mask Dataset which is used to simulate current set of data on a given dataset which is available publicly for better testing and training of images. Image augmentation is used so that we can remove the data restriction. The proposed model attains 100% accuracy during the testing and 99.9% precision during training [6].

Gupta et al. [7] proposed a model named Smart City and Intelligent Transport System which uses IoT devices and different sensors to enforce the social distancing so that people can follow government guidelines promptly. For monitoring real-time movements of objects their model describes deploying sensors in different parts of the city and it also offers data sharing. Overall, the system architecture has the ability to store and share data and information to the relevant central cloud facilities and also has the ability to record the data at real time and exchange messages with nearby sensors.

Sonn et al. [8] the authors in this paper have discussed about the contribution of Smart Cities in containing the spread of COVID-19. They have specifically talked about smart cities of South Korea. Not only they tracked people of wearing a mask in public and maintaining social distance, but it they also tracked the user medical history, purchase history, cell phone location, even the real-time monitoring of people not only on the public places but even in their buildings which helps the government to suppress the COVID-19 infection in the South Korea much better than any other countries.

Singh et al. [9] have talked about how IoT can help to suppress COVID-19 in healthcare sector and can save a lot of life. The developed system gave emphasis on interconnect devices so that hospital can keep track of different cases. They have given major key merits for IoT that can help to fight COVID-19 pandemic such as when the system is operated by IoT device there will be lesser chances of errors as compared to work done by humans. IoT devices will provide effective control and will enhance the diagnosis. They have talked about several application of IoT in the healthcare sectors like Automated treatment process, Telehealth consultation, Wireless Healthcare network to identify COVID-19 patients, Rapid COVID-19 screening, connecting all medical devices through the Internet, accurate forecasting of virus etc.

Sonn et al. [8] has talked about state-of-the art model which helps to suppress the COVID-19 pandemic without having a lockdown in the entire country. The paper titles "Smart city technologies for pandemic control without lockdown" in this research they interview different patients and their past movement has been recorded. They have noticed some of the patients have hidden their past records from the interviewer. But by using the real-time tracking we can get the information accurately.

Jaiswal et al. [10] has proposed a unique way to deal with the pandemic using the position of technology to track infected people. They proposed the use of drones and robot technologies as medical personnel to deal and track the infected patients.

Ponkia et al. [11] has used principal component a(PCA) to detect and recognize face images. They have used small dataset for detection of faces in the images. They worked on the GUI of the model. The model has various buttons and text fields such as selecting an image, running the PCA recognizer to recognize the selected image from the image database. If the user is recognized then the selected image and the image from the database is shown side by side and text field below the image shows the name and grant access to the system and if the person is not in the database, then the text field will show access denied, until the user clicks on register them as a new user.

3 Existing System

The existing system deals with Mobile_NetV2 algorithm for classification and prediction. The system used 20% images for test and rest for 80% training purpose. The existing system uses face recognition library from python to recognize faces and identity of the person [12]. The system helps to identify the person wearing mask or not but it fails do it efficiently and sometimes returns ambiguous results.

3.1 Issues in Existing System

The existing systems does not allow the system to easily scale and maintain real-time inference capability when the number of users increases.

3.2 Drawbacks in Existing System the Major Limitations of Existing Schemes Are as Follows

- Mobil_NetV2 algorithm is used to train and test the data in the existing which is slower as compared to new generation models.
- The existing system does not have the inference capability and does not use powerful GPU to compute the results.
- The facial recognition system used in the existing system is old and does not cover complete face.

4 Proposed System

In the proposed work, we have incorporated a new deep learning neural network named as InsightFace in place of old face recognition library. The InsightFace used MXNET (DNN Framework) as its inference backend which allows our system flexibility, fast model training with high scalability which helps in maximize productivity and efficiency.

The InsightFace framework uses face detectors with Additive Angular Margin Loss for Deep Face Recognition (ArcFace) algorithm to obtain highly discriminative features for face recognition. The ArcFace uses modified SoftMax loss classification that makes predictions between the feature and weights (Fig. 1).

The proposed model of face recognition implements singleton class which creates a single instance of an object which ensures our model will be time and memory efficient. By implementing the InsightFace bounding box wrapper we allow our model to detect the image positions as well as object of interest in the images and is

Fig. 1 Decision margins of different loss functions under binary classification case. The dashed line represents the decision boundary, and the grey areas are the decision margins

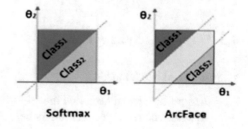

Fig. 2 Bounded box
wrapper with feature
selection

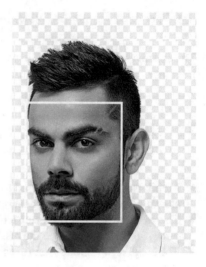

represented by rectangular boxes, and with the help of embedded wrapper, it performs feature selection during the model training (Fig. 2).

For calculating landmarks, we have implemented Is frontal helper function that will return the landmarks if they are true and in between −25 and 25, if not then it will return false for every other value. Finally, for calculating the inference of the proposed model we placed time. Time () statements which returns time in seconds (float value) since the epochs.

4.1 Advantage of Proposed System

- The proposed system will provide faster facial detection and identification as compared to the previous one.
- The InsightFace module will provide lower time complexity while parallel increasing the accuracy of the Model
- The proposed system has increased dimension, i.e., the system will scale as more users are registered with the system.

4.2 Proposed System Pseudocode

- Systems AI Model loaded:

    Insight-face facial detection model
    Insight-face facial recognition model

- Known faces images loaded and facial embedded extracted and saved

- Input feed captured through RGB camera
- Image frame passed through insight-face (arc face) facial. detection model
- Cropped face separated from each frame
- Cropped face passed through insight-face facial recognition model to extract embeddings
- Extracted embedded compared with known embeddings 8. Known face with the highest similarity identified as the person in frame
- Identified person emailed
- Desktop notification generated

4.3 Proposed System Design

When the program detects the user is not wearing the mask it first takes the input from the image, crop the image so that only the facial part will be available to send to the DNN model. After that first, it represents the image by drawing Bounded Box Wrapper and then performs detection to identify the person and in the last it will notify the user by sending an e-mail and also it goes back to iterate through the face mask detection part again.

4.4 Complete System Design

As we have worked on the person detection and identification part in this paper the other half detection Architecture will remain the same as defined in the previous paper and also shown in Fig. 3 Person Identification Model (Fig. 4).

4.5 Formulating the Dataset

For creating the dataset for mask and no mask the data is collected from different free and open-source images websites Kaggle, google images, etc. The images are then classifying in multiple folders as masked and no mask. The first folder labeled as masked consist of 1915 entries, whereas the second folder labelled as no mask consist of 1918 entries.

4.6 Training Process

1. After feature xi and weight W normalization, we get the cos θj (logit) for each class as (Wi)'xi.

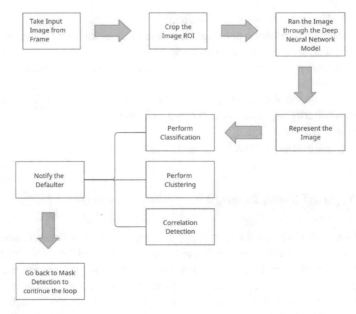

Fig. 3 Person identification model

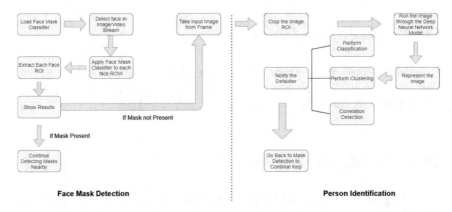

Fig. 4 Face mask and person identification module workflow

2. We calculate the arccosθyi and get the angle between the feature xi and the ground truth weight Wyi.
3. We add an angular margin penalty m on the target (ground truth) angle θyi.
4. We calculate cos(θyi+m) and multiply all logits by the feature scales.
5. The logits then go through the SoftMax function and contribute to the cross-entropy loss

5 Simulation Study

The section provides the benchmarking results of our proposed RFMPI-DNN model and the previous detection model on WIDER FACE dataset.

5.1 Dataset

WIDER FACE is a largest public dataset available which contains 393,703 faces and 32,203 images in its database built on 61 event classes from Internet [13]. The database consists of several human faces, different poses, occlusion, expression, illumination, and low resolution. The images in the WIDER FACE dataset are divided into three classes for training, testing and validation purposes each of them is divided into a ratio of 50%, 10 and 40%, respectively. Depending on the edge box each subset is defined into three difficulty levels that are "Easy", "Medium" and "Hard". The Hard level covers all the detections from easy and medium, meaning hard can show effectiveness of different methods. In our simulation, the proposed RFMPI-DNN model and the previous Detection model both uses WIDER FACE dataset containing 16,102 images from which 196,852 annotated faces are extracted [14].

5.2 Feature Extractor

We have used Resnet-10 as backbone and caffe-model as neck to construct the feature extractor [15]. The combination of both is used in most of detectors, so it would be good for comparison and replication.

5.3 Training Values and Test Size

We have trained both the model using the Stochastic Gradient Decent (SGD) optimizer (momentum $= 0.9$, decay $= 0.01$) having batch size $= 32$ on GeForce GTX 1060ti [16]. The initial learning rate is $1 \times e-4$ as lower the initial learning rate the better the results. WE have used 40 EPOCS so that our training accuracy would be higher. EPOCS are hyper parameter, i.e., it tells us about the number of times the training algorithm will work.

5.4 Comparison on WIDER FACE

Table 1 shows the average precision (AP) of our RFMPI-DNN model and the previous model on WIDER FACE test and validation subset. Our model performs better in all three segments of dataset NAD gets very promising results.

Even when both the backbone is same in the model our model outperforms the previous model in each level in validation as well as in testing phase. The Precision Recall curves (AP) are shown in Fig. 5. Our model achieves best AP in all level faces, i.e., AP = 0.972(Easy), 0.965 (Medium), 0.925 (Hard).

6 Result Discussion and Analysis

The graph is plotted in pythons matplotlib Fig. 6 Precision-Recall curves obtained by our proposed (RFMPI-DNN in red) and the previous detection model All methods trained and tested on the same training and testing set of the WIDER FACE dataset (a): Easy level, (b): Medium level and (c): Hard level. Library which gives us the complete understanding why our RFMPI-DNN model is showing better result than the FACE RECOGNITION model used previously. The graph plots the time inference of four outputs, i.e., Mask, No Mask, Face Detection, and Person Identification. The time taken by both the model is showed as a comparison side by side.

Table 1 Average precision performance

Method	Backbone	Easy	Val medium	Hard	Easy	Test medium	Hard
Previous detection model	ResNet-10	0.969	0.958	0.921	0.965	0.957	0.9211
RFMPI-DNN(OURS)	ResNet-10	**0.972**	**0.965**	0.925	**0.967**	**0.962**	**0.924**

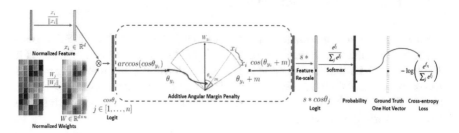

Fig.5 Training diagram

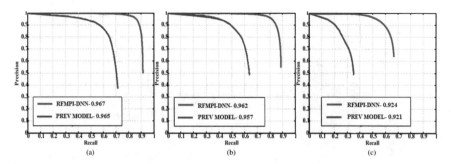

Fig. 6 Precision-recall curves obtained by our proposed (RFMPI-DNN in red) and the previous detection model All methods trained and tested on the same training and testing set of the WIDER FACE dataset **a**: Easy level, **b**: Medium level and **c**: Hard level

6.1 Connectivity

The connectivity between the two models will remain same as in previous paper, i.e., the model first runs the face detection part to detect the face in the frame, if face is not found it will continue until it founds one, when the face is found it will detect the person wearing a mask or not if the person is wearing a mask it will continue detecting other faces (if there are more faces available) or else if the no mask is detected the system will switch to person identification model the model fetched the face ROI from the image and compare it with the face data available in the database and gives out notification that the person is not wearing a mask. Figure 7 Time Comparison of Previous and Proposed Model shows the diagram of connecting two models.

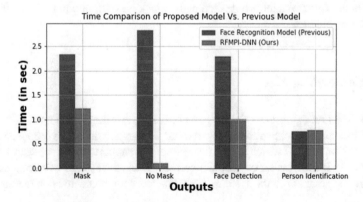

Fig. 7 Time comparison of previous and proposed model

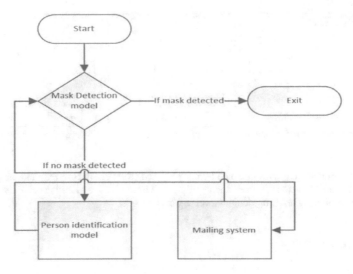

Fig. 8 Connection of both model

6.2 Proposed Model Accuracy Samples

The proposed detection model will tightly bound the complete face while the previous model does not cover the complete face ROI Fig. 8 Connection of Both Model. Shows the proposed model and previous model facial detection accuracy (Fig. 9).

Black Represents the Old Face Recognition Module
White Represents the New Face Recognition Module

White tightly covers the whole face while black is not tightly bound and does not cover the whole face.

6.3 Embedding Size

The old facial recognition model supports only supports dimension up to 128 means system does not have capability to detect more persons at a time. The proposed model has dimensions of 512 which means the system can support more people and will be reliable even if the number of people increases. Thus, the system will scale as more users are registered with the system. The same has been presented in a tabular form in Table 2.

Fig. 9 Face module
comparison

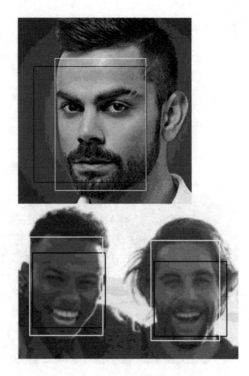

Table 2 Facial recognition
dimensions

	Old dimenstion	New dimentions
Embedding size—facial recognition	128	512

6.4 Evaluation Results

The result of inference time on Face Mask, No Mask, Face Detection, and Person Identification presented in Fig. 5 has been tabulated for better understanding of results in Table 2: Facial Recognition Dimensions. By looking at the table, we can clearly say that our proposed model performed significantly better than the existing model (Table 3).

6.5 Working of Model

Figure 12 shows the Connection of Both Model the model running and successfully detecting the person face mask and Fig. 10 shows the accuracy result around the edges of the box (Fig. 11).

Table 3 Inference time table

Functions	Old system inference time	New system inference time
Detect and predict mask	0.0234	0.0123
Detect and predict no-mask	0.0283	0.0111
Face recognition person	0.230	0.0101
Identification	0.0077	0.0079

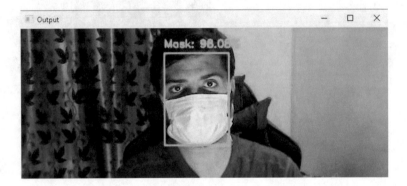

Fig. 10 Model detecting person wearing a mask

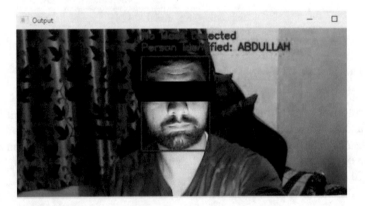

Fig. 11 Working of person identification model

6.6 E-mail Notification

Apart from getting notified on the screen, the default individual which is not covering his/her face with a face mask will be notified by the system by sending an email to

Fig. 12 Person ID and mask detection model combined

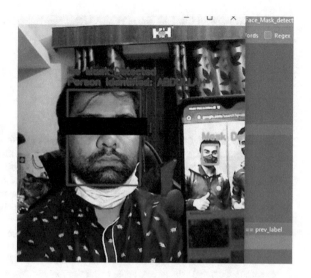

the person from the systems custom email id. This work is done using SMTP module of python.

7 Conclusion and Future Works

7.1 Conclusion

As the pandemic is still going on and currently there is no future prediction regarding when COVID-19 will successfully get over getting vaccinated is the option but wearing face mask is still a need even after getting vaccinated. This proposed model will deal efficiently even the larger groups of people and will detect and identify person more effectively and efficiently. Faster recognition is need of an hour because of how COVID-19 spreads. Apart from faster detection the user will get notification on his/her email address so that they would not forget to cover their face and stick to the terms set by the government for covering the face.

7.2 Future Work

Some of the future works can be considered are given below:

- Detection of Suspicious Person with covered face.
- Adding mobile message system for efficient and fast notification system.
- Adding alarm signal for alerting the person in the real time.

- Performing and adding new classification and detection model for getting better workflow of the system.

References

1. Yadav S, Javad I, for R. in A. Science and undefined (2020) Deep learning based safe social distancing and face mask detection in public areas for covid-19 safety guidelines adherence. researchgate.net. https://doi.org/10.22214/ijraset.2020.30560
2. Loey M, Manogaran G, Taha M, N. K.- Measurement, and undefined (2021) A hybrid deep transfer learning model with machine learning methods for face mask detection in the era of the COVID-19 pandemic. Elsevier, Accessed 10 Oct 2021. [Online]. Available https://www.sciencedirect.com/science/article/pii/S0263224120308289
3. Jiang M, Fan X, Yan H (2020) RetinaMask: a face mask detector, May 2020, Accessed 10 Oct 2021. [Online]. Available http://arxiv.org/abs/2005.03950
4. Rota P, Oberste M, Monroe S, … Nix W, and undefined (2003) Characterization of a novel coronavirus associated with severe acute respiratory syndrome. science.sciencemag.org Accessed 10 Oct 2021. [Online]. Available https://science.sciencemag.org/content/300/5624/1394.abstract
5. Leung N, Chu D, Shiu E, Chan K medicine, and undefined (2020) Respiratory virus shedding in exhaled breath and efficacy of face masks. nature.com, Accessed 10 Oct 2021. [Online]. Available https://www.nature.com/articles/s41591-020-0843-2?fbclid=IwAR1WI0_kJaGqvvm0VvRV1AEitGdgOJ0kAt52NB-Xv_AUY0Ce_C558kuzcbQ
6. Zhu X, D. R. vision and pattern recognition, and undefined (2012) Face detection, pose estimation, and landmark localization in the wild. ieeexplore.ieee.org. Accessed 10 Oct 2021. [Online]. Available https://ieeexplore.ieee.org/abstract/document/6248014/
7. Gupta M, Abdelsalam M, Mittal S (2020) Enabling and enforcing social distancing measures using smart city and ITS infrastructures: a COVID-19 use case, April 2020, Accessed 10 Oct 2021. [Online]. Available http://arxiv.org/abs/2004.09246
8. Sonn JW, Kang M, Choi Y (2020) Smart city technologies for pandemic control without lockdown. Int J Urban Sci 24(2):149–151. https://doi.org/10.1080/12265934.2020.1764207
9. Singh R, Javaid M, Haleem A, R. S.-D. & M. Syndrome, and undefined (2020) Internet of things (IoT) applications to fight against COVID-19 pandemic. Elsevier, Accessed 10 Oct 2021. [Online]. Available https://www.sciencedirect.com/science/article/pii/S1871402120301065
10. Jaiswal R, Agarwal A, Negi R-I. S. Cities, and undefined (2020) Smart solution for reducing the COVID-19 risk using smart city technology. Wiley Online Libr 2(2):82–88. https://doi.org/10.1049/iet-smc.2020.0043
11. Ponkia A, J. C.-I. R. I. & Video, and undefined (2012) Face recognition using PCA algorithm. researchgate.net, 2017, Accessed 10 Oct 2021. [Online]. Available https://www.researchgate.net/profile/Jitendra-Chaudhari/publication/275330326_FACE_RECOGNITION_USING_PCA_ALGORITHM/links/590c366b0f7e9b7fed8f78fd/FACE-RECOGNITION-USING-PCA-ALGORITHM.pdf
12. Lodh A, Saxena U, … A. M.-2020 4th, and undefined (2020) Prototype for integration of face mask detection and person identification model–COVID-19. ieeexplore.ieee.org Accessed 10 Oct 2021. [Online]. Available https://ieeexplore.ieee.org/abstract/document/9297399/
13. Wang Z et al. (2020) Masked face recognition dataset and application, March 2020, Accessed 10 Oct 2021. [Online]. Available http://arxiv.org/abs/2003.09093
14. R. D.-J. of I. I. P. (JIIP) and undefined (2020) Deep net model for detection of covid-19 using radiographs based on roc analysis. irojournals.com vol 02(03):135–140. https://doi.org/10.36548/jiip.2020.3.003

15. S. M.-J. of S. C. P. (JSCP) and undefined (2019) Study on Hermitian graph wavelets in feature detection. irojournals.com. https://doi.org/10.36548/jscp.2019.1.003
16. Zhao Z, Zheng P, Xu S, XW-I Transactions on neural, and undefined (2019) Object detection with deep learning: a review. ieeexplore.ieee.org, Accessed 10 Oct 2021. [Online]. Available: https://ieeexplore.ieee.org/abstract/document/8627998/

An Infrastructure-Less Communication Platform for Android Smartphones Using Wi-Fi Direct

A. Christy Jeba Malar, R. Kanmani, M. Deva Priya, G. Nivedhitha, P. Divya, and T. S. Pavith Surya

1 Introduction

The infrastructure-less communication setup for android devices using Wi-Fi Direct (Wi-FiDi) in a smart home environment can be employed for discovering and advertising themselves. Contract-oriented sensor-based application platform is an infrastructure-less data distribution platform in which sensor data is distributed freely in a Peer-to-Peer (P2P) network. The design aspects of Direct-to-Direct (D2D) communication are discussed by Fodor et al. [1] and Oide et al. [2]. Nowadays, most of the smartphones are equipped with Android 4.0 operating system.

A. C. J. Malar · T. S. P. Surya
Department of Information Technology, Sri Krishna College of Technology, Coimbatore, Tamilnadu, India
e-mail: a.christyjebamalar@skct.edu.in

T. S. P. Surya
e-mail: 20tuit110@skct.edu.in

G. Nivedhitha · P. Divya
Department of Computer Science and Engineering, Sri Krishna College of Technology, Coimbatore, Tamilnadu, India
e-mail: nivedhitha.g@skct.edu.in

P. Divya
e-mail: divya.p@skct.edu.in

M. D. Priya (✉)
Department of Computer Science and Engineering, Sri Eshwar College of Engineering, Coimbatore, Tamilnadu, India
e-mail: devapriya.m@sece.ac.in

R. Kanmani
Department of Electronics and Communication Engineering, SNS college of Technology, Coimbatore, Tamilnadu, India
e-mail: kanmani0808@gmail.com

© The Author(s), under exclusive license to Springer Nature Singapore Pte Ltd. 2023   135
R. Agrawal et al. (eds.), *International Conference on IoT, Intelligent Computing and Security*, Lecture Notes in Electrical Engineering 982,
https://doi.org/10.1007/978-981-19-8136-4_12

This paper focuses on D2D communication among smartphones for device discovery. Wi-Fi P2P technology allows devices with Wi-Fi to connect a camera to a printer, and mouse or keyboard to a computer. Smart devices can connect to one another or in a group. Wi-Fi users can benefit from this to share content and communicate with one another when Wi-Fi Access Points (APs) are available. Nodes cannot communicate with each other directly. All the communications are passed through router or AP. The AP acts as a mediator between networks. This infrastructure acts as a Wireless Local Area Network (WLAN) for all indoor environments [3].

Wi-FiDi allows devices to communicate without using Wi-Fi routers. It facilitates the devices of different manufacturers to connect and communicate at ease. Wi-FiDi is a type of Wi-Fi, wherein any smartphone can connect with another without the need for Internet or AP. There is no need for any infrastructure setting and it is like a P2P network. It does not allow multi-hop communication like ad hoc network [4].

The smartphones have built in soft APs. It is an excellent and secure replacement for Bluetooth-based communication. It works at high speed when compared to Bluetooth though protected with WPA2 encryption. Network connectivities can be established between smartphones and devices in a Wi-FiDi network [4, 5].

Bluetooth and Wi-Fi Direct devices support wireless communication in ad hoc manner [6]. They do not require any infrastructure to work. In both cases, one device act as the owner and others act as clients. Bluetooth permits only seven devices to be connected in a piconet. It is limited to a range of 10 m approximately 33 feet, whereas Wi-FiDi can support a range upto 100 m. So Wi-FiDi is preferable to connect more than 10 devices in a wide physical area. Complete comparative study is performed over Bluetooth and Wi-FiDi by considering education in smart classrooms using Mobile Ad hoc NETworks (MANETs) [7]. Table 1 shows the comparison between Bluetooth and Wi-Fi Direct communications.

Wi-FiDi is faster than Bluetooth which makes it more attractive for sharing huge files.

Table 1 Comparison between bluetooth and Wi-Fi direct	Parameter	Bluetooth 5.0	Wi-Fi direct
	P2P sharing	Yes	Yes
	Speed	1–3 Mbit/s	>54 Mbit/s
	Range	100 m	46–100 m
	Energy consumption	0.01–1.0 W	2–20 W
	Frequency	2.4 GHz	2.4 or 5.0 GHz
	Service discovery	Yes	Yes
	Supported devices	Smartphones, smart TVs, and laptops	

1.1 Wi-Fi Direct Architecture

Wi-FiDi uses the Time of Flight (ToF) approach defined by Fine Time Measurement (FTM) protocol specified in IEEE 802.11–2016. FTM calculates the accurate position of a mobile device from an AP. Each AP in the WLAN network is configured with its location in geospatial coordinates (latitude, longitude, and altitude) and civic addresses. It facilitates precise localization even for multi-floor structures. It supports high accurate localization and data services from service providers within a single network, without the need for separate installations for beacons. One single network within smart home environment provides Wi-Fi connectivity with high accurate services without additional hardware, site surveys and maintenance. Smartphones protect location data by confidential management frame exchange and authenticity in data origin. Once a smartphone is connected with an AP in Wi-Fi network, the location data exchange from smartphone is encrypted using WPA2TM along with some protected management frames. Compared with other location technologies, certain level of privacy is maintained on location data [8].

2 Fine Time Measurement (FTM) Protocol

Wi-Fi fingerprinting-based Indoor Positioning System (IPS) relies on fingerprint or signal strength measurement. Signal strength measurements are variable due to the noise in the environment and signal attenuation or by building structure which leads to an inaccurate positioning system. Due to the limited accuracy of path loss model and imperfect scalability of Received Signal Strength Indicator (RSSI) fingerprint-based positioning systems, industry vendors are in search of techniques which yield better accuracy. Wi-Fi Time of Flight (ToF) is a time-based range management protocol used for getting accurate positioning information [9]. Fine Time Measurement (FTM) protocol along with Angle of Arrival (AoA) is used to determine client positioning with distance and angle [10]. Wi-Fi RTT-based measurements make use of finger-printing and range-based techniques to get optimum accuracy in indoor positioning [11, 12]. In the early release of IEEE 802.11TM, it included techniques for time delay measurement. The measurement was in microseconds (μs), too coarse for any indoor application. Hence, several hardware design modifications were recommended for improving the timing resolution from μs to nanoseconds (ns). The solution given by IEEE 802.11TM involves a time delay in ns. This time delay measurement is done by FTM protocol [13] which is a P2P single user protocol enabling exchange of messages between the initiating station and the responding device. Wi-Fi signals travel in the speed of light. The speed of radio waves is used instead of signal strength. Wi-FiDi localization is rooted on the FTM protocol in IEEE 802.11-2016.

$$\text{Distance} = \text{time of flight} \times \text{speed of light} \qquad (1)$$

The time between transmissions are converted to distance by multiplying with speed of light as given in Eq. 1. For accurate location determination, it is important to synchronize two devices in communication to the same clock reference within an ns. This is done by making a RTT measurement. When a mobile device sends a Wi-Fi frame, the RTT measurement starts and the time is recorded.

When device 1 receives response from device 2, it notes the arrival time. RTT is the difference between device timestamps referred in the same clock. The RTT includes the time taken between sending a frame and receiving its response from the other side. The mobile device can record accurate timestamps to measure the RTT in several ns as illustrated in Fig. 1. The protocol allows many timestamps to be transmitted among mobile devices within the Wi-Fi range and timestamps are collected in one device to measure the RTT. When more than two devices are connected to one mobile device, it should send the measurement to one device, while the other is involved in calculation. Table 2 shows the comparison between time measurement and FTM protocol.

FTM protocol allows a mobile device to request either in scheduled measurement or in immediate measurement mode. In scheduled measurement mode, there is no

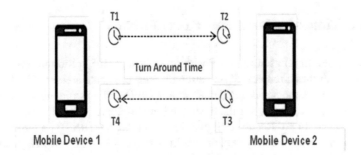

Fig. 1 Round Trip measurement to synchronize time clock

Table 2 Comparison between FTM and time measurement protocol

Parameter	Timing measurement	FTM
Standard	Established with IEEE 802.11–2012	Established with IEEE 802.11-2016
Frame	Class 3 frame	Class 1 frame
Timestamp resolution	In ns (10 ns)	In picoseconds (ps)
When frame sent	No indication of when TM frame will be sent	First frame should be sent at 10 ms after the receipt of initial request
Burst	No concept of burst	Scheduling and operational parameters of an FTM session are formulated
Frame spacing	No control on spacing between two consecutive frames	Min delta FTM

necessity to start an FTM session. The operational parameters of the session are scheduled based on the availability of resources. As Soon As Possible (ASAP), the value is taken as zero (ASAP = 0). In an immediate measurement mode, there is an urgent need to receive location information. The initiating mobile device can request another mobile device to start FTM session ASAP (ASAP = 1).The device captures the timestamps associated with initial frame and sends in the next frame. It involves less overhead but involves more complexity and resources. Figure 2 shows session management in immediate mode. FTM parameter contains a number of fields that are used in FTM session management. FTM frame is helpful in supporting FTM communication to report Time of Arrival (ToA) and time difference of arrival. The first burst instance begins at the time recorded in a partial timer specified in FTM frame format, regardless of ASAP field value. When ASAP value is set to zero by the responding station, the partial time value is set to less than 10 ms. The RTT is computed as shown in Eq. 2.

$$RTT = [(t4 - t1) - (t3 - t2)] \tag{2}$$

The distance is calculated by using the speed of light and various timestamps. Timing offset is calculated as

$$ToF = RTT/2 \tag{3}$$

$$Range = C^*ToF \tag{4}$$

where 'C' represents the speed of light.

2.1 Wi-Fi Direct Simulator

A Wi-Fi Direct simulator WiDiSi is an open source simulator for implementing Wi-FiDi networks. WiDiSi is based on Peersim large-scale simulator for P2P networks. It provides a dynamic large scale environment for Wi-FiDi network, where Android applications can be tested for some scenarios. A configuration test file is available for creating and modifying network parameters. Wi-FiDi Application Programming Interface (API) is similar to Wi-Fi P2P API in Android. Wi-FiDi-enabled smart devices are grouped and one device acts as Group Owner (GO) and plays the role of soft AP. In the discovery phase, device scanning for APs is the same as in traditional Wi-Fi. It also searches for application services available in the higher layer. Devices send probe request frames to establish communication. The GOs stay in the listening state to send response to the response frame.

Devices continuously move around, search and listen between 100 an 300 ms to establish connection. Following the discovery phase, a device can query about higher

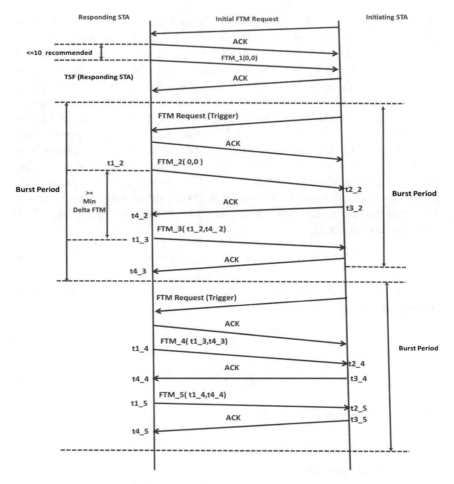

Fig. 2 FTM session management in scheduled mode

level application services. Currently, android supports Domain Name Service (DNS)-based service discovery to exchange higher level services. Based on the intended value in each device, it can negotiate to act as a GO. Intended value is determined based on the remaining battery and computational power decided by the application. The GO invites other unconnected devices to join the group, or unconnected devices can send request to the GO to establish a connection. Multiple smart devices are considered as nodes in the simulator. The nodes run in the node container which is responsible for including permanent nodes. Each node is capable of executing certain applications and represents the behavior of android devices. Each node includes Wi-FiDi interface that acts as P2P protocol in Android devices. Figure 3 shows the architecture of WiDiSi which is a single threaded simulator used for constructing Wi-Fi Direct infrastructure. Node Initializer is responsible for initializing nodes in environment when it is included. Proximal elements are defined by the proximity

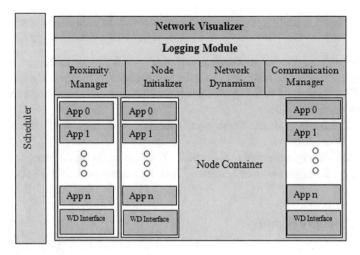

Fig. 3 WiDiSi architecture

manager. Network dynamism is responsible for node mobility, as Wi-FiDi devices are in motion in the network. Message exchange among nodes is handled by node communication manager. The logging component monitors the network behavior with the help of runtime information while the scheduler drives the simulation engine.

The behavior of P2P communication in Wi-FiDi devices are modeled through delays.

- Switching delay occurs while switching the channels in the discovery phase. It is the average time taken by two devices to find a common channel.
- Channel delay denotes the time needed for physical propagation of signals. It is the amount of time taken to exchange a frame from one device to another.
- Authentication delay is dependent on user authentication. It is the time taken for approving the authentication and moving on to the provisioning phase.
- Power management delay denotes the idle state time of a device.
- Internal processing delay is the time taken for processing a basic request.

2.2 Wi-Fi Direct Interface

Wi-FiDi interface has classes that replicate the behavior of Android's Wi-FiDi implementation. They are WifiP2pManager, NodeP2pinfo, and EventListener. WifiP2pManager replicates the behavior of Android's android.net.wifi.p2p.WifiP2pManager. This class is helpful for managing Wi-Fi P2P connectivity. The application can discover and setup peers and query services. WiDiSi is a single threaded simulator where all internal components are synchronized around Peersim. Android APIs are multi-threaded. To overcome this,

Peersim can be used in hybrid mode. The following APIs are supported in Wi-Fi Direct interface.

- **WifiP2pManager.ConnectionInfoListener** - It is similar to that of android.net.wifi.p2p class and is used when connection information is available
- **WifiP2pManager.DnsSdServiceResponseListener** - It is used when connection information is available
- **WifiP2pManager.DnsSdTxtRecordListener** - It is used when Bonjour TXT record is received
- **WifiP2pManager.GroupInfoListener** - It listens to group information
- **WifiP2pManager.PeerListListener** - It is used when peer lists are updated

2.3 Proximity Manager

To keep the network updated, nodes are to be added and removed over time. Node mobility is not possible in Peersim. Node movement is possible in WiDiSi using the following functionalities. The geolocation of each node is kept in the instance of INET coordinate protocol and it is represented as (x, y) coordinates. INET-Initializer is responsible for maintaining the location of each node when it is included in peer group. The node movement control component is responsible for node mobility based on the scenario. It updates the proximity list of each node based on the new location. The proximity list is maintained by the neighbor list linkable protocol installed in each node. The proximity observer checks the neighbor list linkable protocol for changes. If any changes occur in mobility of the node, it is informed to the event listener for further actions.

3 Results and Discussion

Nodes represent smart devices in real Wi-Fi Direct network. The geo-position of each node is maintained as INET coordinates. The proximity controller is responsible for maintaining the node movement. The simulation environment covers an area of 500 m with an average for 100 nodes. All the devices move around within the network range. They move at a speed of 0–15 m/s. The signal strength reduces when the distance between the nodes increases. It is given by

$$PathLoss = 20 \log_{10}(d) + 20 \log_{10}(f) + 32.44 - G_{tx} - G_{rx} \qquad (5)$$

where
f - Channel frequency (MHz)
G_{tx} - Gain in the transmitter
G_{rx} - Gain in receiver

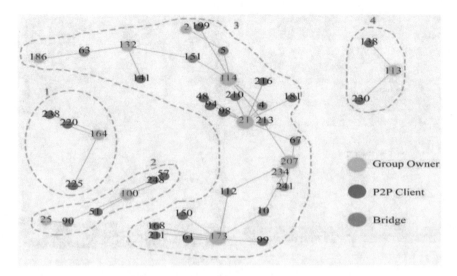

Fig. 4 Node representation and P2P group formation

Figure 4 shows the node representation and peer group formation. GOs in each group are represented in green. Clients are represented as blue color nodes. Nodes are grouped either in standard, autonomous or persistent mode. The nodes can identify nodes positioned at proximity by using APIs running on the node. Each node represents an Android smart device.

3.1 Positioning of a Mobile Device with Wi-Fi RTT

The Android P introduces some new features to measure RTT from APs or other peer group devices. Figure 5a shows the Wi-Fi scan phase in a RTT tool to discover whether the missed smart device is available in the scan list. It lists all the nearby devices in the Wi-Fi Direct range.

The range taken is around 500 m. When a person misplaces his mobile somewhere in a smart home environment, it can be scanned and identified from another smart device by forming a peer group. The device need not connect to an AP to measure RTT. The AP does not maintain user information of mobile devices, and hence, the privacy of users will not be affected. Figure 5b shows the place of a mobile device in an indoor building. It shows the navigation for the mobile device from the search place. If the device's RTT is measured from three different APs, the technique is called multi-lateration where the average positioning accuracy is found to be 1-2 m. With these APIs provided in Android P, new user experiences like indoor navigation, advertisement on offer products and providing fine-grained indoor location services can be created.

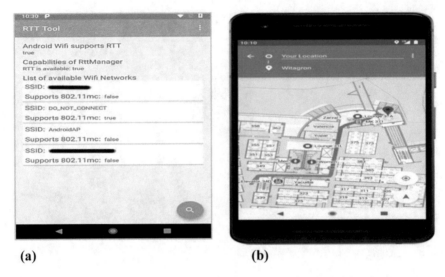

Fig. 5 **a** Wi-Fi scan phase and **b** Navigation of a mobile device from search location

4 Conclusion

Positioning of a smart device in an indoor environment is possible with the help of FTM protocol and RTT measurement in a smart building. D2D communication is possible by forming peer groups in a Wi-FiDi network. In a Wi-FiDi network, each device acts as an AP for connecting with other devices. Neighbouring devices can be discovered and communication can be established at higher speeds than other wireless technologies. The speed of RTT distance measurement is 0.02 s, and the range is also extended to 1 km covering the entire smart building. In the future, this work can be extended to provide location-based services about offers and navigation to target locations to customers in shopping malls.

References

1. Oide T, Abe T (2017) Suganuma: COSAP: contract-oriented sensor-based application platform. IEEE Access, 5:8261–8275
2. Fodor G, Dahlman E, Mildh G, Parkvall S, Reider N, Miklós G, Turányi Z (2012) Design aspects of network assisted device-to-device communication's. IEEE Commun Magazine 50(3)
3. Win HT, Pathan ASK (2013) On the issues and challenges of fiber-wireless (Fi-Wi) networks. J Eng
4. Casetti C, Chiasserini CF, Duan Y, Giaccone P, Manriquez AP (2017) Data connectivity and smart group formation in Wi-Fi direct multi-group networks. IEEE Trans Netw Serv Manag 15(1):249–255
5. Naik S, D'Souza M (2019) Efficient power saving method for WiFi direct devices in IoT based on hidden Markov model. In:11th IEEE international conference on communication systems

and networks (COMSNETS), pp 565–567

6. Hameed A, Ahmed HA (2018) Survey on indoor positioning applications based on different technologies. In: 12th IEEE international conference on mathematics, actuarial science, computer science and statistics (MACS), pp 1–5

7. Endo K, Fujioka G, Onoyama A, Okano D, Higami Y, Kobayashi S (2018) Evaluation of educational applications in terms of communication delay between tablets with Bluetooth or Wi-Fi Direct. Vietnam J Comput Sci 5(3):219–227

8. Huang L, Wang P, Liu Z, Nan X, Jiao L, Guo L (2019) Indoor three-dimensional high-precision positioning system with bat algorithm based on visible light communication. Appl Opt 58(9):2226–2234

9. Banin L, Schtzberg U, Amizur Y (2013) Next generation indoor positioning system based on WiFi time of flight. In: Proceedings of the 26th international technical meeting of the satellite division of the institute of navigation (ION GNSS+ 2013) pp 975–982

10. Rea M, Abrudan TE, Giustiniano D, Claussen H, Kolmonen VM (2019) Smartphone positioning with radio measurements from a single wifi access point. In: Proceedings of the 15th international conference on emerging networking experiments and technologies, pp 200–206

11. Hashem O, Harras KA, Youssef M (2021) Accurate indoor positioning using IEEE 802.11 mc round trip time. Pervasive and Mobile Comput 101416

12. Huilla S, Nigussie E, Thanigaivelan N, Horsmanheimo S (2019) Smartphone-based indoor positioning using Wi-Fi fine timing measurement protocol. MSc Thesis

13. Ibrahim M, Liu H, Jawahar M, Nguyen V, Gruteser M, Howard R, Bai F (2018) Verification: accuracy evaluation of WiFi fine time measurements on an open platform. In: Proceedings of the 24th annual international conference on mobile computing and networking, pp 417–427

Prioritization in Data Warehouse Requirements—Incorporating Agility

Hanu Bhardwaj and Jyoti Pruthi

1 Introduction

New possibilities and markets, dynamic economic conditions, and the creation of rival products and services all require organizations to respond. Software systems have been the backbone of most corporate operations in recent years. A data warehouse is an integrated environment in which data from many systems are combined and presented in a logical manner. Data warehouses contain a variety of components, including data in various formats from various sources, a database, extract, transform, and load (ETL) operations, dashboards, and so on [1].

The most time-consuming and expensive task in establishing a data warehouse is data integration utilizing ETL from heterogeneous sources, which accounts for more than half of the project price [2]. Furthermore, business customer's requirements in data warehouse projects are constantly changing, making the typical waterfall approach difficult to follow [3].

Project management by using data warehouse is distinct from most other types of software project management; actually, a data warehouse is never going to finish truly. Every phase of a data warehouse project has a start and end date; however, the data warehouse will never reach its completion point. Traditional assignment of roles and duties appears to result in excessive rework, and the waterfall methodology does not appear to function for project control.

H. Bhardwaj (✉) · J. Pruthi
Department of Computer Science and Technology, Manav Rachna University, Faridabad, Haryana, India
e-mail: hanu_mru@mru.edu.in

J. Pruthi
e-mail: jyoti@mru.edu.in

© The Author(s), under exclusive license to Springer Nature Singapore Pte Ltd. 2023   147
R. Agrawal et al. (eds.), *International Conference on IoT, Intelligent Computing and Security*, Lecture Notes in Electrical Engineering 982,
https://doi.org/10.1007/978-981-19-8136-4_13

Projects involving data warehouses are always evolving and dynamic [4]. These qualities make data warehouse project management difficult and distinctive; they are also a major reason why agile methodologies are appropriate.

1.1 Agility Principle in Software Development

All the twelve principles of Agile Software Development (ASD) are an umbrella title for a group of concepts/activities which are based on the principles and ideas, and it is mentioned in the ASD's Manifesto as well. It can be said to be a perfect idea to apply the said principles and values and to use them to figure out what to do in a particular scenario [5].

Agile is different from conventional software development methodology as it lays emphasis on the people doing the job and their collaboration. Cross-functional and self-organizing teams collaborate so as to lead to best solutions.

Cooperation and the self-organizing team are highly valued in the ASD community. People from several departments make up teams. Those teams don't need to have separate occupations; instead, they should make sure they have all of the appropriate skill sets when they get together [6].

1.1.1 Prioritization Strategies

Prioritization holds an important place in ASD. Rather than waiting for the whole project to be delivered, it is better to make the most important functionality available to the stakeholder at the earliest [4]. This can be achieved by applying prioritization on the requirements efficiently.

Backlog prioritizing is necessary to categorize the product backlog items (User Stories, defects, spikes, etc.) in order to determine the development and deployment sequence. During sprint planning, the scrum team follows the sequence to identify product backlog items. The influence factors may be as follows [7]:

1. **Customer satisfaction:** It is one of the contributing variables for prioritizing product backlog items. We constantly prioritize stories with the factor of customer satisfaction in mind, allocating a high priority to functionality (backlog item) that has a high possibility of increasing customer satisfaction. Customers may get unsatisfied if those functions are not prioritized first.
2. **Economic Value:** Prioritizing functionality requires a thorough understanding of the company's strategy and goals. As a result, the best person to prioritize product backlog items is the product owner or the business analyst. Prioritize the product backlog item by determining which feature will bring the most value to satisfy the business goal.
3. **Complicatedness:** Functionality that is sophisticated or tough should be started early in the project. Because the team is new, it is easy to assign talented and

experienced members. In addition, if it takes longer than expected, the team gets more flexibility in adjusting the extra time.

4. **Opportunity and Risk:** Prioritizing high-risk functionality at the top of the list helps to maximize early mitigation. When high-risk functionality is postponed until later in the project, the overall risk of implementation rises.

5. **Price:** High costs do not always imply poor priority; we must consider the cost and ROI of each service logically. Less cost–more return will undoubtedly be a top priority, while more cost–less return may be a low priority. However, when other criteria of prioritizing are taken into account, less cost–more return should also be at the top of the list. Cost isn't the sole consideration; we must also consider the functionality's return when prioritizing.

1.1.2 Techniques to Prioritize

There are multiple ways or strategies for prioritization that are mentioned below [8]:

a. **Affinity Analysis:** In the method, the user stories are divided into categories such as high, low, medium, and medium, and select backlog items to plan or construct. This is a set of basketing approaches for prioritizing items of product backlog into groups [7, 8].

b. **Kano Analysis:** By using this technique, the backlog items or functions are being organized, keeping the customer pleasure element in mind. Different aspects of customer satisfaction are shown on the Kano Analysis chart, and we put or arrange the backlog items in particular areas, prioritizing them as per the same [7, 8].

c. **MoSCoW Method:** Another way to prioritize work items in product development is to use MoSCoW. The term comes from the initial letter of each of the four priority categories. Another method of analysis of affinity is to divide the product backlog into 4 category [7]. As the names of each & every category suggest, the teams prioritize must have the list of functionality first, followed by should have, could have & would like to have, in order to maximize the return on investment and avoid losing high-value functionality due to time or money constraints.

d. **Ranking:** A rank value (1,2,3,4,5...) is another means of ranking stack of backlog items. We use this strategy to stack our backlog items one on top of the other. The priority of Rank 1 is higher than that of Rank 2. The higher rank (lower value) or high priority displays above the low rank in our backlog (high value). We prioritize grooming, planning, and constructing the highest-ranking product backlog item before moving on to the lower-ranking items [8].

Determining how to prioritize new functionality and features is one of the most difficult aspects of agile software development. When creating development roadmaps, even the most experienced product owners and project managers may find it challenging to prioritize what to work on first. Prioritization is an important component of product strategy that should be examined and improved on a regular basis [4].

1.2   Agility in Data Warehouse

An agile data warehouse enable a company to progress from reporting and historical analysis to more advanced, predictive analytics. It also encourages the incorporation of new data types and structures into complicated behavioral models, resulting in a level of comprehension previously unattainable. Data warehouses contain a variety of components, including data in various formats from various sources, a database, extract, transform, and load (ETL) operations, dashboards, and so on [1].

The most time-consuming and expensive task in establishing a data warehouse is data integration utilizing ETL from heterogeneous sources, which accounts for more than half of the project price. Furthermore, business customer's requirements in data warehouse projects are constantly changing, making the typical waterfall approach difficult to follow [9].

The goal of agile data warehousing is to reduce time to value. To begin with a specific source or area of data and work on it from that point itself, from the input of the data to visualization, enables the organization to look or get visible output and achieve the desired goals.

With an agile data warehouse, introducing complete new datasets from new data sources, which typically take weeks or months to effectively integrate into the data warehouse and overall analytics workflow using traditional methods, can be done in days or even hours [9]. By supporting open-ended data discovery and self-service analytics, an agile data warehousing system puts data and analysis in the hands of people who need it. The teams can obtain more flexibility in addressing new requirements while also improving the quality and response time to new requirements when they use agile methodology for data warehousing initiatives. Agile provides value to businesses on a regular basis and aids in project management, particularly for data warehouse projects, which are typically vast and complicated [10].

1.3   Prioritization Strategies

As mentioned in the above section, there are many prioritization techniques available in software development but no such fixed techniques or strategies available for prioritization in data warehousing. In the proposed model, we are trying to build a mechanism to prioritize the input that can go in the data warehouse and enable the decision-making process faster and effective. As the agile approach is used in the proposed work that will make the delivery of the work starting from the very beginning rather than waiting for the entire data warehouse to be in place, this will be readily available inputs.

2 Overview of Previous Work

Customer'S desire to purchase software systems is determined by how well the product satisfies their demands. Prioritization provides prospects for good results and client satisfaction [11, 12]. Prioritization of needs is defined as the process of identifying and ordering critical system requirements based on their relevance [13, 14]. Releases or iterations are then used to develop the requirements. The concept is that the most important demand should be implemented first, followed by the others [15, 16].

Several stakeholders decide which requirements should be executed as releases during the requirements prioritization process [16, 17]. Developers are the most powerful stakeholders in the needs prioritization process since they are skilled individuals. Customers with authority over the system, on the other hand, are the best judges of requirements priority. Customers and developers must agree on which requirements should be prioritized [18]. This procedure is challenging, especially when it involves several parties who must achieve an agreement in a frantic and often disoriented environment [15].

The needs are chosen using a specific requirements prioritization technique that must first be determined. The end result is a prioritized project backlog, which is a collection of critical requirements for the project [16, 18]. The most typical criterion for prioritizing needs is commercial value; however, other factors like complexity, stability, and reciprocal interdependencies are also taken into account [19].

Aside from the project, the type of the requirements has an impact on the process of prioritization. Dependencies of requirements increase the complexity of the same, making it more difficult to choose a specific iteration [18]. There are two types of dependencies: chronological and architectural. It is possible that a condition that is deemed complex and hazardous will not be executed [20]. The relevance and an urgent need of the gathering of requirements also have a role, which is heavily influenced by the stakeholder's perspectives [20]. Most important requirements are functions that are required at the early stage of the process and provide the organization with strategic business value [21, 22]. Along with it, the volatility and stability of needs should be considered, as ever changing requirements have an impact on a project's cost and schedule [23].

3 Proposed Algorithm

In the previous work [24], we have elicited indicators for finalizing requirements of the data warehouse. Once indicators have been elicited, an algorithm is being proposed for choosing indicators, from indicator hierarchy for every sprint, so as to develop data warehouse in incremental phases.

We assume that information regarding reporting hierarchy, informs hierarchy, indicator hierarchy, indicator-position pair (consumer/producer) are available with us [24]. Also, information regarding the deadline and frequency of every indicator has been elicited.

Do until sprint capacity is full
A. For each indicator present at leaf level (Left to Right direction)
B. Find the level of position for which indicator is being produced
C. Pick the indicator with highest level/position (indicator hierarchy)
D. In case of tie amongst the indicators at highest level (least distance), check the number of nodes of the tree of indicator hierarchy to which indicator belongs as leaves.
E. Pick the indicator with least number of nodes
F. In case of tie amongst the indicators on account of the same number of nodes, find the number of contributing indicators for each of them, which will go in computing the next level indicator. (Find the siblings of indicator)
G. Choose the sibling set with larger cardinality
H. If cardinalities are the same, check the number of positions that each indicator in the sibling set is being consumed by.
I. Pick the indicator sibling set being used by highest number of positions
J. In case of tie, check the deadline for each indicator
K. Pick the indicator with least deadline period
L. In case of tie, check the frequency of delivering indicator
M. Pick the indicator with highest frequency
N. In case of tie, the choice of choosing indicator is done by product owner, after consulting concerned positions, based on urgency etc.

3.1 Prioritization Strategy in Proposed Algorithm

We have considered following parameters while framing the algorithm for prioritizing and selecting indicators. k and k' represent indicators identified in indicator hierarchy.

1. Given an indicator hierarchy, indicators at leaves must be given priority over the one which is higher in the hierarchy.
2. If the distance of propagation of k is smaller than that of k', then k is preferred.
3. If the number of indicators contributing to compute k is smaller than that for computing k', then k is preferred.
4. If the deadline for delivering k is closer than delivering k', then ki is preferred.
5. If frequency of delivering k is higher than that of k', k is preferred
6. If k is being produced for a position at a higher level than that of the position receiving k', then prefer k.
7. If k being produced is consumed/used by more number of positions than that of k', the k is preferred
8. Product owner to break ties at last.

3.2 Explanation and Example

In our previous work [24], we explored to elicit indicators by exploring reports hierarchy which is obtained from the organization chart. Further from 'Reports Hierarchy' which conveys formal relationship in an organization, 'Informs Hierarchy' was explored which conveys informal relationship other than organization chart but relevant for indicator exchange. From reports and informs hierarchy, we elicited indicators and finally constructed indicator hierarchies which describe the computation structure of every indicator. We also elicit the information regarding the deadline and frequency of each indicator as this information shall help us in prioritizing the indicators.

So, for implementing the proposed algorithm, we start traversing the indicator hierarchies (trees) from leaf level from left to right. Assume that the indicators at leaf level are I1, I2, I3, I4, I5, and I6. We need to check the position for which these indicators are being produced, assume I1 is being produced for P3, and similarly the indicator-consumer position elicited are I1-P3, I2-P4, I3-P4, I4-P4, I5-P4 and I6-P2. Assuming that P4 is the highest level, we observe that there is a tie in I2, I3, I4 & I5 as they are being produced by the highest and same level of positions. So, to break the tie, as per point D in the algorithm, we check the total number of nodes of the tree to which each of these indicators belong.

Assume that I2 belongs to a tree (hierarchy) with 10 nodes, and similarly, I2-10, I3-10, I4-10, and I5-20 information is checked. Here, we again see that I2 and I4 qualify for the next step as they belong to the tree with the least and same number of nodes. Next as per step F of the proposed algorithm, to break the tie between I2, I3, & I4, we check the count of their siblings which contribute with them to calculate the next-level indicator. Assuming that I2 has 5 siblings, and similarly, I2-5, I3-5, and I4-3 information regarding the sibling set is fetched which leads to another tie between I2 and I3 as they have the maximum and same number of sibling indicators. To break the tie at this level, as per point H, we need to check the number of positions which are consuming the siblings. Assuming that siblings of both I2 and I3 are consumed by 10 positions each, we again need to break the tie as per point J. Here, we shall check the deadline of indicators I2 & I3, assuming that both have the same deadline and have to be completed by the same time, we continue having a tie between them. Therefore, we proceed further to point L to break the tie based on the frequency (Daily/weekly/Monthly/Quarterly and so on..) of I2 and I3. Assuming that both have the frequency of usage as weekly, we are not able to break the tie and ultimately the decision to choose between I2 & I3 lies in the hands of the product owner, based on factors like urgency or business needs.

In the above example, we explored till the last step, i.e., N to choose the indicator that should be decided for getting included in the ongoing sprint. The above steps are repeated recursively till a point is reached where sprint capacity is exhausted. Thus, the output of the above algorithm is a set of indicators that need to be handled for phase-wise data warehouse design. This way, as per the various factors used for prioritization, data warehouses are constructed in an iterative manner (Phase-wise) to avoid delay in decision-making.

4 Conclusion

In the proposed algorithm, the priority of choosing indicators has been decided on the basis of different factors. The proposed work can be applicable in designing data warehouse in incremental stages in any of the scenarios or industries wherein an organizational formal or informal hierarchical structure exist, and there is exchange of indicators/key information among the positions. This shall enable the availability of decision-making process quick and effective as there is no need to wait for the entire data warehouse to be in place.

References

1. Taylor D (2021) ETL (Extract, Transform, and Load) process in data warehouse, October 2021. https://www.guru99.com/etl-extract-load-process.html
2. McKnight W (2014) Data warehouses and appliances, information management. Strategies for Gaining a Competitive Advan with Data 2014:52–66. https://doi.org/10.1016/B978-0-12-408 056-0.00006-0
3. Borhan NH, Zulzalil H, Mohd Ali N, Hassan S (2019) Requirements prioritization techniques focusing on agile software development: a systematic literature review. Int J Scientif Technol Res 8(11):2118−2125
4. Olaronke I, Ikono R, Gambo I (2018) An appraisal of software requirement prioritization techniques. Asian J Res Comput Sci 1(1):1–16. Article no. AJRCOS.40763. https://doi.org/10.9734/ajrcos/2018/v1i124717
5. Morlion P (2021) The 12 Agile principles: what are they and do they still matter? June 2021. https://www.plutora.com/blog/12-agile-principles
6. Maarit L, Similä JK, Abrahamsson P (2013) Definitions of Agile software development and agility, communications in computer and information science. In: Conference: 20th European conference, EuroSPI 2013, June 25−27, vol 364. https://doi.org/10.1007/978-3-642-39179-8_22
7. Mahapatra N (2017) Backlog prioritization in Agile software development. https://agiledigest.com/agile-digest-tutorial-2/backlog-prioritization/
8. Omeyer A (2020) 5 Best ways to prioritize your product Backlog, April 20. https://medium.com/agileinsider/5-best-ways-to-prioritise-your-product-backlog-a761fadc8862
9. Herschel RT, Rahman N, Rutz D, Akhter S (2013) Agile development in data warehousing, principles and applications of business intelligence research, Jan pp 286–300. https://doi.org/10.4018/978-1-4666-2650-8.ch020
10. Tobias K, Sebastian O (2015) The impact of in-memory technology on the agility of data warehouse-based business intelligence systems—a preliminary study among experts. In: Wirtschaftsinformatik proceedings 2015. vol 44. https://aisel.aisnet.org/wi2015/44
11. Mulla N, Girase S (2012) Comparison of various elicitation techniques and requirement prioritization techniques. Int J Eng Res Technol (IJERT) 1(3). ISSN 2278-0181
12. Wiegers K (1999) First things first: prioritizing requirements, Software Development, Process Impact. www.processimpact.com

13. Bebensee T, van de Weerd I, Brinkkemper S (2010) Binary priority list for prioritizing software requirements. In: Requirements engineering: foundation for software quality, pp 67–78. https://doi.org/10.1007/978-3-642-14192-8_8

14. Karlsson J, Wohlin C, Regnell B (1998) An evaluation of methods for prioritizing software requirements. Inf Softw Technol 39(14):939–947

15. Paetsch F, Eberlein A, Maurer F (2003) Requirements engineering and agile software development, WET ICE 2003. In: Proceedings. Twelfth IEEE international workshops on enabling technologies: infrastructure for collaborative enterprises 2003. https://doi.org/10.1109/ENABL.2003.1231428

16. Razali R, Anwar F (2011) Selecting the right stakeholder for requirements elicitation: a systematic approach. J Theor Appl Inf Technol 33(2):250–257

17. Svensson RB, Gorschek T, Regnell B, Torkar R, Shahrokni A, Feldt R et al (2011) Prioritization of quality requirements: State of practice in eleven companies. In: IEEE 19th international requirements engineering conference 2011. https://doi.org/10.1109/RE.2011.6051652

18. Bakalova Z, Daneva M, Herrmann A, Wieringa R (2011) Agile requirements prioritization: what happens in practice and what is described in literature. In: Requirements engineering: foundation for software quality, pp 181–195. https://doi.org/10.1007/978-3-642-19858-8_18

19. Jarzębowicz A, Sitko N (2020) Agile requirements prioritization in practice: results of an industrial survey. Proc Comput Sci 176:3446–3455. https://doi.org/10.1016/j.procs.2020.09.052

20. Wohlin C, Aurum A (2005) What is important when deciding to include a software requirement in a project or release?. In: International symposium on empirical software engineering. https://doi.org/10.1109/ISESE.2005.1541833

21. Lehtola L, Kauppinen M, Kujala S (2004) Requirements prioritization challenges in practice. In: Product focused software process improvement, pp 497–508. https://doi.org/10.1007/978-3-540-24659-6_36

22. Lutowski R (2005) In: Software requirements: encapsulation, quality, and reuse: CRC Press. ISBN 9780849328480

23. Lauesen S (2002) In: Software requirements: styles and techniques: Addison-Wesley Professional. ISBN-13: 978-0201745702

24. Bhardwaj H, Prakash N (2016) Eliciting and structuring business indicators in data warehouse requirements engineering. J Expert Syst 33(4):405–413. Wiley Publishing Ltd. https://doi.org/10.1111/exsy.12165

Module Allocation Model in Distributed Computing System by Implementing Fuzzy C-means Clustering Technique

Shipra Singh and Deepa Gupta

1 Introduction

With the advancements within the era dispensed processing structures poses new challenges every day which captivates the attention of many researchers to paintings on this domain. If the proper gain of available resources and heterogeneous processing talents is taken into consideration, then in DCS the overall reaction time of any software can be minimized by way of distributing it over several available computing modules. Module allocation in allotted processing system reveals an extended software in environments where a huge amount of information has to be processed in a quick time or wherein actual-time computations are required. DCS is used to explain a system each time multiple computers are interacting in order that a group of tasks runs on a machine with a couple of processing units. DCS allows speedy calculation that ease the equally aligned execution of application tasks. Partitioning programs into obligations and their assignments are two cornerstone steps in DCS design. Accomplishment of these steps must be achieved effectively, a growth in the tally of processors inside the device may additionally virtually bring about a decrease inside the overall fee because of the saturation effect due to excessive communique among the processors. DCS performance may be near to gold standard best if the disbursed duties and processing modules are mapped well. As an end result, a properly-prepared assignment allocation strategy is needed to properly use processing devices and decrease inter-processor communication when interacting duties are living on distinctive processors. To avail the maximum output with the available assets, it's miles obligatory to reduce the reaction time by using mapping the strategies to the processing modules in step with the inter-module verbal exchange fee, as a way to limit the communique among the modules with the aid of plotting

S. Singh (✉) · D. Gupta
Infosys Ltd., Bangalore, India
e-mail: singh.shipra@infosys.com

the clusters as in step with maximally speaking techniques mapped onto the same processing module. DCS lets in allocation of obligations the usage of any of the two techniques: Dynamic allocation and Static allocation. Static allocation does not allow reassignment of the responsibilities over processors in the course of execution, while dynamic allocation allows reassignment of the duties over processors throughout execution.

The graph-theoretic method has been applied by means of several researchers to the module allocation trouble with the goal of minimizing general execution and conversation costs which includes Bokhari [1], Shen and Tsai [2]. Tindell et al. [3] defined a technique to solving the static module allocation trouble using a way referred to as simulated annealing. Peng et al. [4] have derived premier module assignments to reduce the sum of module execution and conversation charges by way of the usage of the branch and sure method. Attiya and Hamam [5] discuss the trouble of module allocation in a heterogeneous distributed gadget with the aim of maximizing the system reliability then gift a heuristic algorithm derived from a well-known simulated annealing (SA) technique to solve the problem.

Yadav et al. [6] provided an green module scheduling version in a distributed processing device the use of ANN. Martinez et al. [7] taken into consideration the trouble of allocating fixed-priority fork-be a part of allotted real-time modules onto allotted multi-middle nodes linked thru a flexible time-brought on switched Ethernet network. Kumar et al. [8] gift a dynamic module scheduling algorithm for distributed computing systems beneath a fuzzy surroundings. Akbari and Rashidi [9] proposed a set of rules based on a multi-goal scheduling cuckoo optimization algorithm to lessen execution time at the same time as permitting for maximum parallelization. Crespo et al. [10] gift a method to generate a static time table for a multi-core partitioned machine through considering numerous troubles: partitions with specific tiers of severely, hierarchical agenda, allocation of partitions to cores and era of the global agenda. Kumar et al. [11] labeled a brand new heuristic model for the allocation of modules to more than one processors following to obtain most desirable price and highest quality reliability of the gadget. When the comparable data objects are measured via features in ways then it is partitioned into multiple organizations called clusters. The algorithms of clustering have been evolved to remedy a specific problem in a particularized subject. Clustering techniques are categorized into hard clustering (each information belongs to simplest one cluster) and soft or fuzzy clustering (each information belongs to each cluster in keeping with their membership grade). Many researchers have been utilizing the approach of clustering using fuzzy technique. Yadav et al. [12] developed fuzzy membership features for making the clusters of modules with the limitations to maximize the throughput and minimize the parallel execution time of the system. Commutable fault spotting and prognosis based on an evolutionary fuzzy classifier are given via Lemos et al. [13]. Then, Velmurugan et al. [14] have started a performance-based totally analysis among okay-manner and algorithms for C-means clustering for connection-intended telecommunication statistics. Zhou et al. [15] proposed a collaborative fuzzy clustering set of rules in dispensed network environments. Nayak [16] has evolved a fuzzy good judgment-based clustering set of rules for Wi-Fi sensor networks to increase the community lifetime. Kumar et al. [11] designed a model based on mathematical implementation to allocate the modules onto the processors, their effort is to gain best reliability and ideal value of the devices. This paper states a venture undertaking problem and has numerous objectives: minimizing the reaction time, minimizing the machine fee and that too with least number of assumptions.

The stated model utilizes C-means clustering based on fuzzy approach for mapping the cluster of modules to lessen inter-modules verbal exchange value. To estimate the effectiveness of the algorithm said in the paper, the model has been carried out the usage of R Programming and the clusters of tasks has been plotted. This system is impartial of the number of processors and obligations therefore may be applied for any quantity of processing devices and processes. The shape of the paper is as follows: The obligations task trouble is stated in the first location observed by using proposed set of rules structure. And later the effects of the implementation are stated together with the evaluation with other proposed models and subsequently the paper then concludes.

2 Problem Statement and Definitions

Considered a distributed program consisting of a set $P = \{P_1, P_2,…, P_n)$ of 'n' heterogeneous processors and a set $M = \{M_1, M_2,…,M_m\}$ of 'm' tasks with some inter-tasks communication cost, these tasks will be grouped into clusters for execution on these processors. The assignment procedure is executed with the help of function A from the cluster of tasks in set M to the processors in set P:

A: $M \rightarrow P$, where $A(i) = k$ if module M_i is assigned to processor P_k, $1 \leq i \leq m$, $1 \leq k \leq n$.

Aim of the above function A is to achieve AO, i.e., optimal allocation, AO must reduce the response time of the entire system with the help of systematic mapping of processors and tasks. DCS will make the best out of resources available by spreading out the load on all the available processors such that the maximally communicating tasks can be assigned to the same processing unit.

2.1 Definitions

2.1.1 Fuzzy Execution Cost (EC)

The fuzzy execution cost $\tilde{e}_{i,k}$ is the amount of data required for the execution for module M_i on the processor P_k. The execution cost for all the tasks on each processor are fuzzy triangular numbers and the values are considered as matrix of fuzzy execution cost, FECM $= [\tilde{e}_{i,k}]$ of order $m \times n$. For the allocation on each processor the fuzzy execution cost A, is calculated as:

$$FEC(k) = \sum_{\substack{1 \le i \le m \\ i \in MS_k}} \widetilde{e}_{i,A(i)},$$

where $MS_k = \{i: A(i) = k, k = 1, 2 \ldots n\}$

2.1.2 Fuzzy Inter-Module Communication Cost (FIMCC)

The transferable quantity of data needed from module M_i to module M_j is denoted by $d_{i,j}$. The matrix $DT = [d_{i,j}]$ signifies the communication of data among the tasks while the execution. Matrix $DTR = [\widetilde{r}_{i,j}]$ of size n x n is used to store the data transfer rate among the processors. The FIMCC between modules M_i and M_j is defined by:

$$\widetilde{c}_{i,j} = \left\{ \begin{array}{ll} \frac{d_{i,j}}{\overline{DTR}}, & \text{if } i \ne j \\ 0, & \text{if } i = j \end{array} \right\}$$

where $\overline{DTR}$, is the average transfer rate among the processors in the processor's domain. The communication between the modules are taken in form of fuzzy inter-module communication cost matrix, FIMCCM $= [\widetilde{c}_{i,j}]$ of order m × m. The fuzzy inter-module communication cost for each processor for a given allocation A is obtained as:

$$\text{FIMCC(k)} = \sum_{\substack{1 \le i \le m \\ i+1 \le j \le m \\ A(i) \ne k}} \widetilde{c}_{i,j}$$

2.1.3 Response Time of System

The response time of the system is a function of calculating the amount to be performed by each processor and calculation time. This function is defined by considering the processor with the calculation load and communication heavier aggregate. The system response time for a given assignment A is establish as:

$$\text{RT(A)} = \max_{1 \le k \le n} \{\text{PEC(k)} + \text{IPCC(k)}\}$$

2.2 Assumptions

These assumptions have been made to designing the algorithm.

1. Any task may take variable execution instances to execute on exclusive processors and the conversation cost among the tasks can also vary. Also the overall value of the machine is computed the use of the execution and conversation costs. And the execution price of any undertaking is taken as endless in case that venture cannot be executed on a certain processor.
2. Assume that when the processing of an undertaking is completed, the output is stored in the nearby storage by way of the processing unit. And if some other task on the equal processor desires that output, it may be accessed from the nearby memory. Consequently, the communique fee is taken into consideration zero in among the responsibilities from the identical cluster.

3   Proposed Model

In this section, we have proposed a heuristic algorithm for solving modules allocation problem. The model developed in this paper addressed the multiple objectives such as:

i.   Balanced Load assigned on all processors.
ii. Response time of the system must be minimized.
iii. Total communication cost must be minimized.

The developed model has following major steps:

i.   To calculate the fuzzy execution costs.
ii. To calculate the fuzzy inter-module communication costs.
iii. To design the clusters of tasks.
iv. To schedule the clusters of tasks to the processors.

Let 'm' be the total number of modules and the sizes of the modules are defined by vector $MS = [s_1, s_2, \ldots, s_m]^T$. Let n be the number of processors and $\widetilde{er}_k$ (in bites/ second) is the execution rate of the k-th processor. The execution rate $\widetilde{er}_k$ are considered as a triangular fuzzy numbers.

Let $\widetilde{er}_k = (a_k, b_k, c_k)$ represents the fuzzy execution rate of the k-th processor as triangular fuzzy number. The execution rate $\widetilde{er}_k$ of the k-th processor is defined by the following membership function as:

$$\widetilde{er}_k = (a_k, b_k, c_k) = \begin{cases} \frac{x-a_k}{b_k-a_k}, & a_k \leq x < b_k \\ 1, & x = b_k \\ \frac{c_k-x}{c_k-b_k}, & b_k < x \leq c_k \\ 0, & \text{otherwise} \end{cases}$$

The execution cost for the i-th module M_i to the k-th processor P_k will be a triangular fuzzy cost and it is calculated as:

$$\widetilde{e}_{ik} = \left(\frac{s_i}{c_k}, \frac{s_i}{b_k}, \frac{s_i}{a_k} \right)$$

The matrix $DT = [d_{i,j}]$ of order $m \times m$ in this paper represents the communication of data among the interacting tasks during the processing, where $d_{i,j}$ is the notation of transferable quantity of data from M_i to M_j. . The data transfer rate (in bps) in between the processors is stored in the form of a matrix $DTR = [\widetilde{r_{k,l}}]$ of size $n \times n$ and the data transfer rates $\widetilde{r_{k,l}}$ are considered as a fuzzy triangular number.

Let $\widetilde{r_{k,l}} = (l_{k,l}, m_{k,l}, n_{k,l})$ is the representation of the data transfer rate between the processors P_k and P_1 is also in triangular form. The data transfer rate $\widetilde{r_{k,l}}$ between the processors P_k and P_1 is defined by the following membership function as:

$$\widetilde{r_{k,l}} = (l_{k,l}, m_{k,l}, n_{k,l}) = \begin{cases} \frac{x - l_{k,l}}{m_{k,l} - l_{k,l}}, & l_{k,l} \le x < m_{k,l} \\ 1, & x = m_{k,l} \\ \frac{n_{k,l} - x}{n_{k,l} - m_{k,l}}, & m_{k,l} < x \le n_{k,l} \\ 0, & \text{otherwise} \end{cases}$$

The inter-module communication costs between module M_i and M_j is calculated as:

$$\widetilde{c_{i,j}} = \begin{cases} \frac{d_{i,j}}{DTR}, & \text{if } i \ne j \\ 0, & \text{if } i = j \end{cases}$$

Algorithm

Step 1. Input: MS, $ER = [\widetilde{er_k}]$, $DT = [d_{i,j}]$, $DTR = [\widetilde{r}_{k,l}]$.

Step 2. Compute

   (a) Fuzzy Execution Cost Matrix, $FECM = [\widetilde{e}_{i,k}]$

   (b) Average Data Transfer Rate, $\overline{DTR} = \dfrac{\sum\limits_{1 \le k \le n} \sum\limits_{1 \le l \le n} \widetilde{r}_{k,l}}{n}$

   (c) Compute $FIMCCM = [\tilde{c}_{i,j}]$

Step 3. Transform the FIMCCM into $IMCCM = [c_{i,j}]$, by using average defuzzification method.

Step 4. Compute $NITCCM = [\bar{c}_{i,j}]$, the new inter-modules communication matrix, where

$$\bar{c}_{i,j} = \begin{cases} \max\limits_{i} \max\limits_{j} c_{i,j} - c_{i,j}, & \text{if } i \ne j \\ 0 \end{cases}$$

Step 5. Apply fuzzy C-mean clustering:

   (a) Update $t = 0$, and designate membership section to each module M_i in each and every cluster kth such that $\sum\limits_{k=1}^{p} u_{i,k}^{(t)} = 1$, also t is denoted as the count of iterations.

(b) Compute the centers of the clusters with the formula

$$C_k = \frac{\sum_{i=1}^{q} u_{i,k}^m x_i}{\sum_{i=1}^{q} u_{i,k}^m}$$

(c) Update membership grade according to the formula

$$u_{ik}^{(t+1)} = \frac{1}{\sum_{n=1}^{P} \left(\frac{x_i - C_k}{x_i - C_n}\right)^{\frac{2}{m-1}}}$$

(d) If $|u_{i,k}^{(t+1)} - u_{i,k}^{(t)}| \leq \in$

Step 6.

(a) Modify NECM $= [\bar{e}_{k,l}]$ (new execution cost matrix)by appending rows as per the tasks residing in the cluster $\text{Cl}_k = \{M_k, M_l, \cdots\}$.
(b) If the cluster Cl_k has the minimum execution cost on processor P_l then Cl_k will be assigned onto P_l.
(c) Eradicate the k-th row and l-th column from NECM.
(d) If each processor is mapped with a cluster

then jump to step **7**

else go to step **6 (b)**.

Step 7. Compute.

(a) $\text{TCOST}(k) = \text{FEC}(k) + \text{FIMCC}(k)$
(b) $\text{RT}(A) = \max_{1 \leq k \leq n} \{\text{FEC}(k) + \text{FIMCC}(k)\}$

4   Implementation of the Model

Let us consider a distributed program made up of nine modules $\{M1, M2, \ldots, M9\}$ to be divided into clusters and those clusters are executed onto the processor's set $\{P1, P2, P3\}$. The various inputs which are in matrices form are described in Tables 1, 2, 3, 4, 5 and 6. Sizes of the modules are given in Table 1. These modules are to be executed on three processors and the execution rate of these three processors given in Table 2. The data transfer rates between the processors are given in Table 3. Table 4 contains the fuzzy execution cost for each module on each processor and also summarized in Fig. 1. Table 5 represents the data communications between the modules and the fuzzy inter-module communication cost is given in Table 6, also depicted by Fig. 2.

Here DTR = (2.7, 4.7, 7.3) and the fuzzy inter-module communication cost between modules are as follows:

Table 1 Sizes of the tasks

MS =		
	M1	240
	M2	300
	M3	190
	M4	301
	M5	225
	M6	255
	M7	232
	M8	245
	M9	280

Table 2 Execution rate of processors

Processor (P_k)	Execution rate (er_k)
P1	[1.002, 1.709, 1.920]
P2	[0.950, 1.245, 1.565]
P3	[1.525, 1.700, 1.890]

Table 3 Data transfer rates between the processors

		P1	P2	P3
DTR[$\widetilde{r_{i,k}}$] =	**P1**	(0,0,0)	(1, 2,3)	(1,2,4)
	P2	(1,2,3)	(0,0,0)	(2,3,4)
	P3	(1,2,4)	(2,3,4)	(0,0,0)

After the defuzzification of fuzzy inter-module communication cost between modules, the following crisp costs are obtained (Table 7).

After a few iterations of fuzzy C-mean clustering approach the following partition matrix is obtained:

Figure 3, states the base scatter plot for the Inter-Module Communication Cost, on further execution of C-means Clustering following results are obtained.

In Fig. 4 Scatter Plot using the R psych package is represented for the considered data values, the lower diagonal contains Bivariate Scatter Plot, over the diagonal this plot shows histograms and above diagonal values are the Pearson Correlation.

From the Fig. 5, the clusters of modules are formed as:

$$Cl\_1 = \{M2, M3\}$$
$$Cl\_2 = \{M4, M5, M6\}$$
$$Cl\_3 = \{M1, M7, M8, M9\}$$

Table 9 states the minimized costs of the procedure for the optimal allocation of tasks clusters onto the processing units, where Cl_1 → P2, Cl_2 → P1 and Cl_3 → P3. The optimal complexity and time of response for the system are (1247.38,

Table 4 The fuzzy execution cost for each task on each processor

	Modules/ Processors	P1	P2	P3
FECM =	M1	(125.00, 140.43, 239.52)	(153.35,192.77,252.63)	(126.98, 141.18, 157.38)
	M2	(156.25, 175.54, 299.40)	(191.69,240.96,315.79)	(158.73, 176.47,196.72)
	M3	(98.96, 111.18, 189.62)	(121.41,152.61,200.00)	(100.53, 111.76,124.59)
	M4	(156.77, 176.13, 300.40)	(192.33,241.77,316.84)	(159.26, 177.06, 197.38)
	M5	(117.19, 131.66, 224.55)	(143.77, 180.72,236.84)	(119.05, 132.35, 147.21)
	M6	(132.81, 149.21, 254.49)	(162.94, 204.82,268.42)	(134.92, 150.00,167.21)
	M7	(120.83, 135.75, 231.54)	(148.24,186.35,244.21)	(122.75, 136.47, 152.13)
	M8	(127.60, 143.36, 244.51)	(156.55, 196.79,257.89)	(129.63, 144.12, 160.66)
	M9	(145.83, 163.84, 279.44)	(178.91,224.90,294.74)	(148.15, 164.71,183.61)

Table 5 Inter-task communication cost of data

	Modules ↓ →	M1	M2	M3	M4	M5	M6	M7	M8	M9
DT = $[d_{i,j}]$	M1	0	10	21	14	23	16	17	9	9
	M2	10	0	8	11	21	13	16	18	23
	M3	21	8	0	6	9	10	15	19	21
	M4	14	11	6	0	8	9	6	10	3
	M5	23	21	9	8	0	12	13	8	7
	M6	16	13	10	9	12	0	11	9	6
	M7	17	16	15	6	13	11	0	8	7
	M8	9	18	19	10	8	9	8	0	5
	M9	9	23	21	3	7	6	7	5	0

Table 6 Fuzzy triangular values for inter-task communication cost

↓→	M1	M2	M3	M4	M5	M6	M7	M8	M9
M1	(0,0,0)	(1.3699,2.1277, 3.7037)	(2.8767,4.4681, 7.7778)	(1.9178,2.9787, 5.1852)	(3.1507,4.8936,8.518)	(2.1918,3.4043, 5.9259)	(2.3288,3.6170, 6.2963)	(1.2329,1.9149, 3.3333)	(1.2329,1.9149,3.333)
M2	(1.3699,2.1277, 3.7037)	(0,0,0)	(1.0959, 1.7021,2.9630)	(1.5068, 2.3404, 4.0741)	(2.8767,4.4681,7.777)	(1.7808, 2.7660, 4.8148)	(2.1918,3.4043, 5.9259)	(2.4658, 3.8298, 6.6667)	(3.1507,4.8936,8.518)
M3	(2.8767,4.4681, 7.7778)	(1.0959, 1.7021,2.9630)	(0,0,0)	(0.8219, 1.2766, 2.2222)	(1.2329,1.9149,3.333)	(1.3699, 2.1277, 3.7037)	(2.0548, 3.1915, 5.5556)	(2.6027, 4.0426, 7.0370)	(2.8767,4.4681,7.777)
M4	(1.9178,2.9787, 5.1852)	(1.5068, 2.3404, 4.0741)	(0.8219, 1.2766, 2.2222)	(0,0,0)	(1.0959,1.7021,2.963)	(1.2329, 1.9149, 3.3333)	(0.8219, 1.2766, 2.2222)	(1.3699, 2.1277, 3.7037)	(0.4110,0.6383,1.111)
M5	(3.1507,4.8936, 8.5185)	(2.8767, 4.4681, 7.7778)	(1.2329, 1.9149, 3.3333)	(1.0959, 1.7021, 2.9630)	(0,0,0)	(1.6438, 2.5532, 4.4444)	(1.7808, 2.7660, 4.8148)	(1.0959, 1.7021, 2.9630)	(0.9589,1.4894,2.592)
M6	(2.1918,3.4043, 5.9259)	(1.7808, 2.7660, 4.8148)	(1.3699, 2.1277, 3.7037)	(1.2329, 1.9149, 3.3333)	(1.6438,2.5532,4.444)	(0,0,0)	(1.5068, 2.3404, 4.0741)	(1.2329, 1.9149, 3.3333)	(0.8219,1.2766,2.222)
M7	(2.3288,3.6170, 6.2963)	(2.1918,3.4043, 5.9259)	(2.0548, 3.1915, 5.5556)	(0.8219, 1.2766, 2.2222)	(1.7808,2.7660,4.814)	(1.5068, 2.3404, 4.0741)	(0,0,0)	(1.0959, 1.7021, 2.9630)	(0.9589,1.4894,2.592)
M8	(1.2329,1.9149, 3.3333)	(2.4658, 3.8298, 6.6667)	(2.6027, 4.0426, 7.0370)	(1.3699, 2.1277, 3.7037)	(1.0959,1.7021,2.963)	(1.2329, 1.9149, 3.3333)	(1.0959, 1.7021, 2.9630)	(0,0,0)	(0.6849, 1.638, 1.8519)
M9	(1.2329,1.9149, 3.3333)	(3.1507, 4.8936, 8.5185)	(2.8767, 4.4681, 7.7778)	(0.4110, 0.6383, 1.1111)	(0.9589,1.4894,2.592)	(0.8219, 1.2766, 2.2222)	(0.9589, 1.4894, 2.5926)	(0.6849, 1.638, 1.8519)	(0,0,0)

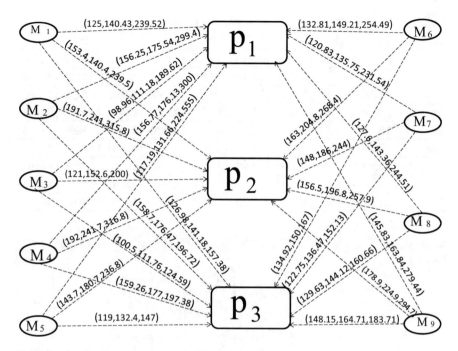

Fig.1 Fuzzy execution cost matrix for each task on each processor

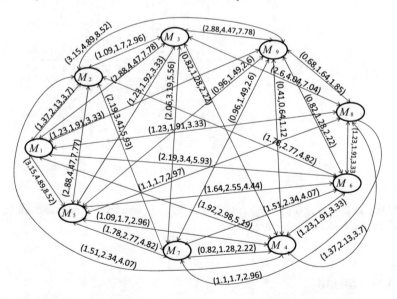

Fig. 2 Fuzzy triangular inter-module communication cost among tasks

Table 7 Crisp value of inter-module communication cost among tasks

Modules	M1	M2	M3	M4	M5	M6	M7	M8	M9
M1	0	2.4004	5.0409	3.3606	5.5209	3.8407	4.0807	2.1604	2.1604
M2	2.4004	0	1.9203	2.6404	5.0409	3.1205	3.8407	4.3207	5.5209
M3	5.0409	1.9203	0	1.4402	2.1604	2.4004	3.6006	4.5608	5.0409
M4	3.3606	2.6404	1.4402	0	1.9203	2.1604	1.4402	2.4004	0.7201
M5	5.5209	5.0409	2.1604	1.9203	0	2.8805	3.1205	1.9203	1.6803
M6	3.8407	3.1205	2.4004	2.1604	2.8805	0	2.6404	2.1604	1.4402
M7	4.0807	3.8407	3.6006	1.4402	3.1205	2.6404	0	1.9203	1.6803
M8	2.1604	4.3207	4.5608	2.4004	1.9203	2.1604	1.9203	0	1.2002
M9	2.1604	5.5209	5.0409	0.7201	1.6803	1.4402	1.6803	1.2002	0

Fig. 3 Scatter plot as per IMCC–I

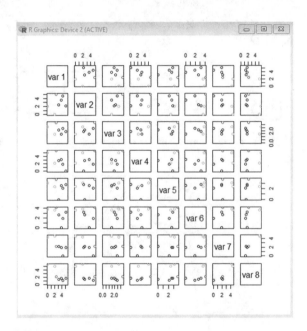

1437.05, 1949.01) and (561.9, 639.89, 746.78), respectively. These allocations are made on the basis of Fig. 5 and Table 8.

5 Conclusion

The effective mission of modules to processors is a critical segment in acquiring DCS's great performance. This manuscript, is a numerical version using R Programming for the project of modules in DCS has been developed via using the C-means

Fig. 4 Scatter plot as per IMCC–II

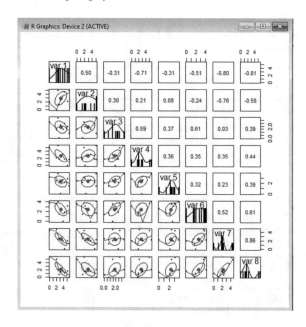

Fig. 5 Cluster plot as per IMCC

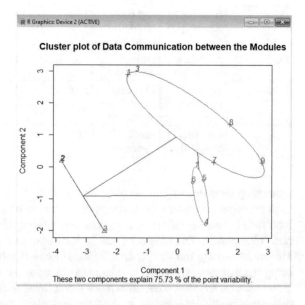

diffuse clustering technique. This version tries to optimize the charge of the device and the reaction time of the gadget. In this version, the wide variety of responsibilities in each cluster is limited with the aid of the inter-module communication cost. This model unearths that the quantity of task groups 'k' is same to the sort of processing units to set up one-to-one correlation among mission businesses and

Table 8 Optimal costs of the program

Processors	Modules	FEC(k) (1)	FIMCC(k) (2)	TCOST(k) (1) + (2)
P1	M4, M5, M6	(406.77,457.0,779.44)	(25.48, 39.59, 71.12)	(432.25, 496.59, 850.56)
P2	M2, M3	(313.1,393.57,515.79)	(25.35, 39.38, 68.52)	(338.45, 432.95, 584.31)
P3	M1, M7, M8, M9	(527.51,586.48,653.78)	(34.39, 53.41, 93.0)	(561.9, 639.89, 746.78)

Table 9 Comparison of the results obtained by proposed model with other existing models

Size of the model (m, n)	Reference model	Technique applied	System time		Response time	
			Reference model	Present model	Reference model	Present model
(4,4)	Shatz et al. [20]	Heuristic algorithm	33	19	32	17
(6,4)	Attiya and Hamam [5]	Simulated annealing approach	145	51	141	49
(5,2)	Daoud and Kharma [17]	List-based scheduling algorithm	134	86	132	86
(8,3)	Akbari and Rashidi [9]	Genetic algorithm	170	111	164	88
(10,3)	Topcuoglu et al. [21]	List-based scheduling algorithm	230	172	220	129
(7,3)	Kopidakis et al. [18]	Heuristic algorithm	285	205	275	198

processing units that improves computing overall performance via the proper use of each processor. To assess the functioning of the put forward model has been examined with the answer of the amazing current fashions which have been quoted from numerous papers. The effects of this look at were in evaluation with the opportunity contemporary models [1, 2, 5, 9, 15, 18] which is probably advanced through using the awesome techniques. It suggests that the put forward model offers even better answers than the other alive models employed to decrease the device cost and reaction time. Figure 6 and seven shows graphical comparison of device charges and reaction instances of the proposed version with special modern fashions are verified, respectively. Graphical outcome displays that the outcomes obtained by the usage of the existing version are higher than existing fashions in respects of each desires, so with the concluding assertion we will use that the proposed version is a terrific healthy for arbitrary finite variety of obligations and processors with fuzzy expenses (Fig. 7 and Table 9).

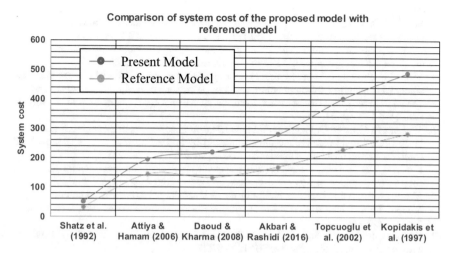

Fig. 6 Comparison of system cost of the proposed model with reference model

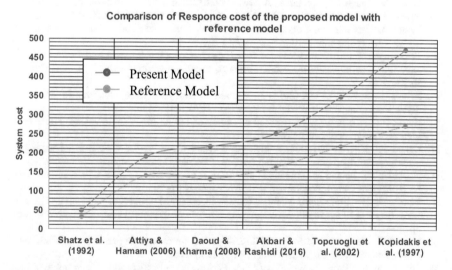

Fig. 7 Comparison of response cost of the proposed model with reference model

References

1. Bokhari SH (1987) In: Assignment problem in parallel and distributed computing. vol. 32(1). Kulwer Academic Publishers, pp 24–156
2. Shen CC, Tsai WH (1985) A group matching approach to optimal task assignment in distributed computing system using a minimax criterian. IEEE Trans Comput 34(3):197–203
3. Tindell KW, Burns A, Wellings AJ (1992) Allocation hard real-time tasks: an NP-hard problem made easy. Real Time Syst 4(2):145–165

4. Peng DT, Shin KG, Abdelzaher TF (1997) Assignment and scheduling communicating periodic task in distributed real-time system. IEEE Trans Software Eng 23(12):745–758

5. Attiya G, Hamam Y (2006) Task allocation for maximizing reliability of distributed system: a simulated annealing approach. J Parallel Distrib Comput 66:1259–1266

6. Yadav PK, Pradhan P, Singh PP (2011) A fuzzy clustering method to minimize the inter task communication effect for optimal utilization of processor's capacity in distributed real time system. In: Proceedings of the international conference on soft computing for problem solving AISC, vol 130. pp 151–160

7. Martinez R, Nelissen G, Ferreira LL, Pinho LM (2015) Allocation of parallel real–time tasks in distributed multi-core architectures supported by an FTT-SE network. In: International conference on architecture of computing system, vol 9017, pp 224–235

8. Kumar H, Chauhan NK, Yadav PK (2016) Dynamic tasks scheduling algorithm for distributed computing system under fuzzy environment. Int J Fuzzy Syst Appl 5(4):77–95

9. Akbari M, Rashidi H (2016) A multi-objectives scheduling algorithm based on cuckoo optimization for task allocation problem at compile time in heterogeneous systems. Expert Syst Appl 60:234–248

10. Crespo A, Balbastre P, Simo J, Albertos P (2016) Static scheduling generation for multicore partitioned system. Inform Sci Appl (ICISA) 376:511–522

11. Kumar S, Kumar K (2019) Neuro-fuzzy based call admission control for next generation mobile multimedia networks. Int J Eng Adv Technol (IJEAT) 8(6) ISSN: 2249–8958

12. Yadav PK, Singh MP, Kumar A, Agarwal B (2012) An efficient tasks scheduling model in distributed processing systems using ANN. Int J Circuits and Syst 1(1–2):53–66

13. Lemos A, Caminhas W, Gomide F (2013) Adaptive fault detection and diagnosis using an evolving fuzzy classifier. Inform Sci: An Int J 220:64–85

14. Velmurugan T (2014) Performance based analysis between k-means and fuzzy C-means clustering algorithms for connection oriented telecommunication data. Appl Soft Comput 19:134–146

15. Zhou J, Chen CLP, Li HX (2014) A collaborative fuzzy clustering algorithm in distributed network environments. IEEE Trans Fuzzy Syst 22(6):1443–1456

16. Nayak P (2016) A fuzzy logic-based clustering algorithm for WSN to extend the network lifetime. IEEE Sens J 16(1):137–144

17. Daoud MI, Kharma N (2008) A high performance algorithm for static task scheduling in heterogeneous distributed computing systems. J Parallel Distrib Comput 63:399–409

18. Kopiddakis Y, Lamari M, Zissimopoulos V (1997) On the task assignment problem: two new heuristic algorithms J. Parallel and Distrib Comput 42:21–29

19. MacQueen JB (1967) Some Methods for classification and analysis of multivariate observations. In: Proceedings of 5th Berkeley symposium on mathematical statistics and probability. pp 281–297

20. Shatz SM, Wang JP, Goto M (1992) Task allocation for maximizing reliability of distributed computer systems. IEEE Trans Comput 41(9):1156–1168

21. Topcuoglu H, Hariri S, Wu MY (2002) Performance-effective and low-complexity task scheduling for heterogeneous computing. IEEE Trans Parallel and Distrib Comput 13(3):260–274

A Soft Computing Intelligent Technique Implication for the Comprehensive Audit of Electric Vehicle

Abhinav Saxena, Rajesh Kumar, Jay Singh, Shilpi Kumari, Mahima Verma, and Priyanshi Kumari

1 Introduction

According to latest censuses, the transport sector is dominated by the use of fuels such as petrol, diesel, CNG, and LPG. The problem with these fuels is that they are exhaustible in nature and due to an increased concern for the greenhouse gasses emissions in recent decades paved the way for finding the new alternatives, and one of them is electric vehicle. A vehicle that is powered totally or partially by electricity is known as an electric car. E-cars, contrast traditional vehicles that rely entirely on fossil fuels, use an electric motor powered by a fuel cell or batteries. These terms "e-vehicle" and "EV" are also acceptable. The term is frequently used to refer to both BEVs and PHEVs. The initials BEV and PHEV stand for battery electric vehicles and plug-in hybrid electric vehicles, respectively. According to electric vehicle news, an e-vehicle is "A vehicle which uses one or more electric motors for propulsion." Depending on the type of vehicle, motion may be provided by wheels or propellers driven by rotary motors, or in the case of tracked vehicles, by linear motors. "Electric cars, electric trains, electrical trucks, electrical lorries, electrical airplanes, electrical boats, electrical motorcycles and scooters, and electric spaceships are all examples of electric vehicles."

Electric vehicles are self-contained and run purely on electricity. It derives its power from an internal battery and to replenish the car's battery, you'll need to link it to a charging point with an EV charging cable. After that, the batteries are recharged for the remaining of the journey. A BEV does not have a gasoline or

---

A. Saxena (✉) · R. Kumar · S. Kumari · M. Verma · P. Kumari
Department of Electrical Engineering, JSS Academy of Technical Education, Noida, U.P, India
e-mail: abhinaviitroorkee@gmail.com

J. Singh
Department of Electrical and Electronics Engineering, G.L Bajaj Institute of Technology and Management, Greater Noida, U.P, India

© The Author(s), under exclusive license to Springer Nature Singapore Pte Ltd. 2023
R. Agrawal et al. (eds.), *International Conference on IoT, Intelligent Computing and Security*, Lecture Notes in Electrical Engineering 982,
https://doi.org/10.1007/978-981-19-8136-4_15

diesel engine. Furthermore, such vehicles have a greater range of electric driving than plug-in hybrid electric automobiles. There isn't even a tailpipe on it. Even if the car does not contaminate the environment directly, the electricity it produces has the capacity to do so. When electricity is generated using renewable energy sources like sun, wind, or hydroelectricity, there is no pollution. Solar power, often known as solar energy, converts the Sun's energy into electricity. Wind power is the process of capturing and converting the energy contained in flowing air into electricity. The Union of Concerned Scientists recommends using electricity instead of fossil fuels: "Not using gasoline or diesel also means that battery electric cars are significantly cheaper to fuel than conventional vehicles. Exact comparisons depend on the vehicle model and fuel prices, but driving a BEV can save drivers over $1000 annually in gasoline money." A plug-in hybrid electric vehicle is a vehicle that runs across both electricity and gasoline. Even before combustion engine comes in, the car's electric motor drives it for a part of the journey. If the journey was not too lengthy, it might just depend on power. A plug-in hybrid electric vehicle (PHEV) isn't just about a hybrid electric vehicle (HEV). The HEV charges its own batteries with the petrol engine when moving. Unlike a PHEV, you don't have to plug it in. In this paper, different intelligent techniques are used for assessing the comparative performance analysis of charging and discharging of electric vehicles.

2 Intelligent Methods of Charging and Discharging

The following intelligent methods were used for charging assessment of electric vehicle as follows.

2.1 Machine Learning Technique

To enable DWC in EV, a hybrid technique based on machine learning is being developed (s). To link all of the participant EVs in the proposed solution, we use the augmented routing protocol in the network topological architecture. EV, CSE, and embedded wireless nodes were also used to analyze the present status of charging, position, vehicle-ID, and distance for each participant. The CSE monitors an EV's charging state in real time and broadcasts the data via a built-in wireless node. Additionally, the administrator can build the collaborating EV(s) remotely. In electric vehicle, a generator was also used to create a magnetic field. Magnetic coupling occurs whenever the magnetic fields of two electric cars (EVs) collide. The connected EV(s) can deliver electricity wirelessly thanks to the magnetic connection interface. The performance of proposed model was verified during single-position charge transfers and up to 5 km/hr movement, which was determined to be extremely significant. The effectiveness of charge transmission between linked EV(s) in a predetermined manner of 1.5 m distance was only around 89%, with moderate MF strength. To

Fig. 1 Network topology [1]

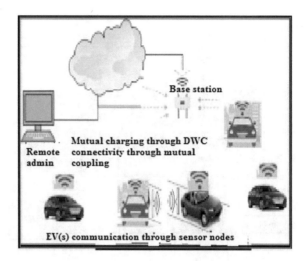

Base station

Mutual charging through DWC
Remote   connectivity through mutual
admin   coupling

EV(s) communication through sensor nodes

confirm the effectiveness, the distance measurements were changed from 60 to 40 cm and then to 25 cm. The power transmission rate was discovered to be 85 and 82% at distances of 40 and 25 cm, respectively, so when distance between both the connected EV(s) was altered. The enhanced routing algorithm was evaluated for performance, latency, and packet loss ratio during hop count communication and BS connectivity. Latency, packet loss ratio, and end-to-end delay were all key communication factors that provided meaningful findings. Our suggested architecture also has the unique feature of being the first to allow each EV in the network to transmit electricity to another EV. In comparison with the existing method, the proposed approach is less expensive to adopt and maintain and produces superior results [1–4]. Topology is shown in Fig. 1.

2.2 Multi-Variant Deep Learning Approach

Two LSTM algorithms were used to predict vehicle speed in this study. The accuracy of the multimodal and basic models was demonstrated. The simulation results demonstrate that the multivariate model greatly outperforms the univariate model due to relative velocity with nonlinear elements that are contained in the dataset used during multivariate model definition. In terms of process, it is preferred to use the multivariate LSTM algorithm in a real-world situation that considers driver behavior as well as weather conditions [5–7]. Experiments in real-sitting scenarios will be conducted using predicting algorithms that have been trained. The efficiency (i.e., execution time) and accuracy of the LSTM algorithm will then be evaluated. Speed prediction has a wide range of applications in the control systems of intelligent transportation systems, particularly in the areas of safety and road efficiency. It is the most dynamic factor in the realm of electro-mobility for optimal online in-vehicle

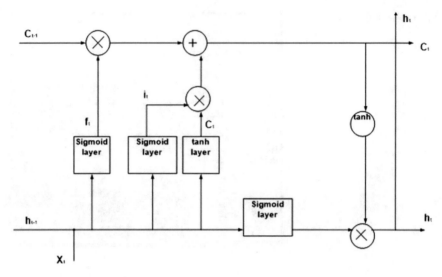

Fig. 2 LSTM gate [11]

energy management [8–10]. Vehicle speed forecasting, on the other hand, is a difficult assignment since it is dependent on a variety of parameters, which can be divided
into two categories: endogenous and exogenous features [11]. The realization of gate
is shown in Fig. 2.

2.3 Dynamic Game Method

How to enable electric vehicles (EVs) perform a full role in shifting the power
grid peak load via demand response regulation has become an urgent problem to
solve (DR) [12–14]. Game theory is frequently employed in the development of
innovative approaches to solving interdecisional problems. The differential game,
as users participate, may show actual changes in moment electricity pricing (TOU
price) on the electricity network and recharge power on electric vehicle [15]. The
structure of differential game model is shown in Fig. 3.

2.4 Artificial Neuro-Fuzzy Control

The paper proposes a smart energy management strategy for a piston hybrid vehicle
in order to minimize total energy consumption. When trying to limit an expended
energy function, which is linked to the amount of electrical energy given by the

Fig. 3 Structure of
differential game model [15]

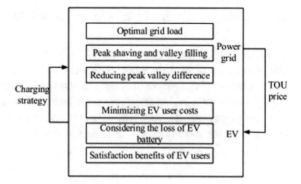

batteries and utilized fuel, it is suggested that you first measure the vehicle's entire
energy consumption.

In specifically, this research presents an intelligent controller that proves its
capacity to increase total vehicle energy efficiency and, as a result, lower total energy
consumption. The proposed technique includes a third-level advanced supervisory
controller that corresponds to a fuzzy system that determines the system's most
optimal operating mode. On the second level, a neuro-fuzzy logic-based intelligent
optimal control technique is devised. Local fuzzy controllers are then used at the first
level to adjust vehicle subsystem set points in order to achieve the greatest operational
performance. The following is a summary of the benefits of the proposed strategy:
It is a web-based solution that minimizes total energy consumption when compared
to many traditional approaches [16]. The developed structure is shown in Fig. 4.

Fig. 4 Developed IHHCS
for BUSINOVA bus
distributed generation
system [16]

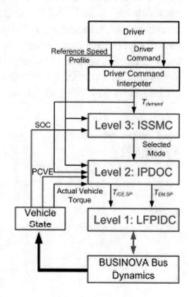

2.5 Bi-Level Event-Based Optimization Method

Connecting renewable energy generation's fluctuating supply with the flexible charging needs of electric vehicles (EVs) can increase renewable energy penetration while reducing load on the state electric power grid. Dimensionality's load, the multiple decision-making steps that must be completed, and the unpredictability on both the supply and demand sides are all concerns that must be handled. EBO is a new method for solving large-scale Markov decision processes [17–21].

2.6 An MPC Method

The hybrid electric storage system is a potential power source for electric vehicles (EVs) that could help them last longer. Battery longevity is influenced by the amount and diversity of power profiles. Unexpectedly increasing power usage during vehicle operation may cause battery degradation, and the super capacitor (SC) in the HESS should be used to compensate. However, due to SCs' limited capacity, the HESS benefit is limited. As a consequence, one of its most significant and challenging duties for a HESS is to properly control the power split between battery and SCs in effort to match the vehicle's driving demand while also reducing battery deterioration rates [22–27].

They evaluate short-term vehicle velocity using a time series forecasting method and estimate future power demand based on the results of the projection. The T-S fuzzy model technique is used to predict system nonlinearity and construct a modeling approach due to the HESS's nonlinear dynamics. Finally, a power management issue is resolved using model predictive control (MPC). To enhance organizational robustness, forecast errors are factored in MPC formulation. When compared with existing methodologies, simulation analysis based on driving profile testing shows that the magnitude and variability of battery voltage are both decreased, and battery life is prolonged by 17.81% [28]. The structure of power management is shown in Fig. 5.

Fig. 5 Power management structure [28]

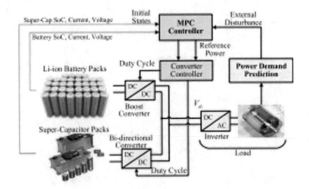

3 Conclusion

This article gives a comprehensive overview and analysis of intelligent electric car monitoring utilizing various soft computing techniques. It all starts with electric car modeling. Following that, many soft computing strategies were developed in order to obtain the best results for EV performance, with machine learning of charging yielding the best results when compared to other ways, for battery charging and discharging in a shorter amount of time while maintaining proper quality.

References

1. Martel F, Kelouwani S, Dubé Y, Agbossou K (2015) Optimal economy-based battery degradation management dynamics for fuel-cell plug-in hybrid electric vehicles. J Power Sources 274:367–381
2. Mi C, Masrur MA, Gao DW (2011) Hybrid electric vehicles: principles and applications with practical perspectives. Wiley, Hoboken, NJ, USA
3. Bohn T, Glenn H (2016) A real world technology testbed for electric vehicle smart charging systems and perverse interoperability evaluation. In: 2016 IEEE energy conversion congress and exposition (ECCE), pp 1–8
4. Adil M, Ali J, Ta QTH, Attique M, Chung T-S (2020) A reliable sensor network infrastructure for electric vehicles to enable dynamic wireless charging based on machine learning technique. IEEE Access 8(6)
5. Chen Z, He F, Yin Y (2016) Optimal deployment of charging lanes for electric vehicles in transportation networks. Transp Res B, Methodol 91:344–365
6. Brecher A, Arthur D (2014) Review and evaluation of wireless power transfer (WPT) for electric transit applications. U.S. Dept. Transp., Fed. Transit Admin., Washington, DC, USA, FTA Rep. 0060
7. Ko YD, Jang YJ (2013) The optimal system design of the online electric vehicle utilizing wireless power transmission technology. IEEE Trans Intell Transp Syst 14(3):1255–1265
8. Lukic S, Pantic Z (2013) Cutting the cord: static and dynamic inductive wireless charging of electric vehicles. IEEE Electrific Mag 1(1):57–64
9. Chopra S, Bauer P (2013) Driving range extension of EV with on-road contactless power transfer—a case study. IEEE Trans Ind Electron 60(1):329–338
10. Jang YJ, Jeong S, Ko YD (2015) System optimization of the on-line electric vehicle operating in a closed environment. Comput Ind Eng 80:222–235
11. Malek YN, Najib M, Bakhouya M, Essaaidi M (2021) Big data mining and analytics. Multivariate deep learning approach for electric vehicle speed forecasting, TUP 4(1)
12. Hwang I, Jang YJ, Ko YD, Lee MS, System optimization for dynamic wireless charging electric vehicles operating in a multiple-route environment. IEEE Trans Intell Transp Syst to be published
13. Xu B, et al (2013) Degradation-limiting optimization of battery energy storage systems operation. M.S. thesis, Power System Laboratory, ETH Zürich, Zürich, Switzerland
14. Omar N et al (2014) Lithium iron phosphate based battery–assessment of the aging parameters and development of cycle life model. Appl Energy 113:1575–1585
15. Zheng Y, Luo J, Yang X, Yang Y (2020) Intelligent regulation on demand response for electric vehicle charging: a dynamic game method. IEEE Access 8
16. Kamal E, Adouane L (2018) Intelligent energy management strategy based on artificial neural fuzzy for hybrid vehicle. IEEE Trans Intell Veh 3(1):112–125. https://doi.org/10.1109/tiv.2017.2788185

17. Green JM, Hartman B, Glowacki PF (2016) A system-based view of the standards and certification landscape for electric vehicles. World Electric Vehicle J 8(2):564–575. [Online]. Available: http://www.mdpi.com/2032-6653/8/2/564

18. Zhang K, Yin Z, Yang X, Yan Z, Huang Y (2017) Quantitative assessment of electric safety protection for electric vehicle charging equipment. In: 2017 international conference on circuits, devices and systems (ICCDS), pp 89–94

19. Falvo MC, Sbordone D, Bayram IS, Devetsikiotis M (2014) EV charging stations and modes: international standards. In: 2014 international symposium on power electronics, electrical drives, automation and motion, pp 1134–1139

20. Freschi F, Mitolo M, Tommasini R (2018) Electrical safety of plug in electric vehicles: shielding the public from shock. IEEE Ind Appl Mag 24(3):58–63

21. Wogan M (2016) Electric vehicle charging safety guidelines. WorkSafe New Zealand. [Online]. Available: https://worksafe.govt.nz/laws-and-regulations/regulations/electrical-reg ulations/regulatory-guidance-notes/electric-vehicle-charging-safety-guidelines/

22. Ghavami M, Esakkiappan S, Singh C (2016) A framework for reliability evaluation of electric vehicle charging stations. In: 2016 IEEE power and energy society general meeting (PESGM), pp 1–5

23. Risk control—fire safety when charging electric vehicles. UK Fire Protection Association

24. Affonso CM, Yan Q, Kezunovic M (2018) Risk assessment of transformer loss-of-life due to PEV charging in a parking garage with PV generation. In: 2018 IEEE power energy society general meeting (PESGM), pp 1–5

25. Naveen G, Yip TH, Xie Y (2014) Modeling and protection of electric vehicle charging station. In: 2014 6th IEEE power india international conference (PIICON), pp 1–6

26. Nie X, Liu J, Xuan L, Liang H, Pu S, Wang Q, Zhou N (2013) Online monitoring and integrated analysis system for EV charging station. In: IEEE PES Asia-Pacific power and energy engineering conference, pp 1–6

27. Xin S, Guo Q, Sun H, Zhang B, Wang J, Chen C, Cyber-physical modeling and cyber-contingency assessment of hierarchical control systems. IEEE Trans Smart Grid

28. Hu Y, Chen C, He T, He J, Guan X, Yang B (2019) Proactive power management scheme for hybrid electric storage system in EVs: an MPC method. IEEE Trans Intell Transp Syst 1–12. https://doi.org/10.1109/tits.2019.2952678

A Review About the Design Methodology and Optimization Techniques of CMOS Using Low Power VLSI

Usha Kumari and Rekha Yadav

1 Introduction

Due to increase, the use of portable systems need of low power consumption is necessary. The low power VLSI techniques are used in portable devices for many applications for decreasing power dissipation and enhancing throughput of circuits such as personal digital assistant, notebook computers, handheld devices or some other communication (portable) devices.

The need for low power design is also becoming a major issue in high performance digital systems, such as microprocessor, digital signal processors and other applications. The power dissipation will increase with the temperature and clock frequency. To remove the difficulties effectively, the temperature of chip should be kept at acceptable level and cooling system is used, so heat dissipation will decrease. The power consumption is reduced by using proper selection of algorithms, proper data processing and by reducing the switching events. Recently, the techniques are used that are based on the constant voltage scaling and field scaling. Some techniques are as VTCMOS, dual threshold CMOS, MTCMOS, dynamic threshold CMOS, glitch reduction and reduction of switching events.

In CMOS, there are three types of power dissipation generally such as

– Dynamic or switching power consumption

  Short circuit power consumption
  Static or leakage power consumption.

1.1 Dynamic/Switching Power Consumption

The power dissipation occurs in the switching period that power consumption is called switching/dynamic power consumption. In CMOS circuit, switching period

U. Kumari (✉) · R. Yadav
Department of Electronics and Communication, DCRUST Murthal, Murthal, India
e-mail: ushatondak@gmail.com

© The Author(s), under exclusive license to Springer Nature Singapore Pte Ltd. 2023   181
R. Agrawal et al. (eds.), *International Conference on IoT, Intelligent Computing and Security*, Lecture Notes in Electrical Engineering 982,
https://doi.org/10.1007/978-981-19-8136-4_16

occurs during charging and discharging phase. In charging phase, the capacitor will charge from 0 to Vdd through PMOS, and in discharging phase, the capacitor will discharge from Vdd to 0 through NMOS as shown in Fig. 1.

The total capacitive load at the output of circuit is

– Gate output node capacitance itself

Total interconnect connected to capacitance
Driven gate input capacitance.

The energy required in switching period for charging the output node from 0 to Vdd and discharge up to Vdd to ground level is

$$\text{Pavg} = \frac{1}{T}.\left[\int_0^{\frac{T}{2}}.\text{Vout}.\left(-\text{Cloadd}.\frac{\text{Vout}}{dt}\right)\right] + \left[\int_{\frac{T}{2}}^{T}.(\text{Vdd} - \text{Vout}).\left(\text{Cloadd}.\frac{\text{Vout}}{dt}\right)\right] \tag{1}$$

The average dynamic power in CMOS logic circuit is

$$\text{Pavg} = \frac{1}{T}.\text{Cload}.\text{Vdd}^2 \tag{2}$$

$$\text{Pavg} = \text{Cload}.\text{Vdd}^2.\text{fclk} \tag{3}$$

$$\text{Pavg} = \alpha.T.\text{Cload}.\text{Vdd}^2.\text{fclk} \tag{4}$$

The mathematical calculation shows that average power dissipation also depends on α that shows switching activities of circuit, so the average power dissipation is shown in Eq. 4.

For diminishing the switching power consumption,

– The Vdd of circuit will reduce.

Fig. 1 Dynamic power dissipation [1]

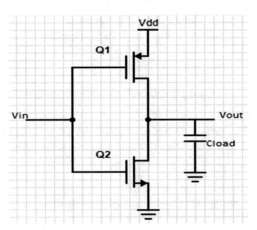

Voltage swing at all nodes should reduce.

Switching activities should reduce.

Load capacitance should reduce.

But there are some other issues that affect the system performance such as delay, as reduction in power supply leads to enhancing in delay parameter.

Switching activities are reduced due to logic optimization, using gated clock signal and also by preventing the hazards. The load capacitance is brought down by using proper design of circuit and using proper sizing transistor.

1.2 Short Circuit Power Dissipation

In switching, power dissipation does not depend on input signal rise time or fall time. In case of CMOS inverter when input is given, voltage waveform should have finite rise time and fall time. During switching time, both *P* and NMOS conduct simultaneously for very short time, so a direct path between ground and power supply is shown in Fig. 2. The power dissipation occurs due to direct path between Vdd and Vss that is called short circuit current. In case of CMOS inverter, PMOS and NMOS are identical/ symmetric, and then rise time and fall time are also equal. In that case, the Kn=Kp=K , $|VTn| = |VTp| = VT$ and also τ.rise = τ.fall =τ having a small capacitive load, so the short circuit current and power consumption are

$$\text{Iavg(S.C)} = \frac{1}{12}\left[\left(\frac{K\tau.\text{fclk}}{\text{Vdd}}\right).(\text{Vdd} - 2\text{VT})^3\right] \tag{5}$$

The S.C power dissipation is

$$\text{Pavg(S.C)} = \frac{1}{12}\left[(K\tau.\text{fclk}).(\text{Vdd} - 2\text{VT})^3\right] \tag{6}$$

Fig. 2 Static power dissipation [1]

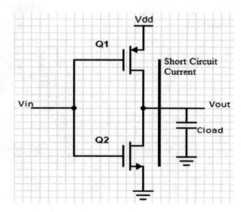

From these equations, it shows that input signal (rise time and fall time) is proportional to the power dissipation and also proportional to transconductance of the transistor.

1.3 Leakage Power Dissipation

In CMOS, the P and NMOS both are present, they generally have nonzero subthreshold or leakage current. The current flows even when the transistor is not in switching mode. The main source of leakage is drain, and bulk terminals are in reverse bios. In CMOS inverter circuit when PMOS is off there is a reverse potential between of VDD between drain and n-well cause a diode leakage through the drain junction. So, another source of leakage current is n-well junction as in Fig. 3. The leakage current of PN junction is

$$\text{Ireverse} = A.Js\left(e^{\frac{qVbois}{KT}} - 1\right) \tag{7}$$

Vbios is reverse bios voltage. Js is current density. A refers to junction area. The reverse saturation current increases with temperature, and the leakage current exits even in standby mode. Another leakage current in CMOS is subthreshold leakage current. By carrier diffusion at source and drain terminal, it causes an existence of weak inversion region. The subthreshold leakage current is

$$\text{ID(Subthreshold)} = \frac{qDnWxcNd}{LB}.e^{\frac{q\Phi r}{KT}}.e^{\frac{q(A.Vgs+BVds)}{KT}} \tag{8}$$

To avoid the subthreshold current, the threshold leakage current should not be low. The total power dissipation in CMOS is

$$\text{Ptotal} = \alpha.T.\text{Cload}.\text{Vdd}^2\text{fclk} + \text{Vdd}(\text{Is}.c + I\text{leakage} + I\text{static}) \tag{9}$$

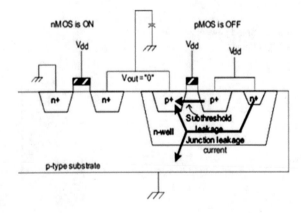

Fig. 3 Reverse leakage current path in CMOS inverter [1]

I s.c shows average short circuit current, α is used for switching activities, I leakage shows the subthreshold and reverse leakage current, and I static shows the DC component from the supply. When the chip contains the conventional CMOS gates, a direct path becomes existing in between input supply and ground, and then a power dissipation occurs that is called the static power dissipation. Static power consumption sources are as follows:

– Leakage current or reverse current

Gate-induced drain leakage (IGIDL)
Subthreshold leakage or weak inversion current (Isub).

Sections first and second show the introduction part about the power dissipation of CMOS circuit. Third section shows literature survey. In this section, the different paper discusses various technologies for reduction of power dissipation. Fourth and fifth sections show mathematical modeling and various optimized technologies which are used in VLSI. The last section shows the comparison between the different technologies and their characteristics.

2 Literature Survey

See Table 1.

3 Techniques of Low Power Design Through Voltage Scaling

To enhance the performance of CMOS circuit, the power consumption decreases by the reduction of input supply voltage.

3.1 Effects of Voltage Scaling on Power and Delay

In scaling techniques, two types of scaling are used that are as follows:

1. Constant voltage scaling
2. Constant field scaling.

In case of constant voltage scaling by reducing Vdd, the dynamic power will be reduced, but it increases the delay. When scaling power is supplied, then the other parameters are kept constant. The speed of circuit also depends on power supply and

Table 1 Literature survey of low power VLSI techniques

S. No and Ref. no.	Author name	Year	Technique used	Advantages
1 [2, 3]	Liqiong Wei	2000	Multiple threshold voltage scaling	Low leakage current, high performance
2 [4]	X. Hao	2011	Power gating technique	Reduces in subthreshold leakage current
3 [5]	A. Jahangir	2012	Transistor stuck technique in standby mode	Leakage current reduces in standby mode
4 [6, 7]	K. Raghava	2016	Glitch reduction technique/pipeline technique	Reduces power consumption, enhances the circuit Performance as well as speed
5 [8]	G. Sravya	2016	Data encoding technique	Transition time decreases
6 [9, 10]	R.Prathiba	2016	Power suppression technique	Increases performance of circuit
				Increases performance of circuit, compact in size
				Low power consumption
7 [11, 12]	Dr. B. T. Geeta	2017	Lector method and stack approach	Decreases complexity in circuit
				Decreases complexity in circuit, reduces power dissipation

also affects the dynamic power of circuit. When circuit is operating in its maximum speed, then switching events will drop and propagation delay becomes increase.

In the second case constant field scaling, the dimensions such as doping densities channel length and gate oxide thickness are scaled down by the same factor and kept constant for both device parameters and load capacitance. But in case of low voltage, it causes the leakage of subthreshold current in standby mode.

3.2   Variable Threshold CMOS (VTCMOS) Circuit

This case uses variable threshold voltage. This is another way to avoid the sub-threshold leakage current by varying bios voltage of substrate in standby mode. In VTCMOS circuit, mosfet body biasing is control with help of substrate bios circuit shown in Fig. 4. In standby mode, reverse bios is applied, so threshold is high and leakage current is reduced. But in active mode, threshold voltage is less, so current is increased. The substrate biasing generally reduces the short channel effects and increases the circuit speed. This technique is effective that automatically controls the threshold voltages and reduces the leakage current, so this approach is known as

Fig. 4 VTCMOS inverter circuit [1]. Bios is usually specified as the direction in which DC from a battery or power supply flows between the emitter and the base

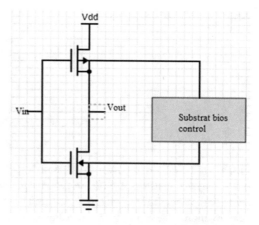

Fig. 5 MTCMOS inverter circuit [1, 13]

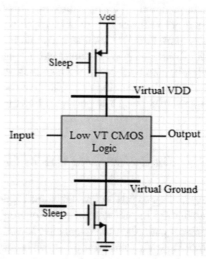

self-adjustable threshold voltage scheme (SATS). But it also has some disadvantages such as it requires triple or twin well CMOS fabrication technology for applying various substrate bios voltage to different circuit parts. The second is to also require an extra substrate bios control circuitry that leads to increase in the circuit area.

3.3 Multiple Threshold CMOS (MTCMOS) Circuit

In this MTCMOS technique reduce leakage current in stand by mode use two transistors with different threshold voltage. Low threshold transistors are use to design logic gate where switching speed is essential. High threshold transistor is used for presentation of leakage power consumption. The MTCMOS structure is shown in

Fig. 6 Dual VT CMOS
circuit [5]

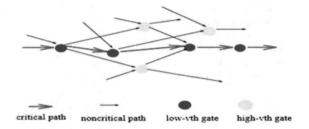

critical path   noncritical path   low-vth gate   high-vth gate

Fig. 5 It is an effective power management technique. In active mode, high threshold voltages transistors are on, and logic operates with the help of low VT. So the switching power dissipation is reduced, and propagation delay becomes small. But in standby mode, the high threshold voltages transistors are off, so there is no conduction path and no leakage current. This technology has benefit over VTCMOS such as it not requires any twin or triple well CMOS technology and also not requires any extra circuitry such as substrate bios control circuit. But it is also having some disadvantages such as due to extra P and NMOS, structure increases the area of the circuit, increases delay and also increases parasitic capacitance. The VTCMOS and MTCMOS technology are effective for the low-voltage and low power logic gates, but for designing low power CMOS circuits, it is not a universal solution (Fig. 6).

3.4 Dual Threshold CMOS Circuits

According to this technique, some higher threshold voltage transistors are designed which are in non-critical path for minimizing the leakage current in sequence of critical path, the transistors have low VT, and circuit performance is determined by those low VT transistors as shown below. The technique is effective for both in active and standby mode. By increasing the threshold voltage of critical path transistors, complexity of circuit will increase and delay will be also increased, by using proper algorithms such as gate level or transistor level for low VT transistor circuits.

3.5 Dynamic Threshold CMOS Circuits

In this technique, a high VT is used in standby mode, so leakage current is low; during ohmic region or in active region, the VT is low, so higher current will flow in the active region. This dynamic CMOS circuit is obtained by attaching the gate and body terminal of transistor. The schematic of DTCMOS is shown in Fig. 7. DTCMOS develops by bulk silicon using triple well technology. To reduce the parasitic capacitance, components need doping mechanism. But it also has advantage like it has good oscillation performance. The supply input voltage control with diode is

Fig. 7 Circuit diagram of
DTCMOS inverter [1]

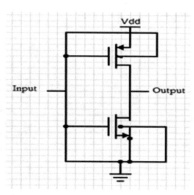

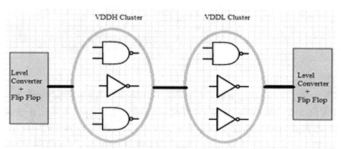

Fig. 8 Cluster voltage dual Vdd scheme [5]

built in potential (source and body in reverse bios). This technique satisfies only for
the low-voltage circuits (0.6 or below).

3.6 Multiple Vdd CMOS Design Technique

The proposed circuit design technique is used only for the low voltage. The high
voltage is given to gate for critical path, while some other gates are connected to the
low voltage (which are in non-critical path). In these two types of circuit clusters
(i) high Vdd and (ii) low Vdd, the high Vdd circuit is placed in front of low Vdd
circuit for avoiding leakage current. These techniques (dual Vth or Vdd) are used for
reducing the leakage power and dynamic power as shown in Fig. 8.

3.7 Standby Leakage Control Using Transistor Stacks

In this technique, the transistors are placed in series connection for leakage current
reduction. The leakage current also diminishes by increasing source voltage Vs which

Fig. 9 Sleep transistor and
stuck transistor [5]

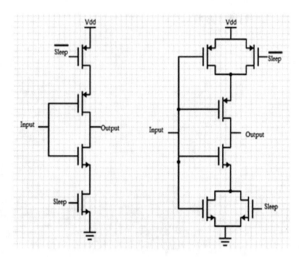

causes (1) Vgs to become negative, (2) Vth to increase due to body effect, and (3) by
decreasing Vds also Vth increases. For reducing the leakage current, a generic-based
algorithm and heuristic search method are used. The technique is effective only for
a single threshold voltage. Stacking transistor will reduce the subthreshold leakage
current. The stacking effect occurs when two or more than two stacked transistors
are off at the same time, which results in decreasing the leakage current or power
dissipation. The stuck effect is the combination of both stuck and sleep transistor
technology. In this stuck effect, the sleep transistor is connected in parallel combination. In another sleep mode, sleep transistors remain in cutoff region and stuck
transistor decreases leakage current. In this technology, each transistor is replaced
with the three sets of logic gates as shown in Fig. 9. In this stuck transistor, the sleep
transistors at sleep mode remain in cutoff region, and one of them from the parallel
transistor with sleep transistor keeps connection with the power supply or Vdd.

3.8 Reduction of Switching Events

For the large circuits, the switching activities become a complicated problem especially for the sequential circuits and feedback loop circuits. In CMOS, the switching
activities are reduced by using optimization algorithm with the proper logic topology or also by circuit level minimization. By reducing the switching activities, the
dynamic power is reduced. The optimization algorithm will depend on data characteristics such as correlation, dynamic range and data transmission statistics. To
reduce the switching activities, proper vector quantization algorithm is also used.
Instead of using full search algorithm, it uses differential tree search algorithm and
reduces the switching events 30 times. Instead of using simple binary codes, it uses

gray code technique. Also in place of two's complement, it uses the sign magnitude representation.

3.9 Glitch Reduction

Switching events are based on delay and also on glitches. By glitch reduction and delay balancing, switching events can be reduced. If the signals at the gate terminal change simultaneously, then no glitch occurs, but when signal changes at different times at the input of circuit, glitch occurs. It is also called the dynamic hazard. In this case, the node voltage not takes fully transition between Vdd and ground voltage terminals. Due to the occurring of glitches and dynamic hazard, it causes the dynamic power dissipation. The glitch occurs due to imbalancing or mismatching in the path length. In this case as shown in Fig. 10, the glitch occurs in the circuit because all inputs do not reach at the same time. For reducing glitch problem the circuit should redesign with balance delay using tree structure. Due to the tree structure, propagation delay will decrease and also the glitches will decrease in circuit. Each signal gave same transition in the tree structure as shown in Fig. 11.

Fig. 10 Signal glitch occurring in multilevel CMOS circuit [1]

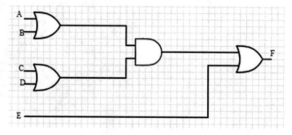

Fig. 11 Tree structure for reducing the glitch transition [1]

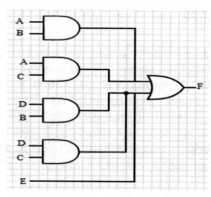

Fig. 12 Lector CMOS gate
[11]

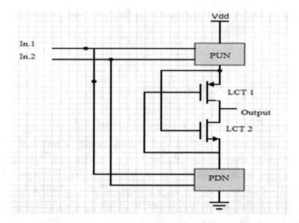

Fig. 13 Power gating circuit
(in ground gating case) [4]

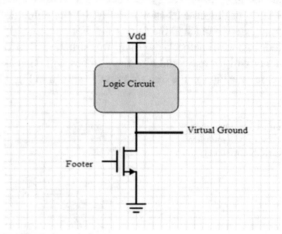

3.10 Modified Lector Technique

In this lector procedure, the leakage current of circuit will control by the stacking of transistors successfully from Vdd to ground as shown in Fig. 12. In this method, more than one transistor becomes off, so it reduces the power consumption. The leakage transistors are controlled by the wellspring of other transistors. So it is also known as self-controlled stacked circuits. Therefore, no outside circuit is necessary to control them. The leakage control transistors (LCT) control the leakage current because there exists a resistance in path (from ground to Vdd), and also the subthreshold leakage and static current will diminish.

3.11 Power Gating Technique

The power gating technique is used for diminishing the effect of subthreshold leakage current. It will work only in transition mode and also provide an effective as well as efficient estimation of power design minimization. Power gating is implemented with the help of supply gating and ground gating as shown in Fig. 13. For practice, the ground gating is widely used. In this, a high Vth (single or multiple) NMOS transistor is placed in between the Vdd and ground terminals. The point in between Vdd and ground is called virtual ground. The working mechanism is when the ground gating is applied, the subthreshold leakage current starts to rise or charge up the internal node of circuit. So voltage becomes decreased at the node. The subthreshold leakage current will also reduce due to smaller Vds of every transistor of circuit. When leakage of circuit becomes equal to the ground and footer, then charging process stops. In case of the power gating, the dynamic characteristic has been ignored during transition mode.

4 CMOS Circuit Characteristics in Low Power VLSI

In the development of high-speed and latest technology circuits, the power and cost are also most important factors. The CMOS technology fulfills all the circuit design techniques and characteristics. In comparison with CMOS technology and other semiconductor devices on the basis of gate length down scaling, the performance of circuit depends on bandwidth, signal-to-noise ratio, dynamic range data rate and inverse power. The clock is a basic unit of circuit which is used for the proper regulation of device. The main aim of the clock cycle in circuit maintains regulation, instructions and the cycle timing. The benefits of CMOS clocks are low power consumption of circuit and also less expensive. The CMOS clock enhances jitter performance as well as phase noise. In the year 2008, CMOS achieves an operating frequency that is between 20 and 30 GHz. The relation between ideal performance and the clock frequency is shown in Fig. 14. In CMOS, another important parameter is gate length, and the relation between the transistor gate length in icons (90–65–45 nm) and years of production is shown in Fig. 15. The trend of high operating speed in CMOS and the production year during the year 2000–2008 is presented in Fig. 16. In CMOS circuits as the technology is developed, (90–65–45 nm) power dissipation will increase. In advance process at 65 nm technology, the CMOS circuit I/O achieves power consumption 10 mW. Due to leakage effect (in reverse bios) in CMOS, it is difficult to minimize the power. The relation between the technology and power consumption is represented in Fig. 17. By gate length shrinkage, two approaches come out first: one for the high-speed applications and second is for low power consumption. The graph shown between the production year and the speed is represented in Fig. 18. In this graph shown, the highest operating frequency is 24 GS/s. The CMOS circuit having main advantages is very low production cost, high reliability as well

Fig. 14 Comparison between the performance and clock frequency for high-speed IC [11]

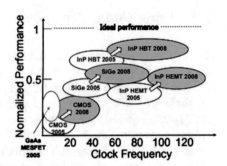

Fig. 15 Comparison between the performance and clock frequency for high-speed IC [11]

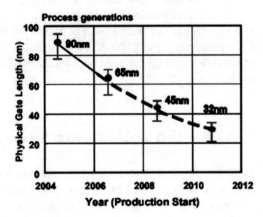

Fig. 16 Relation between production years and the CMOS speed [14]

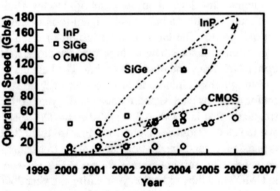

as low power dissipation. The CMOS circuit fabrication is based on silicon wafer, so it is used for production in system on chip, DSP logic circuits and also suitable for VLSI circuits.

CMOS circuits also have some limitations such as low driving force devices and weak frequency response.

Fig. 17 Graph between the technology and power consumption [14]

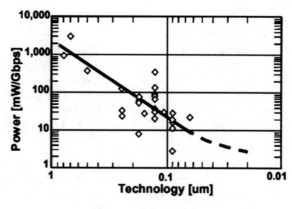

Fig. 18 Graph between the operating speed and production year [14]

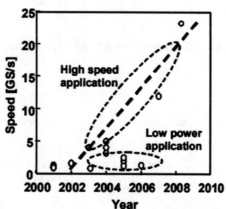

Table 2 Comparison table with different technologies

S. No.	Technique used	Advantages	Disadvantages
1	Voltage scaling	a. Power dissipation reduce	a. Delay increases
2	VTCMOS	a. Reduces overall power dissipation	a. Needs extra circuit (area increase)
		b. Increases the speed or performance	b. Requires twin or triple well CMOS technology
3	MTCMOS	a. Prevents leakage power dissipation	a. Extra transistors are added to increase the overall circuit area
		b. Not requires twin or triple well CMOS technology	b. Delay increases
4	Dual threshold CMOS	a. Leakage current will reduce	a. Complexity of circuit will increase
			b. By increasing, the VT of critical path delay increases
5	Dynamic threshold CMOS	a. Leakage will reduce	a. Needs triple well CMOS technology
			b. Needs doping mechanism
			c. Only used for the low power applications
6	Stuck at transistor	a. Reduces subthreshold leakage current, so overall power dissipation reduces	a. Only effective for the single threshold voltage technology
7	Power gating	a. Diminishes the subthreshold current effect	a. It will work only in transition mode

5 Comparison Table with Different Techniques

See Table 2.

6 Conclusion

This paper describes the various technologies for reducing power consumption and also describes briefly the static dynamic and short circuit power. These also show that the leakage power is the main source of static power dissipation in circuit. Reducing of static power consumption is an important concern for the low power VLSI circuits. For reducing this power dissipation, different technologies are used in VLSI design which is described along with their advantages and their limitations. In these technologies, two approaches such as low threshold and high threshold voltages with suitable critical path are used. The circuit design with any technology is used according to their gain performance and their power consumption. It shows a brief description about multiple Vth and also multiple Vdd technologies for controlling the power dissipation and also for increasing the performance of the circuit.

References

1. Kang SM, Leblebici Y (2003) CMOS digital integrated circuits. Tata McGraw-Hill Education
2. Shikata T, Kondou S, Nose R, Kuniyasu Y, Naitoh M, Suzuki H (1998) Proceedings of the IEEE 1998 custom integrated circuits conference (Cat. No. 98CH36143). IEEE, pp 123–126
3. Wei L, Roy K, De VK (2000) VLSI Design 2000. Wireless and digital imaging in the millennium. Proceedings of 13th international conference on VLSI design. IEEE, pp 24–29
4. Xu H, Vemuri R, Jone WB (2009) IEEE Trans Very Large Scale Integrat (VLSI) Syst 19(2):237
5. Chowdhury AJ, Rizwan MS, Nibir SJ, Siddique MRA (2012) 2012 2nd international conference on power, control and embedded systems. IEEE, pp 1–4
6. Ramirez-Angulo J, Carvajal RG, Lopez-Martin A (2009) 2009 22nd international conference on VLSI design. IEEE, pp 26–27
7. Katreepalli R, Chemanchula H, Haniotakis T, Tsiatouhas Y (2016) 2016 IEEE computer society annual symposium on VLSI (ISVLSI). IEEE, pp 367–372
8. Prathiba R, Sandhya P, Varun R (2016) 2016 international conference on electrical, electronics, and optimization techniques (ICEEOT). IEEE, pp 794–798
9. Jalaja S, Prakash AV (2016) 2016 international conference on microelectronics, computing and communications (MicroCom). IEEE, pp 1–6
10. Sravya G, Sailaja M (2016) 2016 international conference on signal processing, communication, power and embedded system (SCOPES). IEEE, pp 1683–1687
11. Geetha B, Padmavathi B, Perumal V (2017) 2017 IEEE international conference on power, control, signals and instrumentation engineering (ICPCSI). IEEE, pp 1759–1763
12. Zhang Z, Mertens K, Tiebout M, Marsili S, Matveev D, Sandner C (2008) 2008 IEEE international conference on ultra-wideband, vol 2. IEEE, pp 69–72
13. Chen W, Hwang W, Kudva R, Gristede G, Kosonocky S, Joshi RV (2001) ISLPED'01: proceedings of the 2001 international symposium on low power electronics and design (IEEE Cat. No. 01TH8581). IEEE, pp 263–266
14. Ikeuchi T, Cheung T, Onaka H (2008) 2008 international symposium on applications and the internet, IEEE, pp 397–400

Characterization of SPEC2006 Benchmarks Under Multicore Platform to Identify Critical Architectural Aspects

Surendra Kumar Shukla and Bhaskar Pant

1 Introduction

With increasing demand for energy constraining small gadgets, multicore systems have covered almost all embedded markets [1]. To support the growth, various attempts in terms of experiments under distinct microarchitectures were carried out in the past to find out and mitigate various performance issues, like energy consumption [2]. Such analysis becomes vital as multicore computing devices exist with a higher diversity [3]. The diversity becomes more critical in case of inter core communication and load balancing aspects.

To address the diversity, compiler-based constructs have been explored, and latent aspects were identified [4]. The analysis is not limited to the performance but is growing toward the energy consumption aspects [5]. To understand such systems behavior, some noteworthy attempts have been made toward the interference analysis for Mi-bench benchmark suite [6]. Additionally, a list of parameters related to distinct benchmarks has also been identified, and their correlation has been established [7]. These studies and investigations are useful, however are limited to the Mi-bench benchmark suite and related architecture considered during the simulation.

In this research, an attempt has been made to characterize the SPEC2006 benchmark for the multicore systems [8]. The analysis of these benchmarks was carried out for distinct parameters such as IPC, CPI, cache hit rate, and time taken to execute the benchmark [9, 10]. The noteworthy contribution of this research work is that

- The correlation of the parameters has been established for relatively less number of instructions.

S. K. Shukla (✉) · B. Pant
Department of Computer Science and Engineering, Graphic Era Deemed to Be University, Dehradun, India
e-mail: surendrakshukla21@gmail.com

© The Author(s), under exclusive license to Springer Nature Singapore Pte Ltd. 2023   199
R. Agrawal et al. (eds.), *International Conference on IoT, Intelligent Computing and Security*, Lecture Notes in Electrical Engineering 982,
https://doi.org/10.1007/978-981-19-8136-4_17

Table 1 Configuration for simulation

CPU type	Atomic simple CPU	Host architecture
L1-I D cache size	128 KB	Dual core-Intel(R) Pentium(R) CPU N3710 @ 1.60 GHz
L2 Cache size	1 MB	
L1 i cache associativity	2	
L2 cache associativity	1	
Cache line size	64	

- Utilizing the experimental data for devising a new architecture for upcoming state-of-the-art applications.
- Establishing the parameter tuning and trade-off aspects in the presence of powerful application sets.

Rest of the paper is organized as—Sect. 2 includes the system configuration and methodology used for the simulation; Sect. 3 details the simulation results further analysis. Lastly, the paper has been concluded.

2 System Configuration and Methodology

To perform the analysis of SPEC benchmarks, a fixed reference architecture has been selected. The benchmarks were executed under the said architecture, and the results were produced. System configuration used for the simulation is detailed in Table 1. The cache parameters taken for the simulation range from 128 k to 1 MB whereas host architecture is Intel Quad core system. Additionally, gem5 simulator has been executed in the Ubuntu-16.04 operating system. The script for the execution has been prepared in Python and executed in system emulation mode (se.py mode).

3 Simulation Results

Simulation results obtained are detailed in Table 2. It could be noted in the table that, total number of instructions taken for the simulations are 50,000,000, and L2 cache size has been varied from 1 to 2 MB. Total time taken for the simulation is also detailed. It is found that, among all the benchmarks, MFC benchmark has taken the highest time for the simulation. And could be interpreted as that, MFC is a complex benchmark which is essentially a program used for the intelligent parking of the USA. Additionally, most of the instructions of MFC are memory sensitive.

Table 2 Simulation results

Benchmark	l1d_size (kb)	l1i_size (kb)	l2_size (MB)	l1d_assoc	l1i_assoc	l2_assoc	Cache line size	Total_Inst_num	Time	L1 miss penalty	L2 miss penalty	CPI
401.bzip2	128	128	1	2	2	2	64	500,000,000	30 m 24.512 s	6	50	1.32263202
429.mcf	128	128	1	2	2	2	64	500,000,000	35 m 22.012 s	6	50	1.777607164
456.hmmer	128	128	1	2	2	2	64	500,000,000	27 m 10.580 s	6	50	1.003113984
458.sjeng	128	128	1	2	2	2	64	500,000,000	32 m 40.100 s	6	50	1.939774552

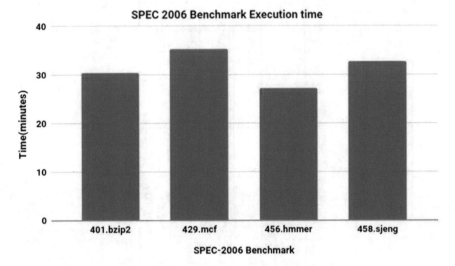

Fig. 1 Execution time analysis of SPEC2006 Benchmark (Multicore Arch.)

4 Result Analysis

4.1 Total Time for Execution

After the simulations, it is identified that among all the benchmarks selected from the SPEC2006, benchmark suite, Hmmer has taken less time for the execution, and MCF had taken large time.

Total time taken during the execution is depicted in Fig. 1.

4.2 CPI Analysis

In Fig. 2, CPI of the benchmark was illustrated. CPI value is essentially the total number of cycles required to execute each instruction [11]. The CPI value here is higher for Sjeng and MFC benchmarks whereas Hmmer benchmark enjoys the less number of cycles incurred for the execution. The reason for such behavior is that if the benchmarks have higher memory-sensitive instructions, they consume more cycles, and further, cycles count increases. Small CPI value is a pointer toward better performance whereas the large CPI value indicates higher execution time [12]. In Fig. 2, the Sjeng benchmark has a higher CPI value (Fig. 3).

The CPI value and execution time parameters are analyzed in Fig. 4 (Figs. 5 and 6).

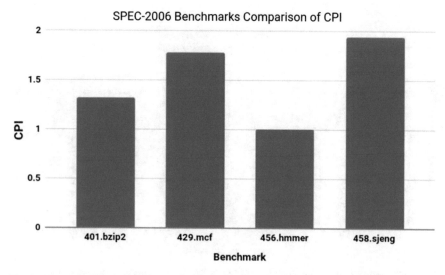

Fig. 2 CPI analysis of SPEC2006 benchmark under multicore arch

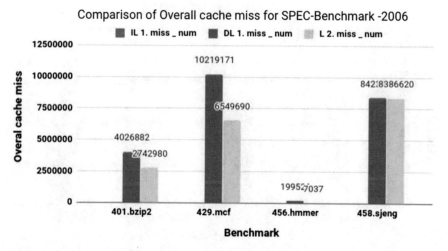

Fig. 3 Overall miss analysis of SPEC2006 benchmark under multicore arch

5 Conclusion

Multicore systems are becoming the backbone of modern computing devices. Analysis of such systems is a mandatory research aspect to understand the latent behavior. Understanding the architectural parameters of these systems for available benchmarks could provide the better architectures. In this study, a detailed analysis of distinct parameters of SPEC2006 benchmarks was achieved, and the correlation

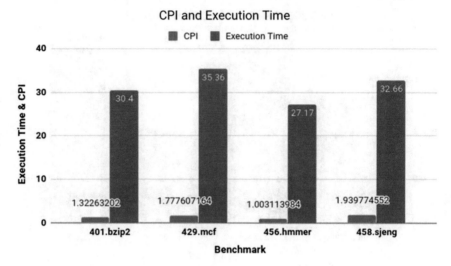

Fig. 4 CPI and execution time analysis of SPEC2006 benchmark under multicore arch

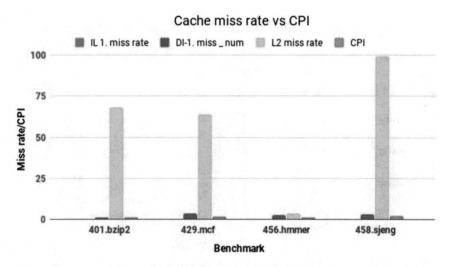

Fig. 5 Miss rate and CPI analysis of SPEC2006 benchmark under multicore arch

among the parameters was established. After the simulation, it is identified that benchmarks which involve higher memory-sensitive characteristics are prone for the performance degradation. Additionally, state-of-the-art devices are coming with a lot of memory operations. Tuning the related design time parameters like cache size could mitigate the performance degradation; however, such steps must be adaptive one and to be taken at the runtime for tuning the performance. The study has used a limited number of instructions for the analysis, which could be extended for the full

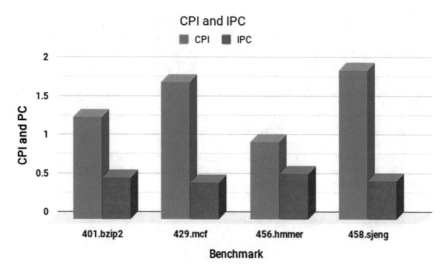

Fig. 6 CPI and IPC analysis of SPEC2006 benchmark under multicore arch

simulation in near future. In future, an innovative adaptive mechanism to tune such parameters could be identified for the performance improvement.

References

1. Das A, Kumar A, Jose J, Palesi M, Revising NoC in future multi-core based consumer electronics for performance. In: IEEE consumer electronics magazine. https://doi.org/10.1109/MCE.2021.3062001
2. Papadimitriou G, Kaliorakis M, Chatzidimitriou A, Magdalinos C, Gizopoulos D (2017) Voltage margins identification on commercial x86–64 multicore microprocessors. In: 2017 IEEE 23rd international symposium on on-line testing and robust system design (IOLTS), pp 51–56. https://doi.org/10.1109/IOLTS.2017.8046198
3. Chiang M-L, Tu S-W, Su W-L, Lin C-W (2018) Enhancing inter-node process migration for load balancing on linux-based NUMA multicore systems. In: 2018 IEEE 42nd annual computer software and applications conference (COMPSAC), pp 394–399. https://doi.org/10.1109/COMPSAC.2018.10264
4. Kasahara H, Kimura K, Adhi BA, Hosokawa Y, Kishimoto Y, Mase M (2017) Multicore cache coherence control by a parallelizing compiler. In: 2017 IEEE 41st annual computer software and applications conference (COMPSAC), pp 492–497. https://doi.org/10.1109/COMPSAC.2017.174
5. Kaliorakis M, Chatzidimitriou A, Papadimitriou G, Gizopoulos D (2018) Statistical analysis of multicore CPUs operation in scaled voltage conditions. IEEE Comput Architect Lett 17(2):109–112. https://doi.org/10.1109/LCA.2018.2798604
6. Shukla SK, Chande PK (2019) Parameter analysis of interfering applications in multi-core environment for throughput enhancement. Int J Eng Adv Technol (IJEAT) 9(2):1272–1286
7. Shukla SK, Chande PK (2019) Investigating policies for performance of multi-core processors. Int J Comput Sci Eng 7(2):964–980

8. Li S, Cheng B, Gao X, Qiao L, Tang Z (2009)Performance characterization of SPEC CPU2006 benchmarks on Intel and AMD platform. In: 2009 first international workshop on education technology and computer science, pp 116–121. https://doi.org/10.1109/ETCS.2009.288

9. Shukla SK, Murthy CNS, Chande PK (2015) A survey of approaches used in parallel architectures and multi-core processors, for performance improvement. In: Proceedings of the twenty-third international conference on systems engineering. Springer, Las Vegas, USA, pp 537–545

10. Shukla SK, Murthy CNS, Chande PK (2015) Parameter trade-off and performance analysis of multi-core architecture. In: Proceedings of the twenty- third international conference on systems engineering. Springer, Las Vegas, USA, pp 403–409

11. Ji S, Dong J, Wang Y, Liu Y (2020) Research on CPI prediction based on space-time model. In: 2019 6th international conference on dependable systems and their applications (DSA), pp 377–382, https://doi.org/10.1109/DSA.2019.00058

12. Patel R, Kumar S (2018) Visualizing effect of dependency in superscalar pipelining. In: 2018 4th international conference on recent advances in information technology (RAIT), pp 1–5. https://doi.org/10.1109/RAIT.2018.8388992

Design of Buck Converter with Modified P&O Algorithm-Based Fuzzy Logic Controller for Solar Charge Controller for Efficient MPPT

Prashant, Abhinav Saxena, Jay Singh, Amit Kumar Sharma, and Nitin Kumar Pal

1 Introduction

Photovoltaic power systems have recently become one of the quickest adopting renewable source due to the benefits they offer over all other options: reliability, ease, availability, environmental friendliness, and renewable energy [1]. PV systems, on the other hand, have two key disadvantages: a high initial cost and nonlinear features that are dependent on climatic circumstances [2]. MPPT is crucial in PV systems [3]. Many ways have been proposed to boost the output power of the solar module. They have advantages and disadvantages that the others lack [4]. All of these improvements will be incorporated in the DC/DC converter and MPPT control algorithm proposed in this study, with the goal of creating a solar PV system that is highly efficient, low-cost, and reliable [5]. It would be prudent if we could devise some circuits that would allow us to harvest the maximum amount of power from them. As a result, regardless of the climatic circumstances, a DC-DC converter regulated by the MPPT mechanism is required to transmit the maximum power from the PV module to the DC-DC load [6]. A fuzzy logic control system is capable to transform difficult information of panel into control signals of the converter [7]. An FLC works

Prashant · A. Saxena (✉)
Department of Electrical Engineering, JSS Academy of Technical Education, Noida, U.P, India
e-mail: abhinaviitroorkee@gmail.com

J. Singh
Department of Electrical and Electronics Engineering, G.L Bajaj Institute of Technology and Management, Greater Noida, U.P, India

A. K. Sharma
Department of Electrical and Electronics Engineering, Galgotia College of Engineering and Technology, Greater Noida, U.P, India

N. K. Pal
Government Polytechnic College Puranpur, Puranpur, U.P, India

© The Author(s), under exclusive license to Springer Nature Singapore Pte Ltd. 2023
R. Agrawal et al. (eds.), *International Conference on IoT, Intelligent Computing and Security*, Lecture Notes in Electrical Engineering 982,
https://doi.org/10.1007/978-981-19-8136-4_18

on the error and change in error as inputs in the controller. The design is complex for FLC as of its rule base is difficult to build and proper fuzzification and defuzzification is required [8]. In this regard, we have implemented three techniques in this work for the best results. First, we have implemented the conventional P&O technique, secondly, we have gone for modified P&O, and last technique used in this work is fuzzy logic controller. All the techniques used will have same basic configuration parameters. Our proposed and implemented modified P&O is found better working with the conventional P&O method. The rest of the work describes as the working of photovoltaic system on different temperature and irradiances and calculation of buck converter presented in Sect. 2. In Sect. 3, the different MPPT techniques were discussed and the simulation model has been demonstrated and noted the results getting from the simulation. The obtained results have been discussed and explained in the Sect. 4, and the conclusion of this research paper is presented in Sect. 5.

2 Photovoltaic System Electrical Configuration

This methodology is based on a photovoltaic system, as shown in Fig. 1. PV panel, buck converter regulated by MPPT mechanism, and DC/DC load are the three primary components [9].

2.1 PV Panel Model

To investigate the impact of various photovoltaic cell variables, many mathematical models have been devised, but the simplest basic implemented is a single diode and its equivalent for the PV cell, as shown in Fig. 1. The I_{pv} equation can be determined using the Kirchhoff law.

$$I_{PV} = n_p I_{ph} - n_p I_s \left(\exp \left[\frac{q \left(V_{pv} + Rsipv \right)}{kTA} \right] - 1 \right)$$

Fig. 1 PV cell electric model

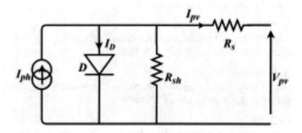

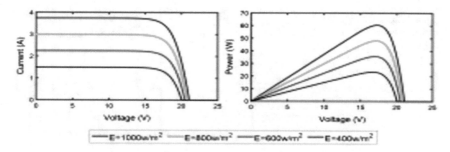

Fig. 2 Impact of irradiance at constant temperature ($T = 25$ °C)

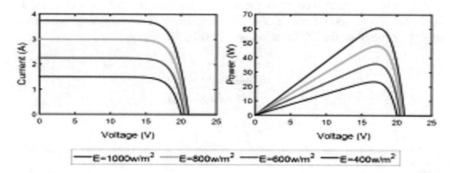

Fig. 3 Impact of temperature at constant irradiance ($E = 1000$ W/m^2)

$$-\frac{V_{pv} + ipvRs}{Rsh} \tag{1}$$

The cell shunt and series resistances are Rsh and Rs, respectively. The cell saturation current and light-generated current, respectively, are Is and Iph. The Boltzmann constant and cell temperature are represented by k and T, respectively. The ideal factor, electron charge, and number of parallel PV cells are depicted as A, q, and np, respectively [10–12].

The electrical (PV and IV) properties under varied irradiation are shown in Fig. 2. It is obvious that these features are influenced by variations in solar radiation. Figure 3 shows the electrical properties as a function of temperature. We can see that when the temperature rises, the voltage and power drops.

2.2 Buck Converter

The buck converter is a switched mode power supply with a linear output voltage response that may be regulated digitally by PWM of a switch [13–15]. A buck converter has two switches, a diode, and transistors, as well as an inductor to keep

Fig. 4 Circuit diagram for buck converter [1]

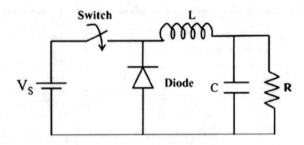

the output capacitor at a constant voltage. It is typically utilised when the input voltage is too high and needs to be decreased to acceptable limits. A buck converter is similar to a step down converter in that it produces an output voltage with lower average values than the DC input voltage Vs (Fig. 4).

$$L = \frac{V_{out}(V_{in} - V_{out})}{F_{sw} * I_{ripp} * V_{in}} \tag{2}$$

$$C = \frac{I_{ripp}}{8 * F_{sw} * V_{ripp}} \tag{3}$$

where

L—inductance value of buck converter
C—capacitance value of buck converter
V_{out}—output voltage
V_{in}—input voltage
F_{sw}—switching frequency
I_{ripp}—10% of Imax
V_{ripp}—1% of Vout

3 Maximum Power Point Algorithm

Varied PV panels have different maximum power points, much like in the photovoltaic model. The maximum power point algorithm is a method for computing the duty ratio in order to place the system in the maximum power point area. The duty ratio gives operating time to the MOSFET and maintain the required output by changing the duty ratio as per changing conditions of PV panel. Here, we implement some of the MPPT techniques like conventional P&O method, modified P&O method, and fuzzy logic controller and simulation has been done and the given outputs has been discussed.

3.1 Conventional P&O Algorithm

The P&O method is often used in MPPT because it is simple to construct and has fewer measured parameters. Tracking the MPP under the impact of weather changes includes perturbing the system by reducing or boosting the output voltage and measuring the effect on the output power, as illustrated in Fig. 5. If the PV output P_{pv} grows, the output V_{pv} is controlled in the same manner as in the previous cycle, as shown in Figure. The output V_{pv} is affected in the inversion direction as the output P_{pv} drops. The output V_{pv} will revolve around the maximum operation voltage after the MPP is found. The given power and voltage values will decide the duty ratio by the algorithm and send it to the PWM converter which has fixed switching frequency. The output of the PWM converter sends the signal to the MOSFET of buck converter. The buck converter is connected to a controller which controls the overcharging of the battery and also controls the required nominal voltage condition to charge the battery. The Simulink models of the buck converter and the MPPT controller have been implemented, respectively. The duty cycle of the gating signal for the MOSFET switch of the buck converter is adjusted by the MPPT Algorithm, in such a way that impedance on both the load and source sides are matched and maximum output power is obtained (Fig. 6).

Based on Fig. 4, we have designed the code for the MATLAB function block which is been used in Fig. 5 which is the Simulink model for the conventional P&O technique. In this method, by generating a maximum power point, we tracked the time for battery charging. Keeping the initial SOC of lead acid battery to be 40, it takes 13.2 s to charge the battery to 40.1. This method was charging the battery quiet slow so we shifted to new MPPT method.

Fig. 5 Flowchart for conventional P&O technique

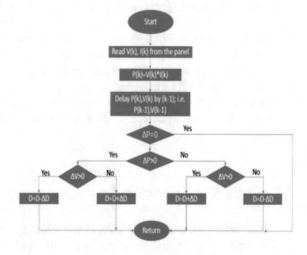

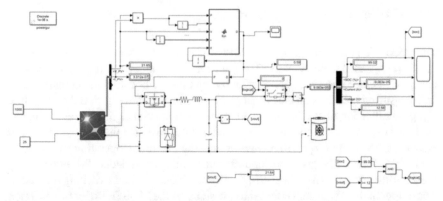

Fig. 6 Simulink model for conventional P&O [2]

3.2 Modified P&O Technique

The fundamental P&O method was devised to improve tracking speed and eliminate oscillations in the vicinity of MPP. To address the shortcomings of the traditional P&O approach, a new P&O methodology is provided to minimise oscillations around MPP on the one hand and enhance tracking speed on the other. Based on a comparison of dP/dV with a given error, the algorithm decides the next suitable step; if the real point is far from the maximum power point, the software offers a large step value. Aside from that, the technique returns a little step value. A little step reduces the number of modifications surrounding the MPP, but a large step accelerates the process of reaching the MPP. Figure 7 shows the flowchart of the improved P&O method.

Based on flowchart illustrated Fig. 7, we have designed the code for MATLAB function block which is been used in Fig. 8 for the Simulink model of modified P&O technique. The code will include the condition of the current also for accurration. In this method, by generating a maximum power point, we tracked the time for battery charging. Keeping the initial SOC of lead acid battery to be 40, we charged the battery and it takes around 3.82 s to charge till 40.1.

In this method, we find that the battery is charging fast as compared to the conventional method. Thus, the modified P&O method is better than conventional method.

3.3 Fuzzy Logic Controller

The three phases of fuzzy logic control are fuzzification, rule basis lookup table, and defuzzification. Fuzzification converts numerical input variables to language variables using a membership function. An error E and a change in error E are generally the two inputs to an MPPT fuzzy logic controller. The user has complete

Fig. 7 Flowchart for modified P&O technique

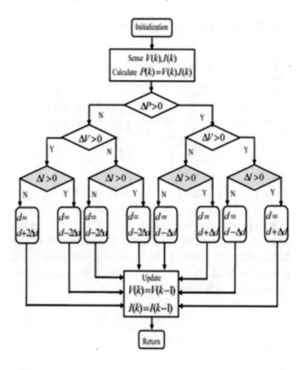

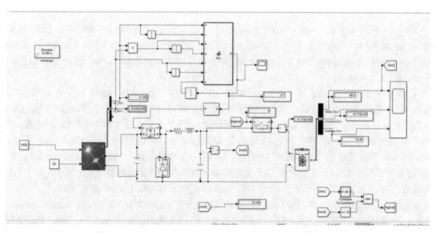

Fig. 8 Simulink model for modified P&O method

freedom in calculating E and E in whatever way he or she thinks suitable. The fuzzy logic controller output is changed from a linguistic fuzzy logic-based MPPT controller variable to a numerical variable using a membership function during the defuzzification stage. The MPP, which regulates the power converter, will receive an analogue signal. The capacity to operate with imprecise inputs, the absence of a

Table 1 Rule base for fuzzy logic controller

S ΔP_{PV}	NB	NS	Z	PS	PB
NB	PB	PS	Z	NS	NB
NS	PB	PS	Z	NS	NB
Z	PS	PS	Z	NS	NS
PS	PB	PS	Z	NS	NB
PB	PB	PS	Z	NS	NB

perfect mathematical model, and the ability to manage nonlinearity are all positives of fuzzy logic controllers, but the primary downside is the high cost of implementation. This approach has the benefit of making determining whether the working point is on the right or left of the MPP straightforward, enabling the converter's duty cycle to be raised or lowered. PPV can also be used to detect changes in radiation and accelerate tracking for low radiation levels. ΔP_{PV} and ΔV may be described using equations, respectively.

$$S(n) = \frac{\Delta P_{pv}(n)}{\Delta V_{pv}(n)} = \frac{I_{pv}(n).V_{pv}(n) - I_{pv}(n-1).V_{pv}(n-1)}{V_{pv}(n) - V_{pv}(n-1)} \tag{4}$$

$$\Delta P_{pv}(n) = P_{pv}(n) - P_{pv}(n-1) \tag{5}$$

The specific criteria for tracking the maximum power point by adjusting the duty cycle must be specified in order to develop the fuzzy logic controller. The guidelines we have established are for voltage and power. The fuzzy logic controller's rule base is listed in Table. 1.

The two inputs, one being a change in voltage and the other being a change in power, and specified the duty cycles based on the conditions. The membership functions of fuzzy logic control were created with the use of a rule basis. The range and features of input and output are defined by membership functions. We must export the FIS file in the fuzzy control block after defining the function. The duty cycle is delivered to the PWM generator as a result of the fuzzy logic's output. The duty cycle of the MOSFET pulses will be changed by the PWM generator.

Figure 9 shows the result of implementing fuzzy logic control for maximum power point tracking in the supplied Simulink model. The charging time, i.e. time taken to charge the battery has been noted. The starting state of charge (SOC) of the battery will be 40%, and the time taken to reach that level will be 40 min (Fig. 10).

Fig. 9 PV characteristics of
PV panel for MPPT

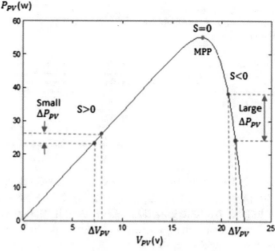

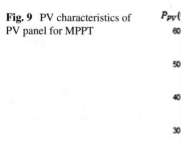

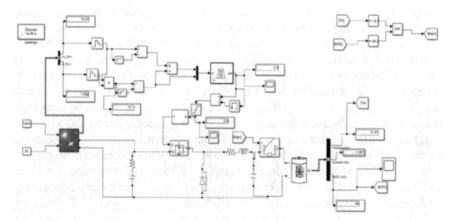

Fig. 10 Simulink model for fuzzy logic technique [4]

4 Discussion and Results

The different MPPT techniques have been simulated successfully. The modelling has been done as taking PV array from the Simulink and takes parameters of the array as discussed above. Then, we take constant block for giving constant irradiance and temperature to the array. The buck converter has been connected as shown in the simulation figure and parameters have been given as mentioned. The bus selector has been connected with the m output of the PV array, and then, the different outputs of bus selector are connected to the display and the measure the generated voltage and current. Then, product block is used to measure the power getting from PV array by giving the input as current and voltage in the block. The AND gate is also

Table 2 Comparison between different methods for MPPT tracking

%SOC	Charging time of battery (in seconds)		
	Conventional P&O	Modified P&O	Fuzzy logic
40.02	1.50	0.58	0.56
40.04	3.50	1.39	1.24
40.06	5.80	2.13	2.02
40.08	8.50	2.98	2.79
40.10	11.32	3.82	3.48

used in the simulation as to overcome the overcharging of the battery and to give minimum required voltage for charging of the battery. If battery is charging at low voltage or overcharging takes place, then it decreases the lifetime of the battery and requirement of changing of the battery takes place which increases the expenses on the renewable energy source of photovoltaic model. From here, different MPPT techniques have been used and illustrated above. The MATLAB function block is designed by using the flowchart in which algorithm has been given and duty ratio has been calculated. In the fuzzy logic controller system, the memory block is used which gives the previous value in the subtraction block and the fuzzy logic block is used in which the fuzzification and defuzzification takes place to calculate the duty ratio. The battery has been connected with the buck converter, and a bus selector is connected with the battery output to record the changes and to note the state of charge, voltage, and current getting from the simulation. The simulation has been run for different time in the different techniques to get proper results and outputs to compare the different techniques we used. The results have been obtained by keeping all the conditions same and varying the MPPT techniques we have found that the proposed MPPT method is giving comparable results to the fuzzy logic controller. The state of charge of battery is taken 40–40.1 and the time taken to charge the battery by different techniques has been shown in Table 2.

Based on the above methods by comparing all the technique, the results are presented in Table 2.

5 Conclusion

In this research, MPPT is utilised to capture maximum output from a photovoltaic panel using a suggested P&O method. The power slope of the PV panel curve is used as an input, and the change in the new P&O step size is outputted by the proposed control approach. For maximum power point tracking, the devised algorithm and fuzzy logic controller resulted in faster and more efficient maximum power tracking and higher voltage. The use of the fuzzy controller's duty ratio output signal to trigger the value of the estimated reference maximum power improved power tracking and increased the maximum power range.

From the results, it can be analysed that the fuzzy logic controller with buck converter performs significantly better than the conventional P&O MPPT approach. Fuzzy logic is also quite close to the modified P&O MPPT approach for charging batteries. The fuzzy logic control is favoured above all other approaches since it is simple to implement.

References

1. Jiang A et al (2005) Maximum power tracking for photovoltaic power systems. Tamkang J Sci Eng 8(2):147–153
2. Rezk H, Eltamaly A (2015) A comprehensive comparison of different MPPT techniques for photovoltaic systems. Sol Energy 112:1–11
3. Verman D, Nela S, Shandilya AM, Soubhagya KD (2016) Maximum power point tracking (MPPT) techniques: recapitulation in solar photovoltaic systems. Renew Sustain Energy Rev 54:1018–1034. https://doi.org/10.1016/j.rser.2015.10.068
4. Husseinali A, Abdulwahhababdulrazzaq A (2018) Evaluating the performance and efficiency of MPPT algorithm for PV systems. Int J Eng Technol 7(4):2122–2126
5. Yang Y, Blaabjerg F, Modified P&OMPPT algorithm for singlephasepv systems based on deadbeat control, Department of Energy Technology, Aalborg University, DK-9220 Aalborg East, Denm
6. Ba A, Ehssein C, Mouhamed M, Hamdoun O, Elhassen A (2018) Comparative study of different DC/DC power converter for optimal PV system using MPPT (P&O) method. Appl Sol Energy 54:235–245
7. Chiang SJ, Shieh HJ, Chen MC (2009) Modeling and control of PV charger system with sepic converter. IEEE Trans Ind Electron 56(11):4344–4353
8. Blorfan A, Sturtzer G, Flieller D, Wira P, Mercklé J (2014) An adaptive control algotithm for maximum power point traking for photovoltaic energy conversion systems—a comparative study. Int Rev Electric Eng (IREE) 9(3):559–565
9. Seguel JL, Seleme SI Jr (2021) Robust digital control strategy based on fuzzy logic for a solar charger of VRLA batteries. Energies 14:1001. https://doi.org/10.3390/en14041001
10. Kuo Y et al (2001) Maximum power point tracking controller for photovoltaic energy conversion system. IEEE Trans Ind Electron 48:594–601
11. Subudhi B, Pradhan R (2013) A comparative study onmaximum power point tracking techniques for photovoltaic power systems. IEEE Trans Sustain Energy 4:89–98
12. Batzelis IE (2017) Simple PV performance equations theoretically well founded on the single-diode model. IEEE J Photovoltaics 7(5):1400–1409
13. Lei P, Li Y, Seem JE (2011) Sequential esc-based global mppt control for photovoltaic array with variable shading. IEEE Trans Sust Energy 2(3):348–358
14. de Brito MAG, Galotto L, Sampaio LP, Melo GDAE, Canesin CA (2013) Evaluation of the main MPPT techniques for photovoltaic applications. IEEE Trans Indus Electron 60(3):1156–1167
15. Rezkallah M, Hamadi A, Chandra A, Singh B (2018) Design and implementation of active power control with improved P & O method for wind-PV-battery-based standalone generation system. IEEE Trans Ind Electron 65(7):5590–5600

Security in Smart Computing Environment

DDoS Attack Detection Using Artificial Neural Network on IoT Devices in a Simulated Environment

Ankit Khatri and **Ravi Khatri**

1 Introduction

IoT is a technology which has found applications in a variety of areas [1] and has transformed industries too. It has revolutionized the industries such as food production, health, manufacturing, etc., and deeply influencing our lives too. Apart from these industries, it has also become a crucial part of what we call the critical infrastructure of a country. Due to its such vast applications, it has also attracted the threat of attacks which immensely impact the socio-economic security of the nation and also intrude the privacy of its citizens.

IoT basically consists of a network of devices interconnected with each other, the cloud servers and the internal system. These devices collect data which are then sent further for various analysis. With the rise of the Internet, the count of such interconnected devices over the network is also on a high and thus becoming an influential part of our daily lives. According to some reports [2], the count of such devices is increasing continuously and expected to rise by a significant amount in coming years. As the IoT devices use batteries as their source of power [3], they have low processing capabilities. It leads to a lack of security and the required defense against cyber-attacks. The number of attacks on IoT has increased immensely in the recent past. The attackers generally prefer IoT for performing the assault for a variety of reasons which puts him/her in a position of advantage.

The attacker may have different motives. It may be performed for personal gains, state sponsored assault, business rivalry, etc. The attackers may belong to the same network which is being compromised or may belong to an external network. The

A. Khatri (✉) · R. Khatri
Dr B R Ambedkar National Institute of Technology Jalandhar, Jalandhar, Punjab, India
e-mail: ankitk.cs.21@nitj.ac.in

R. Khatri
e-mail: ravik.cs.21@nitj.ac.in

© The Author(s), under exclusive license to Springer Nature Singapore Pte Ltd. 2023   221
R. Agrawal et al. (eds.), *International Conference on IoT, Intelligent Computing and Security*, Lecture Notes in Electrical Engineering 982,
https://doi.org/10.1007/978-981-19-8136-4_19

difference being, the internal ones [4] have the access and privileges to easily carry out such an attack, whereas the external ones [4] usually carry out the attacks by exploiting the system using malware. There are a variety of attacks that can be performed on IoT devices which can paralyze the functioning of critical sectors of a nation and can even lead to a war among the conflicting parties. However, these attacks can be detected using a variety of means [5]. The assaults on such devices [6] can be user privacy infringement, malware attacks such as Mirai, DoS attacks, physical tampering, ransomware attacks, etc. In this paper, we tackle the menace caused by DDoS attacks which are targeted upon the IoT equipment by providing an IDS centered on the use of ML. In DDoS attacks, the attacker floods the target host with a massive amount of traffic [7] with an aim to restrict its access for the legitimate users. Various DDoS attacks have occurred in the recent past which have immensely impacted various big MNC's [8]. For example, the GitHub attack of 2015 [9], Dyn attack [10], the attack on University of Minnesota [11], websites getting crashed during an uprising in Turkey [12], attack on the White House website [13], etc. To detect and mitigate such attacks, IDS are deployed. These systems have existed since the late twentieth century and perform their intended task of monitoring the networks for such attacks using various methodologies that evolve continuously with time such as ML-based IDS as proposed by [14].

We have discussed an ANN-based IDS in this paper. Our proposed algorithm sniffs the incoming traffic packets which is then sent to the trained model of ANN for classification of these packets into safe or hostile.

In this paper, we have discussed various concepts and terminologies in Sect. 2. In Sect. 3, we have mentioned the related works. Our proposed methodology is discussed in Sect. 4, followed by the observations and results in Sect. 5. Finally, we have concluded this paper is Sect. 6.

2 Basic Concepts and Terminologies

2.1 Artificial Neural Networks

In ANN [15], we try to replicate the working of a real human brain. A real human brain contains a huge network of neurons which are interconnected to each other. These neurons are responsible for transferring the impulses from the brain to different parts of the body and vice-versa. In ANN, we replicate this process with the use of several layers like an input layer, one or many hidden layers, and at the end there is an output layer. The first layer, i.e., input is responsible for taking the inputs either from the user or the environment and passing these inputs onto the next layer. The hidden layers are responsible for performing a mathematical computation of the inputs and the weights associated with the inputs. Then, the weighted inputs are passed to an activation function which basically transforms the output as per the function and also tells which node would fire and which node would not fire. Finally, we get the desired output through the output layer. ANNs can be used for various problems like regression, classification, image recognition, etc. We have several advantages of using an ANN, like it can perform multiple tasks simultaneously, node failure does not result in failure of the whole network, whole of the data is stored within the network, they have the ability of giving the results even with incomplete information, etc.

2.2 Intrusion Detection System

Intrusion to any system refers to gaining unauthorized access to it by a perpetrator with a motive to intrude on the privacy of the network. There may be various types of such intruders which are broadly categorized into three classes. First one is the masqueraders who do not have any privileges to access a particular system but gain access by illegitimate means. Second, the misfeasors, who may be some internal user who misuse the privileges and gains unauthorized access. Third ones are the clandestine users who use the credentials of the privileged access people of the same network to steal confidential information [16].

IDS [5] can be categorized into two categories.

- Host-based IDS (HIDS)—The host system is equipped with software-based tools that can analyze and observe all the activities in the system and also for possible intruders.
- Network-based IDS (NIDS)—The traffic flow is monitored on the complete network to check for possible malicious users trying to intrude in the network by illegitimate means.

The comparison of these IDS based on certain performance measures is shown in Table 1. These systems can detect potential threats by identifying patterns which are based on misuse and anomalous intrusion. Various methodologies are known for intrusion detection. Few most widely used ones are based on statistical analysis [4], rule-based detection [17], ANN, protocol verification [18], etc. In this paper, we have proposed an ANN-based IDS which is trained to effectively classify safe and hostile packets based on the patterns detected in flooding attacks.

As the IoT devices have limited capacity in terms of storage and computational power, standard IDS may not be suitable for them which are otherwise suitable for highly complex environments. Recent works on the dataset used in the KDD competition have shown the potential of using ML for intrusion detection.

2.3 DDoS Attack

In such an attack, numerous systems are utilized to carry out the attack on the target host. It is faster than the DoS attack and difficult to block and trace due to

Table 1 Comparison of performance of HIDS and NIDS [19]

Host-based IDS	Network-based IDS
Performs well at analyzing the traffic to perceive possible threats	Performs well at analyzing the traffic to perceive possible threats
It performs better for the detection of internal perpetrators	It performs better for the detection of external perpetrators
Performs good in analyzing the magnitude of the damage	Performs poorly in analyzing the magnitude of the damage
Poor reaction to real-time threats, but works better for persistent ones	Strong reaction to real-time threats

its distributed nature, i.e., packets are coming simultaneously from multiple devices and from different geographical locations. The volume of such attacks is also huge when compared to DoS attacks. These attacks can be classified as application layer, protocol-based and volumetric-based attacks. Some of the most common DDoS attacks are SYN flood, DNS amplification, UTP flood, etc.

3 Literature Review

The term IoT was coined by Kevin Ashton of MIT in 1999. However, it is yet to reach its maturity level. Still, there is a high risk of attacks that can be performed on IoT devices due to its vulnerabilities. To detect and mitigate such attacks, numerous researches have already been done. A few of the literature surveys are as follows.

Kasinathan et al. [20], in his work, explored flooding attacks in an IoT environment. They have proposed a scheme which uses an IDS to detect the attacks. In their paper, they have focused on detecting the UDP flood attacks in the IoT environment. The solution which they proposed was implemented for creating a network-based IDS. They designed a network manager which detected hostile packets based on the parameters defined.

Misra et al. [21] developed a learning automata-based solution to detect DDoS attacks on IoT networks. They developed an architecture that was service-oriented to shape their solution. The framework was defined to automatically select the best choice. The drawback of this solution was that no real implementation was done.

Hodo et al. [22] in their study about threat analysis of IoT networks using ANN-based IDS have shown how ANN could be used in IDS. The model showed a significant accuracy of 99.4%. It demonstrated that the ANN could successfully be used in IDS. However, it lacked any insight on how to prevent such an attack if detected.

Tseung et al. [23] proposed a solution which used the power of ML along with an auto-learning Bloom filter (BF) to identify and mitigate the DDoS attacks targeted on a network. They have used multiple feature selection and extraction algorithms of ML like linear SVM, ANOVA to assess the incoming data packets and extract important features from them. Then, the extracted features are kept in a customized feature list.

The input data packets and the extracted information associated with them are given to the auto-learning BF which identifies and classifies the incoming data packets as malicious or not. In this way, both ML and BF are used to identify such attacks and defend them. There is an issue associated with their approach, i.e., the solution proposed by them is not cost-efficient because the feature selection approach used is computationally expensive, hence may not be suitable for an IoT environment.

Zekri et al. [24] have proposed a ML-based solution to identify DDoS attacks in a cloud computing environment. They have used the C4.5 algorithm to design an IDS to identify the attack on a cloud-based environment. They have also used the combination of C4.5 algorithm with signature identification methods for efficient identification of signatures attacks. A decision tree is generated to automatically and effectively detect such attacks targeted on the cloud environment.

A SVM-based detection methodology was proposed by Liao et al. [25]. Their proposed scheme focuses on the similarity among the devices that are a part of the zombie network carrying out the attack. The request patterns of users are stored and analyzed to detect similarities.

A KNN-based DDoS attack detection framework was proposed by Xiao et al. [26]. Their methodology is based on the assumption that similar bots and softwares will generate identical traffic. But, one restriction of their study is that non-identical traffic flows can also be generated by similar bots.

ANN is used for detection of DDoS attacks in the study conducted by Jie-Hao et al. [27]. They have compared its performance with various other ML models. The user traffic is checked for aberrations by the neural network.

The detection methodology proposed by [28] is based on the analysis of traffic by radial basis function neural networks. It categorizes the traffic into safe or hostile. The source IP addresses of the hostile traffic are then sent to a different module for filtering. A FCM cluster algorithm-based methodology is proposed by [29]. A decision tree-based solution is put forward by [30], which observes the traffic flow to classify safe and hostile packets. Moreover, it also describes the packet flow pattern.

An intrusion prevention system is developed by Li et al. [31], which uses SVM in addition with SNORT so that the accuracy of the overall system is improved.

A different type of neural network is used by Liu et al. [32] which finds applications in areas such as classification, data compression, pattern identification, etc. Here, the data is fed to the neural network in numeral form which then classifies the packets as safe or malicious and takes further actions. The information about the protocols that are present in the packets are analyzed by a data mining algorithm in the detection scheme proposed by [33].

4 Proposed Methodology

We shall emulate the scenario shown in Fig. 1 where IoT devices are connected. A bot shall be created to carry out DDoS attack on these devices. The server device shall be equipped with network IDS to detect the DDoS attacks. The IDS would be

Fig. 1 Depiction of attack
on IoT devices in a simulated
environment

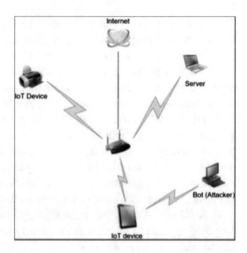

based on ANN, a multilayer perceptron. Our discussed approach uses a simple ANN, the structure of which is depicted in detail below. The server will keep on analyzing the incoming traffic. The ANN will receive the input data through Wireshark with different parameters (highest layer, transport layer, source IP address, destination IP, source port, destination port and packet length, target). Once the attack has been carried out on the device, the input data will be analyzed by the trained model of ANN. The input data packets will be analyzed and classified as hostile or normal packets that will define whether the traffic is DDoS traffic or normal traffic. The parameters have been fine-tuned multiple times to achieve a better accuracy. Further, details are discussed in the implementation section.

4.1   Detailed Configuration of Our Model

A regular perceptron is being used by us which is developed using TensorFlow. We have increased the width of our model to achieve better results. Our proposed model has a configuration which contains, layer (hidden 1): [50 neurons] fully connected layer, layer (hidden 2): [50 neurons] fully connected, layer (hidden 3): [50 neurons] fully connected, output layer: [2 neurons] one hot encoding layer, initialization of the weights: random/normal. The activation functions used are ReLU (hidden layers) and Softmax (output layer).

4.2 Workflow and Algorithms of Our Proposed Methodology

The workflow of our proposed system includes the following parts: sniffing the incoming traffic, collecting the data (real time), training the ANN, viewing the data, analyzing the real-time traffic and graphical presentation of our model's architecture. We have trained our model in 30 epochs of 15,000 iterations each. Our proposed system includes the following phases.

- Sniffing the Incoming Traffic: This model captures the packets in real time and shows detailed information about each of them. We have included the necessary details that needs to be displayed in order to analyze the traffic. The algorithm for that is depicted in Fig. 2.
- Neural Network Trainer: This algorithm deals with collecting the data at real time and then using that data to train our neural network. Data collected is saved in

---

Algorithm 1: Capturing Packets (Sniffing)

This model captures the packets in real-time and shows detailed information about each of them.

Start

```
1. For pkts in capture
2.    If (pkts.TopLayer != 'ARP'):
3.        ip_add = None
4.        ip_layer_name = get_ip_layer_name(pkt)
5.        If (ip_layer_name == 4):
6.            ip_add = pkts.ip_add
7.        End If
8.        Else If (ip_layer_name == 6):
9.            ip_add = pkts.ipv6addr
10.       Display the details about each packet captured
11.          pkt.Top_Layer
12.          packet.trnsprt_layer_name
13.          Time Taken
14.          Layers
15.          Source IP
16.          Destination IP
17.          Length of the Packet
18.       Try displaying Source and Destination Port
19.          Source Port
20.          Destination Port
21.       Except AttributeError
22.          Source Port
23.          Destination Port
24.       End If
25.    Else
26.       ARP = pkt.arp
27.       Display the packet information
28.    End Else
```

Stop

Fig. 2 Algorithm for packet sniffing

a file named data.csv on which the ANN gets trained. The algorithm for that is depicted in Fig. 3.

- Real-Time Prediction: This algorithm reads the live data.csv file and classifies the traffic as malicious (DDoS) or normal based on our already trained model. This model then prints the results/observations using a confusion matrix and classification report. The algorithm for that is depicted in Fig. 4.

Algorithm 2: Data collector

This trains our Artificial Neural Network(ANN).

Input: [Already Saved model, if user wishes to load existing model]

Start
1. User inputs name of Saved Model file
2. Ask user if they want to train a saved algorithm, if no, the new model is created and trained
3. If(*load _model*== *'Y'*):
4.   train= Load_model_saved()
5.   **End If**
6. **Else**
7.   Imports the Neural Network Class from Sci-kit learn
8.   train= Classifier(size_of_hidden_layer,activation_function,iter_max,verbose,tol)
         # Setting the parameters of our classifier
         #size_of_hidden_layer = layers that are hidden in the neural net, (6) = single layer 6, (6,6) = Two layers, each having 6 nodes
         #activation_function = activation function, 'logistic' is equivalent of the sigmoid activation function
         #iter_max = maximum amount of iterations that the model will do
         #Verbose = whether the model prints the iteration and loss function per iteration
         #tol = the decimal place the user wants the loss function to reach
9. Read CSV

#Neural Net Training
12. X =Input (Independent Variables)
13. Y = Dependent Variable
14. Train the model
15. Display predictions
16. malicious = 0, secure = 0
17. For c in predicted_results:
18.   **If**(*c == 1*):
19.   increment malicious packets
20.   **End If**
21.   **Else**:
22.   increment secure packets
23.   **End Else**;
24 Display secure Packets and malicious Packets
25. Display confusion matrix
27. Store model?
28. If true: then Store Model

Stop

Fig. 3 Data collection and training of the neural network

Algorithm 3:Real Time Prediction

#This model is used for classifying the traffic as DDoS or normal traffic.This is not used for training purposes.

Input: [Already Saved model, if user wishes to load existing model]

Start

1.User inputs name of the Data File.
2.Process the Data (Live Label Encoding)
3.Print the data for viewing
4.Load model and model coefficients
5.predicted_result = model.predict(Input Data)
6.malicious,secure = 0
7.For c in predicted_results :
8. **If** c == 1: # *replace it as zero so that DDoS attack can be forced*
9.   **increment** malicious packets
10. **Else:**
11.   **increment** secure packets
12.   **End If**
13.   **print** secure and malicious packets
14.   If No.of malicious Packets > Sum of Both/2 :
15.   **print Attack Detected**
16.   **Else:**
17.   **print Normal Activity Detected**
18. **End If**
19. **print** Confusion Matrix
20. **print** Classification Report

Stop

Fig. 4 Real-time prediction of DDoS attack by an ANN

4.3   The Process of DDoS Attack Classification by the ANN Model

Our proposed algorithm first captures the incoming traffic which is used to create a dataset for training our model. Once the model is trained it can be used to classify DDoS attacks in real time. Real-time data is gathered which is then provided as input to the ANN model for classification. The trained model classifies the real-time packets as hostile or safe based on its learning on previous data. If the number of hostile packets is greater than the average of the count of both malicious and safe packets, then a DDoS attack is notified, otherwise normal activity is reported.

5   Observations and Results

The study consists of two phases: First is the DDoS attack which is performed on an IoT device in a simulated environment and second is its detection using deep learning. The DDoS attacks that we have performed are TCP SYN, Slow Loris and UDP flood, each attack was able to take our server down and make the services unavailable to the user. We have achieved satisfactory results in this phase. The confusion matrix, model evaluation, and the test results are shown in Figs. 5, 6, and 7, respectively. Figure 8 shows the data flow diagram of our proposed methodology.

Fig. 5 Confusion matrix

```
Safe Packets:   162533
Hostile Packets:   99578
Time Taken: 11.340712799999991
Confusion Matrix:
 [[162383    162]
 [   150  99416]]
```

Fig. 6 Model evaluation results

Classification Accuracy	0.993665724
Initial Loss Value	0.049470
Final Loss Value	0.033887
Total Loss	0.015583
Precision (Safe)	0.99926
Precision (Hostile)	0.983123724
Precision (Average)	0.991191177
Recall (Safe)	0.991119463
Recall (Hostile)	0.983123724
Recall (Average)	0.987121593
F1 Score (Safe)	0.995172405
F1 Score (Hostile)	0.983123724
F1 Score (Average)	0.989148064

	Control	Normal	TCP	UDP	Combined (TCP)	Combined (UDP)
Total Packets	0	5171	260	21963	11748	15482
Safe	0	5167	0	1	11624	6667
Hostile	0	4	260	21962	124	8815
Percentage	0	0.077354	100	99.99545	1.055498808	56.93708823
DDoS detected	0	0	1	1	0	1
DDoS in Action	0	0	1	1	1	0

Fig. 7 DDoS detection test results

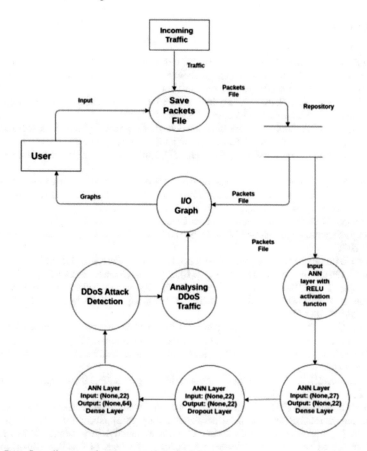

Fig. 8 Data flow diagram of our proposed methodology

6 Conclusion and Future Scope

The methodology we have proposed in this work successfully identifies DDoS attack in a simulated environment. The ANN-based IDS accurately detects attacks such as SYN flooding, UDP flood, etc., with 99.3% accuracy by using a criterion based on the count of hostile and safe packets. The future plans could include modifying this algorithm to identify other DDoS attacks. Moreover, the simulated environment can also showcase methods to filter hostile IP addresses and redirect them to other/fake addresses. Apart from that, mitigation strategies can be used, such as blocking the hostile IP addresses using probabilistic data structures like Bloom filter. Although we have detected DDoS attacks using an ANN in a simulated environment, a real implementation can easily be extracted from this emulation and could be applied in real scenarios. Moreover, the use of CNN and RNN for this problem can also be explored.

References

1. Jain D, Krishna P, Saritha V (2012) A study on internet of things based applications
2. Evans D (2011) The Internet of Things—how the next evolution of the internet is changing everything. CISCO white Pap., no. April, pp 1–11
3. Iqbal A, Suryani MA, Saleem R, Suryani MA (2016) Internet of things (IoT): on-going security challenges and risks. Int J Comput Sci Inf Secur 14(11):671
4. Shun J, Malki HA (2008) Network intrusion detection system using neural networks. In: 2008 Fourth international conference natural in computing, vol 5, pp 242–246
5. SANS institute (2001) InfoSec reading room TU, application of neural networks to intrusion detection
6. Paganini P (2015) Internet of Things: how much are we exposed to cyber threats?—Infosec resources. [online] Infosec Resources. Available at: https://resources.infosecinstitute.com/topic/internet-things-much-exposed-cyber-threats/. Accessed 6 Nov 2021
7. Sharwood S (2015). GitHub wobbles under DDoS attack. [online] Theregister.com. Available at: https://www.theregister.com/2015/08/26/github_wobbles_under_ddos_attack/. Accessed 6 Nov 2021
8. Garber L (2000) Denial-of-service attacks rip the internet. Computer 33(4):12–17
9. Nast C (2018). A 1.3-Tbs DDoS Hit GitHub, the Largest Yet Recorded. [online] Wired. Available at: https://www.wired.com/story/github-ddos-memcached/. Accessed 6 Nov 2021
10. Ademola O (2021) Dyn DDoS cyber attack: a position paper. 13–20. https://doi.org/10.22624/AIMS/MATHS/V9N1P2
11. Kessler GC (2002, 2000) Defenses against distributed denial of service attacks. SANS Institute
12. Simpson S, Shirazi SN, Marnerides A, Jouet S, Pezaros D, Hutchison D (2018) An inter-domain collaboration scheme to remedy DDoS attacks in computer networks. IEEE Trans Netw Service Manage 1–1
13. Evans D, Larochelle D (2002) Improving security using extensible lightweight static analysis. IEEE Softw 19(1):42–51
14. Van NTT, Thinh TN (2015) Accelerating anomaly-based ids using neural network on GPU. In: 2015 international conference on advanced computing and applications (ACOMP), pp 67–74
15. Mishra M, Srivastava M (2014) A view of artificial neural network. In: 2014 international conference on advances in engineering & technology research (ICAETR—2014), pp 1–3. https://doi.org/10.1109/ICAETR.2014.7012785
16. Wallen D (2020) Intrusion detection systems: a deep dive into NIDS & HIDS. [online] Security Boulevard. Available at: https://securityboulevard.com/2020/03/intrusion-detection-systems-a-deep-dive-into-nids-hids/. Accessed 6 Nov 2021
17. Bellekens XJA, Tachtatzis C, Atkinson RC, Renfrew C, Kirkham T (2014) A highly-efficient memory-compression scheme for GPU-accelerated intrusion detection systems. In: Proceeding 7th international conference security information networks—SIN '14, pp 302–309
18. Verwoerd T, Hunt R (2002) Intrusion detection techniques and approaches 25(15)
19. Kozushko H (2003) Intrusion detection: host-based and network-based intrusion detection systems, Sept., vol 11
20. Kasinathan P, et al (2013) Denial-of-Service detection in 6LoWPAN based Internet of Things. In: 2013 IEEE 9th international conference on wireless and mobile computing, networking and communications (WiMob). IEEE
21. Misra S, Krishna PV, Agarwal H, Saxena A, Obaidat MS (2011) A learning automata based solution for preventing distributed denial of service in Internet of Things. In: 2011 international conference on internet of things and 4th international conference on cyber, physical and social computing, Dalian, pp 114–122
22. Hodo E, et al (2016) Threat analysis of IoT networks using artificial neural network intrusion detection system. In: 2016 international symposium on networks, computers and communications (ISNCC). Yasmine Hammamet, pp 1–6. https://doi.org/10.1109/ISNCC.2016.7746067

23. Tseung C, Chow K, Zhang X (2017) Anti-DDoS technique using self- learning bloom filter. In: Intelligence and security informatics (ISI), 2017 IEEE international conference. IEEE, pp 204

24. Zekri M, Kafhali SE, Aboutabit N, Saadi Y (2017) DDoS attack detection using machine learning techniques in cloud computing environments. In: 2017 3rd international conference of cloud computing technologies and applications (CloudTech), pp 1–7, https://doi.org/10.1109/CloudTech.2017.8284731

25. Liao Q, Li H, Kang S, Liu C (2015) Application layer DDoS attack detection using cluster with label based on sparse vector decomposition and rhythm matching. Secur Commun Netw 8(17):3111–3120

26. Xiao P, Qu W, Qi H, Li Z (2015) Detecting DDoS attacks against data center with correlation analysis. Comput Commun 67:66–74

27. Jie-Hao C, Feng-Jiao C, Zhang (2012) DDoS defense system with test and neural network. In: IEEE international conference on granular computing (GrC). Hangzhou, China, pp 38–43

28. Karimazad R, Faraahi A (2011) An anomaly-based method for DDoS attacks detection using RBF neural networks. In: Proceedings of the international conference on network and electronics engineering, pp 16–18

29. Zhong R, Yue G (2010) DDoS detection system based on data mining. In: Proceedings of the 2nd international symposium on networking and network security. Jinggangshan, China, pp 2–4

30. Wu Y-C, Tseng H-R, Yang W, Jan R-H (2011) DDoS detection and traceback with decision tree and grey relational analysis. Int J Ad Hoc Ubiquitous Comput 7(2):121–136

31. Li H, Liu D (2010) Research on intelligent intrusion prevention system based on snort. In: International conference on computer, mechatronics, control and electronic engineering (CMCE), vol 1. IEEE, pp 251–253

32. Li J, Liu Y, Gu L (2010) DDoS attack detection based on neural network. In: 2nd international symposium on aware computing (ISAC). IEEE, pp 196–199

33. Gao N, Feng D-G, Xiang J (2006) A data-mining based dos detection technique. Jisuanji Xuebao (Chinese J Comput) 29(6):944–951

ABBDIoT: Anomaly-Based Botnet Detection Using Machine Learning Model in the Internet of Things Network

Sudhakar and **Sushil Kumar**

1 Introduction

The development of new technologies and growth of internet-connected devices provides a better space for cybercriminals to launch a large scale attack including click fraud, cryptocurrency mining, malspam campaign, DDoS (Distributed Denial-of-Service) etc. on internet infrastructure with the help of huge bot-armies. The bots are the internet-connected devices which has its control to its bot-master. Cybercriminals uses vulnerabilities and malicious programs to compromise and build the large army of bots or network of bots e.g. botnet. The bot-master controls its botnet by issuing commands using command and control servers [31, 32]. Now, everything is connected to the internet with the emergence of IoT (Internet of Things), IIoT (Industrial IoT), OT (Operational Technology),[1] and Industry 4.0 such as refrigerator, microwave oven, smart lights, smart fan, air conditioner, health devices, manufacturing machines, and more [16, 25]. According to the IoT business news, the

---

[1]Amnesia:33 (https://www.forescout.com/company/resources/amnesia33-how-tcp-ip-stacks-bre
ed-critical-vulnerabilities-in-iot-ot-and-it-devices/).

Sudhakar (✉) · S. Kumar
School of Computer and Systems Science, Jawaharlal Nehru University, New Delhi, New Delhi
110067, India
e-mail: sudhak82_scs@jnu.ac.in

S. Kumar
e-mail: skdohare@mail.jnu.ac.in

R. Agrawal et al. (eds.), *International Conference on IoT, Intelligent Computing
and Security*, Lecture Notes in Electrical Engineering 982,
https://doi.org/10.1007/978-981-19-8136-4_20

expected growth in internet-connected devices will be 125 billion by 2030.[2] IoT devices will generate numerous amount of data which presents the challenges of managing the data and security Further, these technologies are used to implement in many other sectors to ease the human life like agriculture, health sector, manufacturing, food supply management, pharma supply management, environment monitoring using various sensors, surveillance, smart homes and so on [6]. Some of IoT botnet attacks in the past are listed in the following points which shows the importance of securing IoT networks and its devices.

- In year 2016, Mirai botnet has devastated major infrastructures and shutdown the network such as NetFlix, Twitter, CNN, and many including one whole country Liberia. The attacker build the botnet by exploiting the unsecured devices, devices working on default passwords and some old vulnerabilities of security cameras. The botnet further being used to launch DDoS attacks and illicit mining of crypto currencies. Since, developer has open the source code of Mirai, many variants has been coming with nefarious capabilities [2].
- In year 2017, Reaper botnet came with more capability than Mirai using some of the code of it but different methods of compromising the IoT devices (approximately one million). This malware uses combination of nine vulnerabilities of IoT devices to exploit and recruit the devices. The malware became more damaging than Mirai due to tremendous capability and complexity. The IoT devices such as IP cameras, routers, network attached storage devices and servers are been targeted by this malware.

In the IoT network architecture has many layers, and each layers has its security issues that could be exploited by the attacker to infect the devices with the malware and further make it a part of a botnet that can be used in the large scale attacks. The layered-wise security issues presents in the IoT devices shown in Fig. 1. It clearly indicates that every layer is susceptible to malicious in the IoT devices which makes it vulnerable by design. (1) *Five layer IoT architecture*; (2) *Four layer IoT architecture*, and (3) *Three layer IoT architecture*.

The security researchers are apprehensive about the detection of IoT botnet because IoT has diverse nature of devices which generate enormous amount of data. The attacker has to take control of the device by compromising it with malware and recruiting the device for the botnet. This process has to be communicated via some channel and packet need to be exchange between command and control (C2) and victim device. The exchange of packets leaves traces of indicator of compromise (IoC); hence, to detect the botnet one has to detect the pattern of packet exchange between victim and C2.

The detection of IoT botnet is broadly cover with host-based detection and network-based detection as shown in Fig. 2. Over the time researchers has proposed many solutions for automatic detection of IoT botnet by analyzing and extracting the features of IoT network using machine learning and deep learning techniques. The

---

[2] Comprehensive Guide to IoT Statistics You Need to Know in 2021 (https://www.vxchnge.com/blog/iot-statistics/).

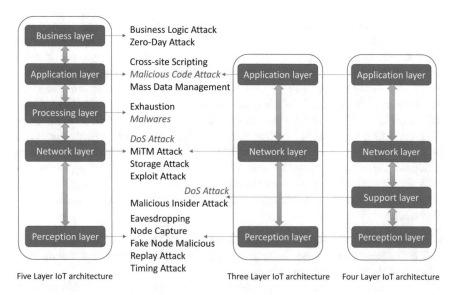

Fig. 1 Layer-wise security issues in the IoT architectures

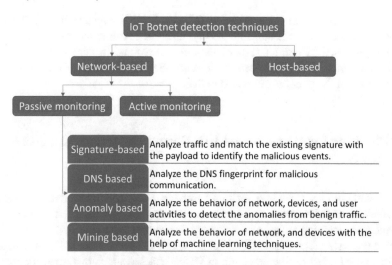

Fig. 2 Taxonomy of IoT botnet detection techniques

botnet can be detected at two levels; one at the host-level and another at the network-level. However, the IoT devices has small memory and computational power to be able to install any security solutions for malicious event detection. Therefore, the working behavior of the IoT devices needs to monitor for any suspicious activity in the network. At the network level it is possible to implement a security solutions that managed to identify the malicious behavior of the connected devices in this case botnet. The identification of malicious traffic from benign can be refer as binary classification problem and in some specific type of botnet can launch multiple type of attacks that can be concluded as multi-class classification. The machine learning and deep learning techniques can be deployed for the classification purpose. In these model can learn and identify the discerning network features of malicious traffic as well as the benign traffic using varieties of classification model architectures like RF (Random Forest) [1], SVM (Support vector machine) [22], DNN (Deep neural network) [13], RNN (Recurrent neural network), LSTM (Long short-term memory) [24]. The different classification algorithms can possibly classify the malicious events by analyzing the different aspect such as DNS footprint [29], signature of an existing malicious payload [12], anomaly based [26], and with the help of mining techniques [3, 5, 20, 21, 33] of the network communication occur among the connected-devices in the network. The main contributions of this research are highlighted in the following points -

- The filter-based chi-square scoring technique is used to find relevant features.
- A technique based on over-sampling is used to balance the class distribution of highly imbalance intrusion dataset.
- A lightweight machine learning model has been trained for the classification of attacks in the Bot-IoT dataset.

The rest of the paper is organized as follows. Section 2 reviewed the IoT Botnet detection related research papers based on machine learning techniques. Section 3 has the proposed model, analysis of dataset and results. In Sect. 4, we compared the performance of our proposed model with the existing state-of-art. Finally, the conclusions of the paper is discussed in Sect. 5.

2   Machine Learning Based IoT Botnet Detection—Related Work

The researchers has proposed various approaches to classify and detect malicious activities in the IoT network using different learning algorithms. The machine learning and deep learning techniques are used to construct these solutions using various open source available datasets like Bot-IoT [14]. Odusami et al. [23] constructed a model using LSTM for the detection of DDoS attacks launch by botnet. Popoola et al. [24] also proposed a model using layers of RNN in the form of cascading network for detection of botnet in the network.

Tyagi and Kumar [34] performed a comparative study with many machine learning techniques and suggested a random forest model due to it performs better than others like k-Nearest Neighbour (kNN), Logistic Regression (LR), SVM, Multi-Layer Perceptron (MLP) and Decision Tree (DT) models. Lo et al. [19] proposed a model based on Edge-based Graph Sample and Aggregate (E-GraphSAGE) and compared the performance with XGBoost and decision tree technique. Chauhan and Atulkar [4] used a different approach and train the proposed model with LGBM. The performance then compared with other techniques like RF, Extra Tree (ET), Gradient Boost (GB) and XGBoost.

Idrissi et al. [10] proposed a CNN based model and compared the results with other techniques like RNN, LSTM and GRU. The proposed model performs better than the other techniques.

Huong et al. [8, 9] proposed a DNN model based on edge-cloud and outperformed the other models like kNN, DT, RF and SVM. Lee et al. [18] and Shafiq et al. [28] employed RF, Bayes Network (BN), C4.5 DT, Naive Bayes (NB), RF and Random Tree (RT) to identify the botnet command & control communication to classify the attack category in the IoT network.

Sriram et al. [30] established in the article that due to not having feature engineering overhead the DL models achieves significantly better as compared to conventional ML models. Kunang et al. [17] and Ge et al. [7] presents a model by combing Autoencoder (AE) and DNN structures in a cascading manner. AE for feature extraction and DNN for classification whereas Ge et al. [7] combined the concept of transfer learning with DNN.

Samdekar et al. [27] experimented with the feature engineering methods and concluded that the Firefly Algorithm is performing better in feature dimensionality reduction than the Chi-Square, ET and Principal Component Analysis when SVM was used for classification. Also, Kumar et al. [15] has proposed the combination of the correlation coefficient, RF mean decrease accuracy and gain ratio for selection of the important features. Kunang et al. [17] and Injadat et al. [11] proposed the Bayesian Optimisation Gaussian Process (BO-GP) method to optimise the hyper-parameters of the AE-DNN and DT models, respectively.

3   Proposed Work

In this section, we analyses the dataset and its features. Further, we will present our proposed model architecture, hyper-parameters and performance evaluation in the subsequent section.

3.1   Description and Analysis of IoT Botnet Dataset

The dataset used in this paper is Bot-IoT [14] published by University of New South Wales Canberra, Australia to facilitate the research community to develop an efficient machine learning or deep learning model for the botnet detection or intrusion detection or malicious traffic in the IoT environment. The network traffic is captured by botnet attacks through internet-connected devices in the network. The network is simulated in the real-time testbed consisting of five IoT devices; first, weather monitoring devices which provide the data like temperature, humidity, and atmosphere. Second, smart cooling refrigerator used to control and adjust the temperature of the

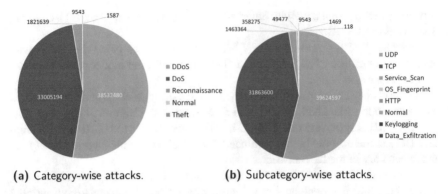

(a) Category-wise attacks.   (b) Subcategory-wise attacks.

Fig. 3 Type of attacks

Table 1 Label instances in the dataset

Label		Instances
Attack	1	73360900
Benign	0	9543

refrigerator. Third, smart light is automatically tern on when motion is detected works on the random signal of pseudo. Fourth, a smart door works according to the probabilistic input. Fifth, intelligent thermostat which is used to control the temperature of entire house by adjusting the temperature of air conditioner. These IoT devices are monitored and controlled by a smartphone. In the testbed, these IoT devices are compromised with botnet malware to generate malicious network traffic. Further, all the extracted features are labelled with type of attacks along with the supporting network features. The labels are in categories and subcategories that can be used for multi-class classification problem. The distribution of dataset towards categories and subcategories are shown in Fig. 3.

In the dataset, the attack labels are defined which is 1 for attack and 0 for benign. The distribution of label values in the dataset which shows the imbalance nature in Table 1. The model might be biased towards the majority class so we have performed random oversampling of minority class to balance the dataset and then train our proposed model.

3.2 Model Architecture

In this section, we present our proposed model that can efficiently classify the IoT botnet attacks in the network using machine learning model. We have used an optimized lightGBM machine learning algorithm for classification. The complete architecture can be divided in three component as presented in Fig. 4. First, the data preparation

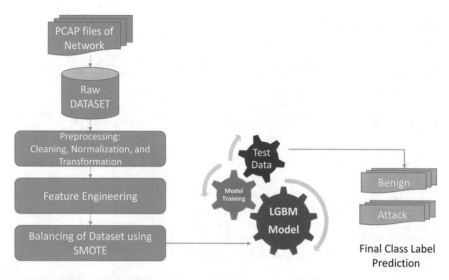

Fig. 4 Architecture of proposed model

phase, in this we will clean the dataset with unknown values. Then, normalize and transform the dataset. Further, the important features are selected using chi-square scoring method is used to select the most important features among all the features as given in Eq. 1. Feature engineering process helps model to improve the accuracy and computational overhead by reducing the dimensionality and selecting important features. We have already discussed about the imbalance nature of dataset and how we have addressed the problem in the previous section.

$$\chi^2 = \sum \frac{(O - E)^2}{E} \tag{1}$$

Second, the LightGBM machine learning algorithm we have used to train with dataset. The algorithm is significantly fast for training with the dataset and uses less memory. The hyper-parameter tuning plays an important role to further optimize the algorithm. The hyper-parameters along with its values that we have uses to train our model are 'num_leaves': 170, 'min_data_in_leaf': 230, 'objective':'binary', 'max_depth': −1, 'learning_rate': 0.07, 'boosting': "gbdt", 'feature_fraction': 0.8, 'bagging_freq': 1, 'bagging_fraction': 0.8, 'bagging_seed': 1009, 'metric': "auc", 'lambda_l1': 0.1, 'random_state': 1009, 'verbosity': −1. Third, in this phase of the architecture the model is tested with the test dataset and obtain the classification results in the form of benign and attack array. This array then used to produce the other results like confusion matrix, instances and percentage of true-negative, true-positive, false-negative, and false-positive.

Table 2 Confusion matrix for botnet detection

Actual class	Predicted class	
	Normal	Malicious
Normal/benign	True Negative (TN)	False Positive (FP)
	Instances	Instances
	Percentage	Percentage
Malicious/attack	False Negative (FN)	True Positive (TP)
	Instances	Instances
	Percentage	Percentage

Table 3 The detailed evaluation metrics

Description	Formula
The overall percentage of correctly predicted instances	Accuracy $= \frac{(TP+TN)}{(TP+FP+TN+FN)}$
The ratio of correctly predicted instances and all instances	Recall $= \frac{TP}{(TP+FN)}$
The relevant prediction among all the predicted values	Precision $= \frac{TP}{(TP+FP)}$
We can calculate the F-measure with the help of precision and recall	F-measure $= \frac{2*((Precision*Recall))}{((Precision+Recall))}$

3.3 Model Training and Evaluation Metrics

The performance of the proposed model can be evaluated by analyzing the performance metrics like confusion matrix, accuracy, precision, recall, and F-measure as described in Tables 2 and 3.

3.4 Result Analysis

In this section, we discuss the performance of the proposed model. The model is successfully able to classify the malicious and benign events in the network. The malicious events are generated by the bots, receiving commands from the bot-master via command and control server. The model achieves 99.99% accuracy (true-positive) with only 0.01% true-negative. The performance of the proposed model with confusion matrix is presented in Fig. 5.

Fig. 5 Performance of the proposed mode

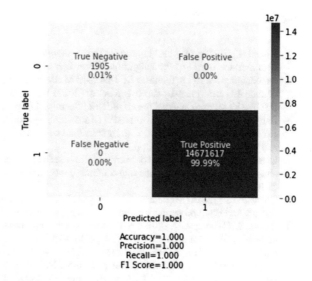

Accuracy=1.000
Precision=1.000
Recall=1.000
F1 Score=1.000

Table 4 Performance comparison with existing research

Techniques	Accuracy	Precision	Recall	F1-score
SVM [14]	0.99	0.99	1	0.99
RNN [14]	0.97	1	0.97	0.98
LSTM [14]	0.98	1	0.98	0.99
RF [34]	0.99	0.99	0.99	0.99
RT [28]	0.99	1	1	–
Our model	0.9999	1	1	1

4 Comparison with Existing Results

Here, we compare the performance of our model with existing state-of-art research available. Our results shows the dominance in all metrics as shown in Table 4.

5 Conclusion

In this paper, an anomaly-based model, ABBDIoT, was proposed to classify the malicious events among benign in the network. The model is trained with highly imbalance dataset. First, the dataset is prepared by cleaning the unknown and missing values then normalize and transform the data. Second, synthetically generate minority samples to balance the dataset using SMOTE oversampling technique. Finally, ABBDIoT model was trained and tested with the Bot-IoT dataset. The model performed shows only 0.01% true-negative with 99.99% true-positive.

References

1. Ajdani M, Ghaffary H (2021) Introduced a new method for enhancement of intrusion detection with random forest and PSO algorithm. Secur Priv 4(2):e147
2. Antonakakis M, April T, Bailey M, Bernhard M, Bursztein E, Cochran J, Durumeric Z, Halderman JA, Invernizzi L, Kallitsis M et al (2017) Understanding the mirai botnet. In: 26th {USENIX} security symposium ({USENIX} Security 17). pp 1093–1110
3. Bahşi H, Nõmm S, La Torre FB (2018) Dimensionality reduction for machine learning based IoT botnet detection. In: 2018 15th International conference on control, automation, robotics and vision (ICARCV). IEEE, pp 1857–1862
4. Chauhan P, Atulkar M (2021) Selection of tree based ensemble classifier for detecting network attacks in IoT. In: 2021 International conference on emerging smart computing and informatics (ESCI). IEEE, pp 770–775
5. Chawathe SS (2018) Monitoring IoT networks for botnet activity. In: 2018 IEEE 17th international symposium on network computing and applications (NCA). IEEE, pp 1–8
6. Dange S, Chatterjee M (2020) IoT botnet: the largest threat to the IoT network. In: Data communication and networks. Springer, pp 137–157
7. Ge M, Syed NF, Fu X, Baig Z, Robles-Kelly A (2021) Towards a deep learning-driven intrusion detection approach for internet of things. Comput Netw 186:107784
8. Huong TT, Bac TP, Long DM, Thang BD, Binh NT, Luong TD, Phuc TK (2021) Lockedge: low-complexity cyberattack detection in IoT edge computing. IEEE Access 9:29696–29710
9. Huong TT, Bac TP, Long DM, Thang BD, Luong TD, Binh NT (2021) An efficient low complexity edge-cloud framework for security in IoT networks. In: 2020 IEEE eighth international conference on communications and electronics (ICCE). IEEE, pp 533–553
10. Idrissi I, Boukabous M, Azizi M, Moussaoui O, El Fadili H (2021) Toward a deep learning-based intrusion detection system for IoT against botnet attacks. IAES Int J Artif Intell 10(1):110
11. Injadat M, Moubayed A, Shami A (2020) Detecting botnet attacks in IoT environments: an optimized machine learning approach. In: 2020 32nd International conference on microelectronics (ICM). IEEE, pp 1–4
12. Ioulianou P, Vasilakis V, Moscholios I, Logothetis M (2018) A signature-based intrusion detection system for the internet of things. Inf Commun Technol Form
13. Kang MJ, Kang JW (2016) Intrusion detection system using deep neural network for in-vehicle network security. Plos One 11(6):e0155781
14. Koroniotis N, Moustafa N, Sitnikova E, Turnbull B (2019) Towards the development of realistic botnet dataset in the internet of things for network forensic analytics: Bot-IoT dataset. Future Gener Comput Syst 100:779–796
15. Kumar P, Gupta GP, Tripathi R (2021) Toward design of an intelligent cyber attack detection system using hybrid feature reduced approach for IoT networks. Arab J Sci Eng 46(4):3749–3778
16. Kumar S, Dohare U, Kumar K, Prasad Dora D, Naseer Qureshi K, Kharel R (2019) Cybersecurity measures for geocasting in vehicular cyber physical system environments. IEEE Internet Things J 6(4):5916–5926. https://doi.org/10.1109/JIOT.2018.2872474
17. Kunang YN, Nurmaini S, Stiawan D, Suprapto BY (2020) Improving classification attacks in IoT intrusion detection system using Bayesian hyperparameter optimization. In: 2020 3rd International seminar on research of information technology and intelligent systems (ISRITI). IEEE, pp 146–151
18. Lee S, Abdullah A, Jhanjhi N, Kok S (2021) Classification of botnet attacks in IoT smart factory using honeypot combined with machine learning. Peer J Comput Sci 7:e350
19. Linda O, Vollmer T, Manic M (2009) Neural network based intrusion detection system for critical infrastructures. In: 2009 International joint conference on neural networks. IEEE, pp 1827–1834
20. McDermott CD, Petrovski AV, Majdani F (2018) Towards situational awareness of botnet activity in the internet of things. Inst Electr Electron Eng

21. Meidan Y, Bohadana M, Mathov Y, Mirsky Y, Shabtai A, Breitenbacher D, Elovici Y (2018) N-Baiot-network-based detection of IoT botnet attacks using deep autoencoders. IEEE Pervasive Comput 17(3):12–22

22. Mohammadi M, Rashid TA, Karim SHT, Aldalwie AHM, Tho QT, Bidaki M, Rahmani AM, Hoseinzadeh M (2021) A comprehensive survey and taxonomy of the SVM-based intrusion detection systems. J Netw Comput Appl, p 102983

23. Odusami M, Misra S, Adetiba E, Abayomi-Alli O, Damasevicius R, Ahuja R (2019) An improved model for alleviating layer seven distributed denial of service intrusion on webserver. J Phys: Conf Ser, vol 1235. IOP Publishing

24. Popoola SI, Adebisi B, Hammoudeh M, Gacanin H, Gui G (2021) Stacked recurrent neural network for botnet detection in smart homes. Comput Electr Eng 92:107039

25. Rani R, Kumar S, Dohare U (2019) Trust evaluation for light weight security in sensor enabled internet of things: game theory oriented approach. IEEE Internet Things J 6(5):8421–8432. https://doi.org/10.1109/JIOT.2019.2917763

26. Sajjad SM, Yousaf M (2018) Ucam: usage, communication and access monitoring based detection system for IoT botnets. In: 2018 17th IEEE international conference on trust, security and privacy in computing and communications/12th IEEE international conference on big data science and engineering (TrustCom/BigDataSE). IEEE, pp 1547–1550

27. Samdekar R, Ghosh S, Srinivas K (2021) Efficiency enhancement of intrusion detection in IoT based on machine learning through bioinspire. In: 2021 Third international conference on intelligent communication technologies and virtual mobile networks (ICICV). IEEE, pp 383–387

28. Shafiq M, Tian Z, Sun Y, Du X, Guizani M (2020) Selection of effective machine learning algorithm and bot-IoT attacks traffic identification for internet of things in smart city. Future Gener Comput Syst 107:433–442

29. Singh M, Singh M, Kaur S (2019) Detecting bot-infected machines using DNS fingerprinting. Digital Invest 28:14–33

30. Sriram S, Vinayakumar R, Alazab M, Soman K (2020) Network flow based IoT botnet attack detection using deep learning. In: IEEE INFOCOM 2020-IEEE conference on computer communications workshops (INFOCOM WKSHPS). IEEE, pp 189–194

31. Sudhakar, Kumar S (2019) Botnet detection techniques and research challenges. In: 2019 International conference on recent advances in energy-efficient computing and communication (ICRAECC). IEEE, pp 1–6

32. Sudhakar, Kumar S (2020) An emerging threat fileless malware: a survey and research challenges. Cybersecurity 3(1):1–12 . https://doi.org/10.1186/s42400-019-0043-x

33. Sun P, Li J, Bhuiyan MZA, Wang L, Li B (2019) Modeling and clustering attacker activities in IoT through machine learning techniques. Inf Sci 479:456–471

34. Tyagi H, Kumar R (2021) Attack and anomaly detection in IoT networks using supervised machine learning approaches. Rev d'Intell Artif 35(1):11–21

A Hybrid Mechanism for Advance IoT Malware Detection

Aijaz Khan, Gaurav Choudhary, Shishir Kumar Shandilya, Durgesh M. Sharma, and Ashish K. Sharma

1 Introduction

IoT is the future of this planet. It works by interconnecting several devices like sensors, routers, and webcams with the help of the Internet. IoT devices are helping in transforming every individual in their daily tasks improving the interaction between mankind and the rest of the world. The complexity of IoT devices is relatively high. This makes it vulnerable to cyberattacks. IoT devices are extensively used in industries, offices, research laboratories, educational centers, etc. IoT devices can integrate themselves with other devices and increase their functionality. The vast domain of IoT devices' usability increases the chances of cyberattack much more. In October 2016, major US DNS servers were infected with Mirai IoT malware, which directly impacts popular online services like Google and Amazon [1].

IoT devices are majorly affected by distributed denial of service (DDoS) attacks. The most famous and widely spread Ddos was Mirai. It was used to initiate the most significant DDoS attack ever on IoT devices. This attack targeted the domain name servers, and an order of 1.2 Tbps was created. Major Internet services such as Twitter, Amazon, and Netflix were shut down since this attack targeted several categories of IoT devices. Mirai created a botnet where several devices performed DDoS attacks, leading to a massive downfall. This led to the rise of other samples of malware which were a version of Mirai.

A. Khan · S. K. Shandilya
VIT Bhopal University, Bhopal, Madhya Pradesh, India

G. Choudhary
DTU Compute, Technical University of Denmark, 2800 Kgs., Lyngby, Denmark

D. M. Sharma (✉)
Shri Ramdeobaba College of Engineering and Management, Nagpur, Maharashtra, India
e-mail: durgesh_sharma54@yahoo.com

A. K. Sharma
Bajaj Institute of Technology, Wardha, Maharashtra, India

© The Author(s), under exclusive license to Springer Nature Singapore Pte Ltd. 2023
R. Agrawal et al. (eds.), *International Conference on IoT, Intelligent Computing and Security*, Lecture Notes in Electrical Engineering 982,
https://doi.org/10.1007/978-981-19-8136-4_21

247

Mirai's source code was leaked in 2017, which created havoc since people were now able to create a new IoT malware by using Mirai's source code. The new malware uses a similar technique of brute force by scanning a random IP address in search of open Telnet port to bypass it using dictionary attacks of common credentials. The more complex variant of Mirai uses IoT vulnerabilities for remote code execution, phishing, and spamming. These variants of Mirai can cause network downtime for a more extended period and may lead to substantial financial losses and leak of confidential data. In 2017, it was reported that 100,000 devices were infected with Mirai or similar kinds of IoT malware. Kaspersky Laboratory [2] said that there were 121,588 IoT malware samples captured in the first half of 2018.

Therefore, the security of IoT devices is an essential aspect of the current generation. The infected IoT devices are expected to remain infected for an extended period, and the chain of infection proliferates. So the need to stop this chain of infection is necessary. Researchers have used static, dynamic, machine learning, neural networking, opcode analysis, and image analysis for IoT malware detection [3].

In summary, the major contributions of this paper can be summarized as follows:

- We present a summary of recent works done in this field of research.
- We provide a background overview of IoT malware concerning two famous malware samples Mirai and Darlloz.
- We present a comprehensive discussion of current implementation with their limitations.
- We present a hybrid model solution to solve the limitations in the current work.

2   Related Works

Many studies have been made to classify, detect, and defend against IoT malware. Su et al. [4] discussed converting the IoT malware image to an 8-bit binary sequence to convert it into the grayscale image. They used ML classifiers to classify these malware images further. Malware files were divided into four categories: "Linux.Gafgyt.1" , "Linux.Gayft.2", "Trojan.Linux", and "Mirai". They used a neural network to automate malware classification. Wang et al. [5] analyzed two famous malware in IoT devices: "Mirai" and "Darlloz.Mirai". They found that this malware spreads through brute force attacks and used a weak password policy in IoT devices. Mirai uses a command and control server to execute its functionality. Running a defensive service on port 48101 kills Mirai every 10 seconds. Darlloz takes advantage of CVE-2012-1823 to exploit IoT devices. Costin and Zaddach et al. [6] analyzed 60 malware families to find out information about their vulnerabilities, exploits, and defensive rules. They found the CVSS score for all malware samples analyzed and computed the mean, which came out to be 6.9. Alhanahnah et al. [7] used the technique of static analysis and utilized string, statistical, and structural features for classifying malware. They used IoTPOTS [14] to collect malware. Hisham et al. [8] compared IoT and Android malware together. They extracted CFGs from samples and created graphs based on factors like nodes, edges, betweenness, and density. Kumar

Table 1 Existing research studies on advance IoT malware detection approaches

Authors	Contributions	P1	P2	P3	P4	P5	P6
Jiawei et al. [4]	Classification of malware using image recognition	No	No	Yes	Yes	No	No
Aohui et al. [5]	Study of malware Mirai and Darlloz	No	No	No	Yes	Yes	No
Andrei et al. [6]	Static analysis of 60 malware families	No	Yes	No	Yes	Yes	No
Mohannad et al. [7]	Signature generation for malware classification	Yes	No	No	Yes	No	No
Hisham et al. [8]	Graph-based comparison of IoT and Android malware	No	No	No	Yes	No	Yes
Ayush and Teng [9]	IoT malware network traffic analysis	Yes	Yes	No	No	No	No
Quoc et al. [10]	Comprehensive study of IoT malware detection using static techniques	Yes	No	No	Yes	No	No
Ahmed et al. [11]	Graph-based deep learning model against adversarial machine learning attacks	Yes	No	No	Yes	Yes	Yes
Jueun et al. [12]	Dynamic analysis of IoT malware using convolution neural network model	Yes	Yes	No	No	Yes	No
Hamid et al. [13]	Opcode-based polymorphic IoT malware detection	Yes	No	No	No	No	No

P1: machine learning, P2: network traffic, P3: neural network , P4: static analysis, P5: dynamic analysis, P6: graph analysis

and Lim [9] used machine learning algorithms to propose a modular solution for analyzing IoT malware traffic. Ngo et al. [10] divided the detection techniques into two groups: "non-graph-based" and "graph-based methods". Graph-based malware detection seems more prominent in detecting novel malware than non-graph methods. Abusnaina et al. [11] used graph-based deep learning models against adversarial machine learning attacks. They used control flow graph to analyze malicious binaries considering multiple factors such as nodes, edges, shortest path, and density. A total of eight different adversarial techniques were used to force the model to misclassify. Jeon et al. [12] proposed a dynamic analysis detection model of IoT malware using neural network model. They trained the model using the CNN algorithm. The feature data was then converted to images to reduce computation problems. A visualization technique was used to process the data further. Darabian et al. [13] used the opcode technique to differentiate between good and malware. They collected the dataset and applied a sequential mining algorithm to classify polymorphic malware. This resulted in the detection of 36 features with an F-score of 99%.

3   Market Perspective of IoT and Malware Impact

Internet of Things is a device that is connected to the Internet. Sensors and software are embedded in IoT for connecting and exchanging data with other devices. IoT devices are present in homes, transport, organizations, military places, etc. It is predicted that there will be 60 billion IoT devices by 2022. IoT devices typically have eight components, i.e., external interfaces, analytics, additional tools, data visualization, processing and action management, device management, and connectivity [15]. Malware is a piece of malicious software that is created for unethical practices. IoT malware shares some similarities with PCs malware, but its features are infrequent. The malware tries to open the Telnet port and use a common wordlist of passwords for brute forcing into the device. IoT devices are vulnerable to various attacks since it is vulnerable to attacks like zero-day attacks and remote code execution. IoT malware takes advantage of the devices' vulnerability and carries on their malicious process. The IoT malware families include Mirai, Bashlite, muBOT, Hajime, PNScan2, etc. The majority of malware is supposed to be performing DDoS attacks on IoT devices. The Mirai and Darlloz botnet working is shown in Figs. 1 and 2.

In the case of the Mirai botnet, a device is initially compromised. It will select a random IP address to scan. If the chosen machine has port 23 or 2323 open, the malware attacks the device using brute force. After successful login, the device's data such as IP address, username, and password is transferred to the CB server. The CB server is then responsible for downloading the bot program. Mirai disables other functions to gain complete control of the device.

The next most common IoT malware is Darlloz malware. It is a low-level threat and works on the CPU. This malware looks for executable and linkable format(ELF) files and expands its reach to ARM architectures. It exploits a device and scans a random IP address to search for port 23 to open. The request is sent in HTTP POST format. Initially, it performs brute force on the target. If a brute force attack fails, then it goes for CVE-2012-1823. This CVE is linked with PHP. Those devices with older web-based versions of PHP are vulnerable to this attack. It allows an attacker to execute arbitrary code in the query string.

Another IoT malware is Bashlite. It is used for large-scale DDoS attacks and exploits Shellshock to gain a foothold into vulnerable devices. It does not depend

Fig. 1 Mirai botnet working

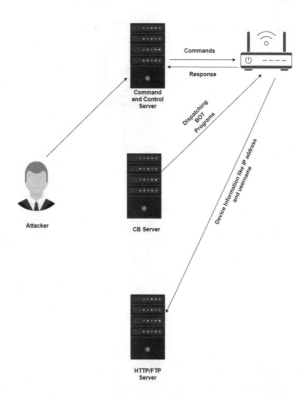

on any specific vulnerability. Instead, it relies on an openly available remote code execution Metasploit module. Bashlite also has an updated version. OWASP suggests that the top vulnerabilities found in IoT malware are buffer overflow, denial of service, and poor encryption implementation. With increasing time, IoT malware is supposed to be upgrading itself. Researchers fear a ransomware attack on IoT devices since IoT devices are used widely in the military, power grid, industries, etc. Recently, IBM Xforce found malware similar to Mirai at their enterprise IoT devices. This malware was supposed to be dropping cryptocurrency miners and backdoors on affected devices.

Static analysis of IoT malware is a technique where the malware is not run in real or simulated environments for analysis purposes. In this method, the malware file is analyzed externally. The fundamental static analysis of IoT malware can be classified under two broad categories: non-graph-based and graph-based techniques. The non-graph-based methods further constitute string, opcode, ELF header, and image-based detection. The graph-based detection includes function call graph and opcode graph-based detection.

Opcode analysis is trendy among researchers. The opcode is a single instruction that is executed by the CPU. The opcode is used to determine the behavior of files. The string is a fixed sequence of characters. String obtained from a malware sample using static analysis can either be useful or gibberish. Useful string tells a lot of information about the malware like the URL it is connecting to, IP address it is

Fig. 2 Darlloz botnet
working

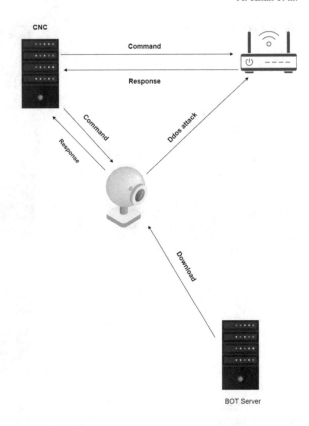

using, information about the command and control server, API calls it is making, etc. An executable and linkable format (ELF) file format contains many valuable details which can be used to detect malware. A usable file that uses the ELF file format has the title ELF header followed by a table in the program header or header table or both. An image-based detection is a way of converting malware files into an image where each pixel of the created image ranges from 0 to 255. In the case of malware detection, the binary files are converted into binary strings of 0 and 1. These binary values are combined into 8-bit vector segments, representing hex values ranging from 00 to FF. The obtained vector gets converted to an image with pixels ranging from 0 to 255. 0 illustrates the black color, and 255 represents the white color. Graph-based malware detection is pretty common nowadays. In malware detection, the control flow graph is widely used. A control flow graph is a directed graph that gives the possibility of every possible execution path taken while a program runs. It is capable of representing the flow inside a program unit. Edges in this graph show that control flow paths and nodes represent basic blocks. There are two designated blocks in this graph. The first one is entry block. This block allows control to enter the control flow graph, and the second is exit block. In this block, the control flow leaves the exit block. A control flow graph can also encapsulate the information per each basic block.

4 Existing Approaches for IoT Malware Detection

IoT is the future of this planet. It works by interconnecting several devices like sensors, routers, and webcams with the help of the Internet. IoT devices are prone to several attacks since these devices do not receive proper updates and have weak login credentials. Several works have been done in this field. Let us discuss each work with its pros and cons in detail.

4.1 Machine Learning-Based Approach for Malicious Traffic Detection

Kumar and Lim [9] used machine learning algorithms for malicious network analysis, which was run at user access gateway for detecting IoT malware based on their scanning patterns. A database was used whose work was to store the malware scanning patterns. A policy module was set up, which decided the further course of action after the gateway traffic was identified as malicious. The detection mechanism also included an optional packet subsampling module that could be implemented at both ends, i.e., physical access and enterprise gateway. The architecture included five modules. The first module was called *ML classifier*. It runs on the access gateway connected to an IoT device at the enterprise end. The gateway level traffic was divided into two levels: *benign* and *malicious*. Benign refers to the type of traffic considered safe, and malicious refers to the unsafe type. Initially, training data samples are created by filtering out traffic sessions and including only TCP packets with SYN flags, then extracting the feature vectors, and lastly retrieving the trained classifier from the second module *ML model constructor*. The following module is *packet traffic feature database*, and it is responsible for storing features that are extracted from traffic samples in the database. The next module is *policy module*, and it is responsible for deciding the course of action to be taken once the traffic is identified as malicious by the ML classifier module. The last module is *Sub-Sampling Module*, which is responsible for sampling the packet traffic from IoT devices. The conclusion was drawn on the basis of 60 traffic sessions. The result came out thoroughly mixed. The feature one graph was distinguishable between benign and malicious samples, whereas features 2 and 3 were not distinguishable. Thus, the machine learning algorithms reduce the complexity involved in analyzing IoT malware. Researchers use both machine learning and graph-based method to automate the whole process of IoT malware detection. However, the extensive use of machine/deep learning techniques in this field has comings. Attackers create malware using obfuscation to bypass normal machine learning detection, limiting this technique. The accuracy of using ML will increase with time since the IoT malware data is scattered [16].

4.2   Graph-Based Analysis

Ahmed et al. [11] discussed possible methods of analyzing IoT malware samples with graph-based analysis methods. The binaries are initially disassembled to generate assembly instruction code and a rough graph. An in-depth analysis of malware binaries is conducted using the CFG method. Parameters like number of nodes, edges, shortest path, betweenness, closeness, and density are considered.Two approaches are used for creating graph embedding augmentation (GEA). GEA helps in generating AEs [17] which helps in maintaining the functionality of IoT software. Eight different adversarial learning techniques were used to force the model toward misclassification, and it was observed that the AEs method yielded a high misclassification rate. The GEA method successfully preserved the original functionality of the sample and achieved a high misclassification rate. Thus, this indicates that a more robust IoT malware detection tool is required. The non-graph method performs well when used against simple and non-obfuscated malware, but its efficiency decreases while analyzing complex IoT malware. The graph-based approach is more effective with complex malware such as the novel and obfuscated ones. The main advantage of using a graph-based approach is that it can even predict features of novel IoT malware up to a certain limit.

4.3   Image-Based Detection

The image-based detection [4] of IoT malware is yet another method of predicting source code. It involves the use of the neural network technique. Initially, a dataset of IoT malware obtained through IoTPOT is used. Each sample is converted into a grayscale image [18]. The image is resized to 64*64 size to use as an input to a convolutional neural network. The experiment was performed five times, and it was found that the accuracy of this system to determine a malicious file is nearly 94%. The effectiveness of this method alone is not enough. The main purpose of using this method is to find out the entropy [19] of malware source code. This method also helped in grouping malware in the same family. It gave an alternative of signature generation using printable strings and statistical features. This detection was proved to be less resource consuming and better against node failures.

4.4   Opcode-Based Detection

The opcode-based analysis is also a way of determining IoT malware. Hamid et al. [13] analyzed IoT malware samples with opcode-based technique. They extracted opcodes from the benign and malware samples. Sequential pattern mining was used to detect maximal frequent patterns (MFP) of opcodes as the best feature for malware classification. MFP of opcodes was used to train a range of machine learning models like K nearest neighbor (KNN), support vector machine (SVM), multilayer perceptron (MLP), AdaBoost, decision tree, and random forest classifiers for detection of IoT malware. Machine learning approaches like True Positive (TP), True Negative (TN), False Positive (FP), and False Negative (FN) were used for evaluating the performance of the approach. IoT malware samples were identified after using a sequential pattern mining algorithm with a machine learning technique. A total of 36 features were identified using the method. These features yielded an F-measure score of 99% in detecting IoT malware from a combination of malware and benign samples. The effectiveness of this method in detecting real-world malware samples is moderate. It performs well with non-obfuscated malware, but when it comes to obfuscated malware, the effectiveness decreases since the bits are modified when a code is obfuscated.

4.5 Static Analysis

Efficient signature generation [7] is a method of clustering malware in their respective families concerning their features for accurate IoT malware detection. The signature detection method uses the code statistics feature for classification and includes five steps. The malware is preprocessed for coarse-grained clustering. *Fine-grain clustering* method is applied next in order to merge the clusters, and finally, signatures are generated. In malware preprocessing, the assembly level instructions are extracted using disassemblers. In *coarse-grained clustering*, the statistical features are normalized. A total of eight features were extracted, like several functions, instructions, etc. K-means clustering was used in order to perform *coarse-grained clustering*. The next phase is *fine-grained clustering*. A binary analysis tool Bindiff is used to compute the similarity between malware samples. The malware samples obtained after *coarse-grained clusters* are partitioned into multiple *fine-grained clusters*. Some of the generated clusters might still share some features; therefore, *cluster merging* method is used in this step. It helps in refining the clustering results. The performance of this technique was quite promising. Compared with API and opcode-based detection, this method showed better results. This printable string clustering and detection method are valid until the IoT malware is not obfuscated or encrypted. This method of detection will completely fail under such circumstances.

5 Hybrid Solution

Much research has been done to overcome the shortcomings in IoT malware detection. The main motive of the advanced word is to solve the issue of novel IoT malware detection. Current research works lack in detecting obfuscated and encrypted IoT malware. Many research works are based only on ARM MIPS-based samples and do not cover a wide range of architecture. Some research work is based on hypothetical

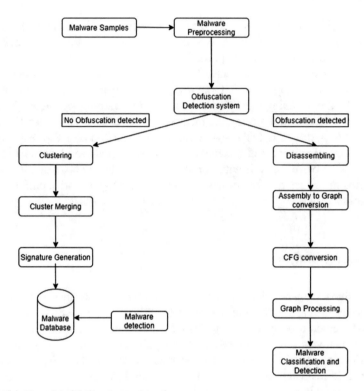

Fig. 3 Hybrid model of IoT malware detection

malware samples. The work which seems promising lacks accuracy. Many works which involve graphs and neural network seem to be time consuming. Therefore, IoT malware is difficult to analyze precisely. We will discuss a hybrid solution for IoT malware detection in this paper. We aim at the most significant issue in malware in recent times, i.e., malware obfuscation. We propose a model which tends to solve this issue. Initially, a malware sample is disassembled to obtain all details. Then it is preprocessed to remove irrelevant data, which would improve accuracy. An image-based classification system is used to determine whether the malware sample is obfuscated or not. If the sample is non-obfuscated, it will go for basic static analysis using machine learning. The non-obfuscated sample will go under the process of clustering, where the statistical features of malware files will be extracted. The average and standard deviation values will also be taken for every IoT malware sample. The function and block-based matching will be performed by removing the graph-based features to determine the incoming and outgoing edges. After clusters creation, we need to combine some of the remaining clusters since they may share some similarities. Thus, we generate refined signatures. These signatures can be clubbed using YARA rules. Detection of IoT malware can be done using a database of malware signatures, either online or offline. The proposed hybrid model of IoT malware detection is presented in Fig. 3.

If malware is obfuscated, then it can be verified using *obfuscation detection system*. It uses image-based classification. The image obtained could give the entropy of the source code, which helps determine whether the IoT malware was compressed/encrypted. After the obfuscation is detected, the assembly code is converted to graph. CFG feature is extracted next to analyze to find valuable data using nodes, edges, density, betweenness, and shortest path. CFG helps determine the code's control flow since the code is unreadable. The obtained result is trained using machine learning algorithm for malware classification and detection. The sample is then trained using a machine learning algorithm.

6 Research Directions

The amount of work in this IoT malware field is limited compared with other malware studies like Windows and Android malware analysis [20]. Current work includes an analysis of the same architecture. No work supports multiple platforms. There is a shortage of IoT malware samples since there are fewer IoT honeypots, and the only known Honeypot is IoTPOT [14]. The analysis of simple IoT malware can be achieved through static analysis with excellent efficiency, but the efficiency decreases with complicated malware. Future research should include points like time complexity, a wide range of supported architecture, novel IoT malware detection, and reducing the verbosity of graph-based analysis. Integration of QFD with IoT is another area for further research as QFD leaves a broader scope being a strong decision-making tool [21–24].

7 Conclusion

Detection of IoT malware is a challenging task for security professionals worldwide since there are different architectures and real malware samples are significantly less when compared with Windows malware. This paper gave a comprehensive review of other IoT malware detection solutions. We discussed the pros and cons of each technique with a depth overview of their detection mechanism. In summary, the IoT malware classification can be classified as opcode-based detection, image-based detection, static analysis, dynamic analysis, machine learning-based analysis, and graph-based analysis. The methods of the non-graphical analysis show promising results when dealing with simple malware, but it loses its accuracy when detecting customized or obfuscated malware. The graph-based approach tended to analyze the program's control flow, which was good enough to explore a more sophisticated malware, or we can say it can detect unseen malware. Dynamic analysis and machine learning were better approaches to analyzing the traffic generated by IoT malware. Based on the mechanism, detection analysis, and efficiency, we present a hybrid model that works on the principle of image-based detection for obfuscation check and decides the analysis approach. If malware is unobfuscated, static analysis for signature generation and detection is performed for higher efficiency. If the malware is obfuscated, then a graph-based approach is used for analysis.

References

1. Sihag V, Choudhary G, Vardhan M, Singh P, Seo JT (2021) Picandro: packet inspection-based android malware detection. Secur Commun Netw
2. Kuzin M, Shmelev Y, Kuskov V (2018) New trends in the world of iot threats. Kaspersky Lab
3. Sinha R, Choudhary VSG, Vardhan M, Singh P. Forensic analysis of fitness applications on android
4. Su J, Vasconcellos DV, Prasad S, Sgandurra D, Feng Y, Sakurai K (2018) Lightweight classification of iot malware based on image recognition. In: 2018 IEEE 42Nd annual computer software and applications conference (COMPSAC), vol 2. IEEE, pp 664–669
5. Wang A, Liang R, Liu X, Zhang Y, Chen K, Li J (2017) An inside look at iot malware. In: International conference on industrial IoT technologies and applications. Springer, pp 176–186
6. Costin A, Zaddach J (2018) Iot malware: comprehensive survey, analysis framework and case studies. BlackHat USA
7. Alhanahnah M, Lin Q, Yan Q, Zhang N, Chen Z (2018) Efficient signature generation for classifying cross-architecture iot malware. In: 2018 IEEE conference on communications and network security (CNS). IEEE, pp 1–9
8. Alasmary H, Anwar A, Park J, Choi J, Nyang D, Mohaisen A (2018) Graph-based comparison of iot and android malware. In: International conference on computational social networks. Springer, pp 259–272
9. Kumar A, Lim TJ (2019) Edima: early detection of iot malware network activity using machine learning techniques. In: 2019 IEEE 5th world forum on internet of things (WF-IoT). IEEE, pp 289–294
10. Ngo Q-D, Nguyen H-T, Nguyen L-C, Nguyen D-H (2020) A survey of iot malware and detection methods based on static features. ICT Express
11. Abusnaina A, Khormali A, Alasmary H, Park J, Anwar A, Mohaisen A (2019) Adversarial learning attacks on graph-based iot malware detection systems. In: 2019 IEEE 39th international conference on distributed computing systems (ICDCS). IEEE, pp 1296–1305
12. Jeon J, Park JH, Jeong Y-S (2020) Dynamic analysis for iot malware detection with convolution neural network model. IEEE Access 8:96899–96911
13. Darabian H, Dehghantanha A, Hashemi S, Homayoun S, Choo K-KR (2020) An opcode-based technique for polymorphic internet of things malware detection. Concurrency Comput Pract Exper 32(6):e5173
14. Pa YMP, Suzuki S, Yoshioka , Matsumoto T, Kasama T, Rossow C (2015) Iotpot: analysing the rise of iot compromises. In: 9th {USENIX} Workshop on Offensive Technologies ({WOOT} 15)
15. Borana P, Sihag V, Choudhary G, Vardhan M, Singh P (2021) An assistive tool for fileless malware detection. In: 2021 World Automation Congress (WAC). IEEE, pp 21–25
16. Sihag V, Vardhan M, Singh P, Choudhary G, Son S (2021) De-lady: deep learning based android malware detection using dynamic features. J Internet Serv Inf Secur (JISIS) 11(2):34–45
17. Papernot N, McDaniel P, Jha S, Fredrikson M, Celik ZB, Swami A (2016) The limitations of deep learning in adversarial settings. In: 2016 IEEE European symposium on security and privacy (EuroS&P). IEEE, pp 372–387
18. Nataraj L, Karthikeyan S, Jacob G, Manjunath BS (2011) Malware images: visualization and automatic classification. In: Proceedings of the 8th international symposium on visualization for cyber security, pp 1–7

19. Lyda R, Hamrock J (2007) Using entropy analysis to find encrypted and packed malware. IEEE Secur Priv 5(2):40–45
20. Moad D, Sihag V, Choudhary G, Duguma DG, You I. Fingerprint defender: defense against browser based user tracking
21. Purohit S, Sharma A (2015) Database design for data mining driven forecasting software tool for quality function deployment. Int J Inf Electron Bus 7(4):39–50
22. Sharma A, Khandait S (2016) A novel software tool to generate customer needs tor effective design of online shopping websites. Int J Inf Tech Comput Sci 83:85–92
23. Sharma A, Khandait S (2017) A novel fuzzy integrated customer needs prioritization software tool for effective design of online shopping website. Int J Operat Res Inf Syst (IJORIS) 8(4):23–42
24. Sharma A, Mehta I, Sharma J (2009) Development of fuzzy integrated quality function deployment software–a conceptual analysis. I-Manager's J Softw Eng 3(3):16–24

A Cloud-Edge Server-Based Cypher Scheme for Secure Data Sharing in IoT Environment

Abhishek Kumar and Vikram Singh

1 Introduction

The present-day Internet is fast transforming in to an Internet of things which will comprise trillions of commonplace Internet-enabled gadgets. These gadgets will produce and consume vast amounts of the data at lightning speeds. The present-day cloud architecture and services will prove poorly insufficient and limiting mainly because of (i) their computation and storage services being housed in a relatively smaller number of data centres and (ii) the comparatively large distance between the IoT gadgets and the remote data centres. To take on these challenges, edge computing is a promising technology with a provision of computing, storage and other resources near to the IoT devices. These transformational proposals may bring up an all-new Internet of things network. Edge computing gives many benefits to Internet of things assisted by cloud computing and helps fulfil the storage and computing requirements by implementing the storage, networking and processing services on servers near to IoT devices, i.e. at the edge of the cloud/networks. Furthermore, because of the limited range of connectivity of the smart devices, the edge servers can mediate the long-haul communications. Technically, edge servers can be described as computing or network devices, personal mobile devices falling within one-hop range of the end-user IoT device. Further, these devices work in strong collaboration with the cloud servers too.

Edge computing allows smart gadgets to share data with higher bandwidth and low latency. But when the smart IoT gadgets share data with other gadgets, serious

A. Kumar (✉) · V. Singh
Department of Computer Science & Engineering, Chaudhary Devi Lal University, Sirsa, Haryana, India
e-mail: abhimonark@gmail.com

V. Singh
e-mail: vikramsingh@cdlu.ac.in

© The Author(s), under exclusive license to Springer Nature Singapore Pte Ltd. 2023
R. Agrawal et al. (eds.), *International Conference on IoT, Intelligent Computing and Security*, Lecture Notes in Electrical Engineering 982,
https://doi.org/10.1007/978-981-19-8136-4_22

security and privacy issues like data integrity and pilferage, and unauthorised reads and writes on data, arise. In this situation, maintaining confidentiality, integrity, availability (CIA) of the shared data at the edge becomes imperative. Currently, there are some solutions to address the challenges of secure data sharing and searching by IoT devices. Of lately and currently, crypto-mechanisms (symmetric, public and homomorphic) are in vogue to ensure confidentiality part of CIA triad, whereas, dynamic attributes and access control lists (ACL) are used for realising availability of shared data. Public and symmetric key-based search-supporting encryption schemes are used to search of shared data. What is novel here? It is the user device that manages the encryption, decryption and access control requirements for securing, searching and sharing of data. In the current IoT architectures, resource-scarce smart gadgets are devoid of computation and storage capabilities required to carry out the data security, confidentiality and access control operations.

2 Related Works

This section presents a panoramic survey on the works done by many researchers in the domain of edge computing and cloud-assisted IoT to examine various security and computation performance related issues, and their efforts to tackle the issues.

Cao et al. [1] have examined that the extremely dynamic nature at the edge of the network makes the network highly vulnerable and tough to protect. Authors have outlined that conventional data encryption and sharing techniques are no longer applicable, because edge computing integrates various trust domains with legitimate entities as trust centres.

Ghosh et al. [2] have evaluated that data and the corresponding network traffic can be decreased even up to 80% without noteworthy loss of accuracy if a large sliding window is utilised in the processing. Authors have examined that the modification in the computation magnitude of the edge nodes would not alter the network traffic, but it must be enough to carry out data encoding.

Nerella et al. [3] have proposed a unique middleware security architecture that makes use of the benefits of cloud computing, fog computing and Internet of things (IoT) to overcome security, privacy, integrity, availability and other issues into the vehicular Internet of things (VIoT).

Hassija et al. [4] have suggested that blockchain can avoid the threat of being a single point of failure due to its distributed architecture in nature.

Javaid et al. [5] proposed an attack resilient cloud-assisted IoT system (ARCA-IoT) to tackle the three issues which are scalability, interoperability and trustworthiness. An innate Naive Bayes approach is used to direct the ARCA-IoT in a way that it computes the chances of the trustworthiness of the operations and then recognise different types of attacks with aid of three proposed algorithms, named as bad mouthing, ballot-stuffing and oscillation attack, bad mouthing algorithm is accomplished by the service consumers.

Santos et al. [6] proposed a lightweight data centre virtualisation approach for edge-cloud computing architecture to deal with the resource limitations of edge devices. Furthermore, authors have presented a data-oriented approach in which the virtual notes are determined based on either raw data or processed data in place of on processing cores or virtual machines. The proposed model supports the data reutilization among various applications with similar requirements in respect of data sources.

Mani et al. [7] have examined the cloud computing solutions depending on the service providers, because many of the international organisations are devoted to developing their specifications of giving a common framework of software and networks. Authors have focused on service providers and operators' problems due to lack of definite specification to follow. Hence, authors have considered the edge computing has potential to tackle the challenges such as data protection, data privacy, bandwidth cost and battery power consumption.

A flexible edge computing (FEC) architecture has been proposed by Sureddy et al. [8] as a flexible structure to perform edge computing. It has been demonstrated that the collaboration of flexible edge computing and deep learning significantly enhances the performance of the system and optimises the task scheduling between the cloud layer and edge layer.

Porambage et al. [9] surveyed on multi-access edge computing (MEC) which aims at enhancing cloud computing abilities to the edge of the radio access network (RAN), consequently providing real time, high bandwidth, low latency access to the radio network resources. Because of leveraging on radio access networks, MEC will increase extensively on bandwidth utilisation and latency, making it simple for both content providers and application developers to get network services. MEC empowers small-scale IoT devices with noteworthy additional computational abilities through computation offloading.

Pravallika et al. [10] proposed an idea based on the potential of the edge nodes to carry out sensing and locally decide through prediction whether to circulate contextual data in the edge company network or to natively reconstruct not delivered contextual data in the light of reducing the required communication interconnection at the cost of precise analytics tasks.

Yu et al. [11] also recognised an issue because of the decentralised management of the edge networks, which cannot provide adequate security and management facilities. For solving this challenge, authors have focused on a service-based solution to secure the network edge, which is called Securebox.

Hussein et al. [12] examined a major challenge around the mobility revolution which is lack of appropriate policies for handling security and privacy in cloud-assisted Internet of things framework. Authors have discussed that the implementation of security with digital encryption standards and dual attestation systems with Internet protocol authentication will enhance the safety and security to operate the systems.

Pan et al. [13] suggested that by the use of network function virtualisation and software defined networking in edge-cloud computing, a small-scale cloud operating platform can be established to direct computing, networking resources and storage to the edge and enable future Internet of things applications. By the use of network function virtualisation at the cloud edge, computing, network control and data storage resources become available and edge gadgets in the vicinity generate ample data and needs to be visible. Software defined networks, on the other hand, reduce the cost and enhances the flexibility and programmability of the virtual network functions in the edge cloud due to the separation of control from data forwarding, usage of centralised network control and configuration.

Abbas et al. [14] have surveyed the emergence of mobile edge computing as an architectural alternative to cloud computing wherein utilities are centrifuged towards the network edge to leverage mobile base stations. However, this architecture is facing major issues, such as security vulnerability, high latency, low coverage and logged data transmission. Authors have examined that the infrastructure where mobile edge computing is nearby integrated with radio access networks, provides a better analysis of network traffic, radio network status and the device location utilities.

Gritti et al. [15] proposed a server-aided, reliable policy-based access control scheme, named cloud-assisted access control for the Internet of things (CHARIOT) that allows an Internet of things platform to verify testimonials of various gadgets requesting access to the data stored within it. CHARIOT allows Internet of things gadgets to authorise themselves to the platform without compromising their privacy by the use of attribute-based signatures. The proposed mechanism permits secure delegation of expensive computational operations to a cloud server, consequently to relieve the workload at the side of IoT devices.

Ni et al. [16] have examined that fog computing is more secure than cloud computing due to some reasons, for example the collected data is briefly managed and analysed on the local fog node nearby data sources, which reduces the dependency on the Internet connections. Local data storage, analysis and exchange make it hard for hackers to obtain access to the data of users. The other reason authors have forwarded is, the exchange of information between the gadgets and the cloud no longer occurs in real time, therefore it is tough for eavesdroppers to catch the sensitive information on a particular user.

Laaroussi et al. [17] investigated the effects of cloud-based and edge-based utilities to provision in vehicular networks. Authors have designed and implemented a testbed, which authorises them to devotedly reproduce a vehicular communication system, by following the recommendations made by foremost standardisation entities.

Abbasi et al. [18] proposed a data management framework to handle the extensive amount of data securely that is being generated by IoT enabled gadgets. Authors have proposed a security layer as a background layer because all remaining layers shall ensure security and privacy of the data. Authors have indicated that current solutions for data management in the Internet of things addresses only limited aspects of the cloud centric Internet of things environment with particular focus on sensor networks, which is only a subgroup of the global Internet of things space.

Kumar et al. [19] suggested the transport layer security (TLS) should be enabled over protocol stacks other than TCP/IP to better suit the needs of things in respect of

complexity and resources requirements. Authors have examined that accurately identification of things and to determine the relevant cloud utilities or tenant applications
is an actual concern. Authors have suggested that cloud deployed policy enforcement
components should be able to dynamically switch between them to allow context
aware coordination, for example, to adjust security levels depending on a perceived
risk.

Shi et al. [20] considered that edge computing is interchangeable with fog
computing, but edge computing focuses more towards the things side, while fog
computing focuses more towards the infrastructure side. Authors have evaluated that
the energy consumption can be decreased by 30–40 per cent by cloudlet offloading.
Authors have surveyed that data can be collected and processed depending on
geographic location without being transmitted to the cloud in edge computing.

Botta et al. [21] claimed that the actual issue towards cloud IoT is the lack of
standards. Authors have outlined that most things are associated with the cloud
through web-based interfaces which are able to decrease the complexity to develop
such applications. However, they are not clearly designed for well organised machine-
to-machine communication and establish overhead in respect of network load, data
processing and delay.

Sharma et al. [22] designed a privacy preserving data aggregation scheme for
mobile edge computing that assists Internet of things applications dependent on the
homomorphic property of Boneh-Goh-Nissim cryptosystem. Authors have claimed
that the introduced scheme can protect privacy and give source authentication and
integrity. In the proposed scheme, there are three participants, for example, edge
server, terminal device and public cloud centre. The terminal devices generate the
data in encrypted format, and then, data is transmitted to the edge server. After
receiving the data, the edge server aggregates the data and submits the aggregate
data to the public cloud centre.

Wen et al. [23] discussed that the cloud mobile media paradigm would authorise the network operators and service providers to give media services to ever
increasing mobile users with much enhanced efficiency. This suggested that leveraging omnipresent clouds could enhance the performance for mobile media networks
notably.

3   The Proposed Cypher Scheme for Secure Data Sharing

This section discusses the proposed cypher scheme—a delicate cryptographic
scheme for sharing and searching the desired data by legitimate users at the edge
of cloud-assisted Internet of things where in entirely security-oriented operations
are offloaded to the within reach edge servers. Furthermore, we added the cache
memory concept, whenever a user issues a query, then the cloud server executes a
search operation on the stored data and sends the result back to the user. If a user
issues the same query again, then the cloud server executes the same operation each
time, resulting in huge computation cost and waste of resources. To handle this issue,

a cache memory module is attached to maintain all previous searches. Whenever the user issues the same query, the cloud server gets results from cache memory instead of frequently searching computation. This technique contributes to save computation cost and wastage of resources. The advanced encryption standard (AES) is used for secret key encryption, and Rivest-Shamir-Adleman (RSA) algorithm is used for public key encryption. SHA-256 algorithm has been to implement the hash function. All security-oriented operations of smart gadgets are processed at a nearby edge server. The proposed scheme, user registration with the edge servers is required to search, share and retrieve the available data. The proposed scheme consists of following four parts:

3.1 Key Generation

In the proposed scheme, on behalf of the owner of the IoT gadgets, two categories of keys are generated by the edge servers. The first category is a randomly created 256-bit key and the second category comprises two types of security keys that are used in searching and sharing of data. The edge server creates both these secret keys uniquely and differently for all the IoT gadgets.

3.2 Data and Keyword Uploading

For uploading the sharable data and the search keywords, firstly, the data owner is required to log into an edge server using an IoT gadget. Thereafter, the data owner can upload the data to connected edge servers. Data owner, additionally, has to upload the search key terms that could be searched by an authentic user who is further authorised to access the search items. A cypher transformation is applied to the data and related search key terms prior to their transmission takes place between the edge server and the cloud storage. Lastly, the encrypted data are digitally signed to validate the data integrity at receiving end. Edge server carries out following steps once it receives the list of IoT gadgets, the data to be shared and the search key terms:

- Encryption of data using a secret transmission key.
- Encryption of search key terms using a searchable secret key. Encryption of secret key using the legitimate recipient's public key to establish a secure link with authorised IoT gadgets.
- Evaluation of hash value of encrypted data using collision resistant hash function to ensure the integrity of the data and then signs the hash value with the private key of the data owner (digital signature).
- Uploading the encrypted data, encrypted keywords, encrypted secret key, signed hash value and digital certificate to the edge storage.
- Verification of the digital certificate and placing the data on cloud storage.

3.3 Data Downloading and Sharing

When an authentic IoT device wants to access the data it is authorised to, it logs in to a nearby edge server using the valid credentials and places its request there. The edge server loads and stores the encrypted data, encrypted secret key, encrypted keywords, signed hash value and digital certificate under the data owner's username from cloud storage or edge server, wherever available. The edge server validates the data owner's digital certificate. If digital certificate is found valid, then the edge server decrypts the secret key, otherwise the process abandons. Once the secret key is decoded successfully, the edge server decodes and retrieves the data. This follows the hash value computation for the data and decryption of the digitally signed hash value. If the both hash values match, then the data integrity test is passed. Thereafter, the data are transmitted to the authentic and authorised recipient.

3.4 Data Search and Retrieval

To search the required data from the encrypted data on edge server or cloud storage, the user, authorised to access the data items, sends the keyword(s) to the edge server after login. Edge server generates a trapdoor using the secret key of the requesting legitimate user. The trapdoor is sent to the storage server with a search request. The request is taken up against the encrypted keyword(s) and the user's name mentioned in the trapdoor. If the keyword is found, then the encrypted data, encrypted keyword, signed hash value, encrypted secret key and digital certificate are handed over to the edge server, which, after verification of the digital certificate, decrypts the secret key. If the requesting user is legitimate, then the edge server decrypts the data and computes the hash value of the data. Edge server also decrypts the digitally signed hash value. If both hash values match, then data integrity stands verified and the data are sent to the requesting legitimate user.

4 Experimental Results

Figure 1 shows that $user_1$ and $user_2$ are registered and secret keys are generated for them.

Figure 2 shows that txt1.txt and txt2.txt files uploaded and shared at the cloud storage. Figure also shows that file has been fetched from cache memory due to repetition of the same query.

Figure 3 shows all the operations were executed on the edge server such as logging in by the authorised users, registration by the new users, execution of the search queries, files uploading and sharing, respectively.

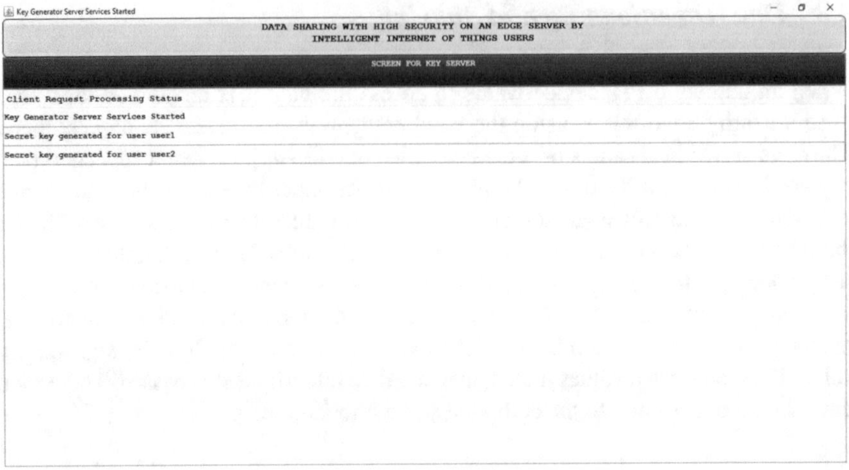

Fig. 1 Screen for key generation server

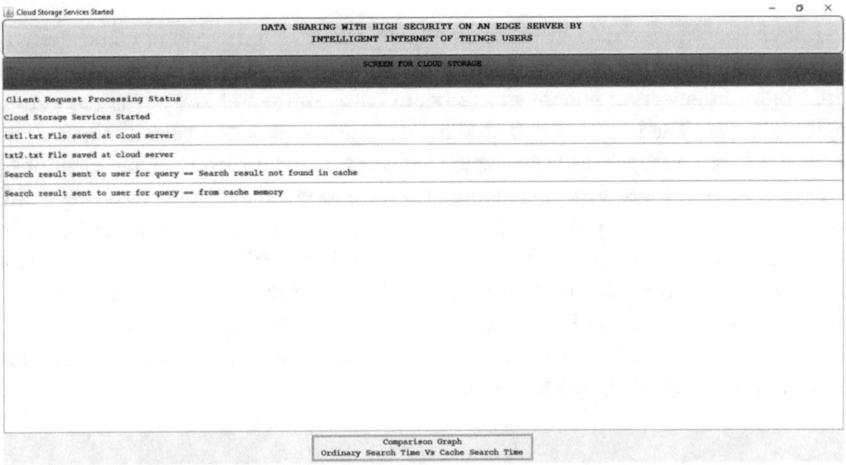

Fig. 2 Screen for cloud server

Figure 4 shows that there is a huge difference between the ordinary search time and cache search time while repeating the same query again.

5 Conclusion and Future Scope

In this paper, a scheme has been proposed to search and share the data among smart IoT devices' users with high security at the edge of cloud. The proposed scheme

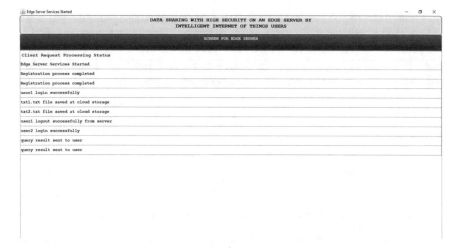

Fig. 3 Screen for edge server

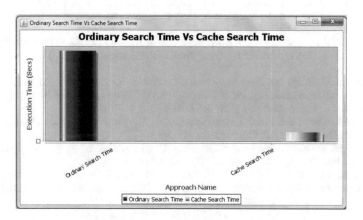

Fig. 4 Comparison graph between ordinary and cache search time

enables the smart IoT gadgets to safely search for some data inside one's own or shared storage. The advanced encryption standard mechanism is used for security that makes the sharing of data more reliable. This scheme not only permits users to perform conjunctive keyword searches over the encrypted data, but also enables encrypted data to be transmitted quickly and many times among several users without needing the "download-decrypt-encrypt" cycle. For reducing the search time, a software cache module has been developed which results in less time than the normal search time and makes the scheme faster. It is hoped that the proposed data sharing and searching scheme is practical and extends a step towards the research in edge-oriented security for cloud-assisted IoT applications. Owing to the limited scope of work and paucity of time, authentication and access control challenges remained to be taken up. Further, the researcher has experimented on only text files, but the system can be enhanced to process PDF, audio, video and other file formats for secure, efficient and faster-search methods. Because of the fact that the future edge-cloud architecture could entail multiple technologies such as network function virtualisation and software defined networks, etc., security issues will be manifold.

References

1. Cao K, Liu Y, Meng G, Sun Q (2020) An overview on edge computing research. IEEE Access 8:85714–85728. https://doi.org/10.1109/ACCESS.2020.2991734
2. Ghosh A, Grolinger K (2020) Edge-cloud computing for IoT data analytics: embedding intelligence in the edge with deep learning. IEEE Trans Ind Inform, pp 1–1. https://doi.org/10.1109/tii.2020.3008711
3. Nerella S, Sateesh G (2019) Confederate process key agreement scheme for cloud assisted vehicular internet of things. Int J Res Advent Technol, Special Issue, March 2019 E-ISSN: 2321-9637
4. Hassija V, Chamola V, Saxena V, Jain D, Goyal P, Sikdar B (2019) A survey on IoT security: application areas, security threats, and solution architectures. IEEE Access 7:82721–82743. https://doi.org/10.1109/access.2019.2924045
5. Javaid S, Afzal H, Babar M, Arif F, Tan Z, Ahmad Jan M (2019) ARCA-IoT: an attack-resilient cloud-assisted IoT system. IEEE Access 7:19616–19630. https://doi.org/10.1109/access.2019.2897095
6. Santos IL, Delicato FC, Pires PF, Alves MP, Oliveira A, Calmon TS (2019) Data-centric resource management in edge-cloud systems for the IoT. Open J Internet of Things, 5(1):29–46. https://www.ronpub.com/ojiot. ISSN 2364-7108
7. Mani AK, Said S, Gokilavani V, Unnikrishnan KN (2019) A review: IoT and cloud computing for future internet. Int Res J Eng Technol 6(5):1098–1102. https://www.irjet.net
8. Sureddy S, Rashmi K, Gayathri R, Nadhan AS (2018) Flexible deep learning in edge computing for IoT. Int J Pure Appl Math 119(10):531–543. http://www.ijpam.eu
9. Porambage P, Okwuibe J, Liyanage M, Ylianttila M, Taleb T (2018) Survey on multi-access edge computing for internet of things realization. IEEE Commun Surv Tutorials 20(4):2961–2991. https://doi.org/10.1109/COMST.2018.2849509
10. Pravallika C, Prakash IB, Sukanya P (2018) Secure data sharing and searching at the edge of cloud-assisted internet of things. Int J Adv Res Comput Sci Softw Eng 8(4):328–332. https://www.ijarcsse.com ISSN: 2277-128X
11. Yu W, Liang F, He X, Hatcher WG (2018) A survey on the edge computing for the internet of things. IEEE Access 6:6900–6919. https://doi.org/10.1109/ACCESS.2017.2778504
12. Hussein IJ, Burhanuddin MA (2018) Internet of things integration with cloud computing. J Adv Res Dyn Control Syst 10(4):616–626. https://www.jardcs.org
13. Pan J, McElhannon J (2018) Future edge cloud and edge computing for Internet of things applications. IEEE Internet Things J 5(1):439–449. https://doi.org/10.1109/jiot.2017.2767608
14. Abbas N, Zhang Y (2018) Mobile edge computing: a survey. IEEE Internet Things J 5(1):450–465.https://doi.org/10.1109/JIOT.2017.2750180
15. Gritti C, Onen M, Molva R (2018) Chariot: cloud-assisted access control for the internet of things. In: 2018 16th Annual conference on privacy, security and trust (PST). https://doi.org/10.1109/pst.2018.8514217
16. Ni J, Zhang K, Lin X, Shen X (2018) Securing fog computing for internet-of-things applications: challenges and solutions. IEEE Commun Surv Tutorials 20(1):601–628. https://doi.org/10.1109/comst.2017.2762345

17. Laaroussi Z, Morabito R, Taleb T (2018) Service provisioning in vehicular networks through edge and cloud: an empirical analysis. In: 2018 IEEE conference on standards for communications and networking (CSCN). https://doi.org/10.1109/cscn.2018.8581855
18. Abbasi MA, Memon ZA, Memon J, Syed TQ, Alshboul R (2017) Addressing the future data management challenges in IoT: a proposed framework. Int J Adv Comput Sci Appl 8(5). https://doi.org/10.14569/ijacsa.2017.080525
19. Kumar S, Karnani G, Gaur MS, Mishra A (2021) Cloud security using hybrid cryptography algorithms. In: 2021 2nd International conference on intelligent engineering and management (ICIEM), pp 599–604. https://doi.org/10.1109/ICIEM51511.2021.9445377
20. Shi W, Cao J, Zhang Q, Li Y, Xu L (2016) Edge computing: vision and challenges. IEEE Internet Things J 3(5):637–646. https://doi.org/10.1109/jiot.2016.2579198
21. Botta A, De Donato W, Persico V, Pescape A (2014) On the integration of cloud computing and internet of things. In: 2014 International conference on future internet of things and cloud. https://doi.org/10.1109/ficloud.2014.14
22. Vaishali, Sharma N (2014) Privacy preserving data aggregation scheme for mobile edge computing assisted IoT applications. Int J Comput Tech 1(1):3–9. http://www.ijctjournal.org
23. Wen Y, Zhu X, Rodrigues JJ, Chen CW (2014) Cloud mobile media: reflections and outlook. IEEE Trans Multimedia 16(4):885–902. https://doi.org/10.1109/tmm.2014.2315596

Attack Detection Based on Machine Learning Techniques to Safe and Secure for CPS—A Review

Durgesh M. Sharma and **Shishir Kumar Shandilya**

1 Introduction

In today's smart world, cyber-physical systems (CPS) connects intelligent control systems and physical components that interact with each other to deliver extensive services including electricity, transportation, building automation, health care, etc. [1]. These interconnected services are designed to enhance the standard of living and to promote technological advances in various domains [2]. System automation brings productivity, controllability, and correctness [3]. Digital and computational technologies play a crucial role in several physical devices and structures. Numerous technological advancements are creating various opportunities in the development of CPS. As a result, the demand for CPS is increasing day-by-day. CPS can be considered as the exploitation of discrete and logical characteristics of computers to oversee and control the continuous and rapidly changing characteristics of physical systems. Controlling unpredictable physical circumstances using existing methods is quite difficult. Thus for handling such circumstances, necessity of advanced technologies arises that are adaptable to dynamic situations and are capable of monitoring real-time status within lesser duration and higher precision. CPS is capable of satisfying these objectives to a wide level. CPS consists of heterogeneous devices linked via several communicational infrastructures. Though CPS still lacks a proper definition, it is substantially considered to be the future generation system that combines control, communication, and computation to attain high performance, efficiency, stability, and robustness. As progressing research emphasizes on achieving these objectives, security within CPS is highly ignored. Compromising security in critical infrastructures

D. M. Sharma (✉) · S. K. Shandilya
Vellore Institute of Technology, VIT Bhopal University, Bhopal, M.P, India
e-mail: durgesh_sharma54@yahoo.com

D. M. Sharma
Shri Ramdeobaba College of Engineering and Management, Nagpur, India

© The Author(s), under exclusive license to Springer Nature Singapore Pte Ltd. 2023
R. Agrawal et al. (eds.), *International Conference on IoT, Intelligent Computing and Security*, Lecture Notes in Electrical Engineering 982,
https://doi.org/10.1007/978-981-19-8136-4_23

may result in catastrophic consequences. Insufficient countermeasures and lack of security breaches will have disastrous impact [4] on system. Moreover, interconnections between CPS and information systems have led to security vulnerabilities owing to legacy of poor information security. Consequently, CPS is subject to numerous threats from cyber-physical attackers like denial of service (DoS) attacks and data integrity threats. The implications of these threats extend beyond data loss.

CPS is distributed across vast geographical regions and typically gather enormous information for decision making and data analysis. The significance of decision making has been reported by [5–7]. Data collection aids in decision making through advanced ML algorithms. Breaches in the data gathering process lead to leakage of sensitive data [8]. Breaches may occur in distinct stages such as data collection, transmission, operation, and storage. In the majority of the existing CPS design systems, data protection is not taken into account. This paper discusses the various aspects of CPS security, ML techniques, and existing challenges.

1.1 Security Objectives of CPS

CPS is basically the integration of several components. Although there exist several numbers data security systems like encryption, firewalls, data erasure, etc., there are some limitations as well [9]. Therefore, safety and security of CPS is quite challenging. The confidentiality, authenticity, integrity, and availability are the basic security objectives of CPS. Confidentiality refers to the capability of CPS to avoid disclosure of sensitive information to unauthorized systems or personnel. Integrity refers to the resources or information that cannot be altered without authorization. Availability of CPS aims to provide service by avoiding control, computation, and communication corruptions as a result of hardware failures, power outages, and system upgrades. Authenticity refers to secure information transmission and communication. For designing a better CPS model, these security objectives need to be considered. In [10], the security objectives, CPS challenges were discussed. In [11], a framework for CPS security was proposed in addition to the discussion of CPS security objectives.

2 Reliability and Security of CPS

Generally, CPSs are deployed in extensive geographical areas wherein they frequently collaborate with continuously varying physical environments. Due to the software implementation or system faults, indiscriminate failures occur in the system. These random failures and uncertainties result in an unreliable system. Furthermore, more security threats are encountered in CPS compared to networked systems due to its widespread heterogeneity. Analogous to the other networks, security threats originate both internally and externally in CPS initiating attacks with reference to every system layer. Numerous fault-tolerance methods derived from intelligent control systems can be applied to CPS. Majority of them work on the basis of inserting duplicate records or redundancy into the systems that are capable of only recovering the indemnities rather than abandoning the damages. The reliability requirement is to guarantee the stability of functional objectives in CPS. Specifically, these functional objectives comprise safety, optimization of function, liveness, and closed-loop stability. The security requirement emphasizes on protecting CPS from being breached by cyber vulnerabilities. The crucial security requirements include better understanding of information availability, security, integrity, and confidentiality [12]. Integrity is concerned with credibility of resources or information. Therefore, to maintain the integrity of the system, the system should be capable of surviving threats with reference to the data received and transmitted by the sensors, actuators, and control units. Availability signifies the system accessibility and its usability upon requisition. DoS is a specific attack occurring due to the insufficient amount of availability. Confidentiality indicates the potential of maintaining privacy of information from illegitimate users [13]. As a whole, confidentiality ensured CPS will efficiently protect against the anomalous behaviors trying to tap the data streams between the layers. Although the confidentiality provision is not the solution to every attack, rather prevention methods are very useful in this context.

In [14], a framework for attack detection in CPS was proposed and performance parameters like privacy, security, and reliability were evaluated whereas an Android-based Pan-Tompkins algorithm application has been proposed to improve cardio-vascular strain and to detect the heartbeat anomalies to find out the variations in the normal patient's. In [15], the improvement of a circulatory strain and detection and notice apparatus as that permits quick discovery of any variations from the norm in a patient's fundamental dependent on the and reports it to the pertinent emergency clinic or clinical staff.

In [16], an approach for CPS threat modeling was proposed that helped in understanding the nature of an attack and its effects on CPS. Krotofil et al. [17] discussed the significance of time in CPS security attacks. In this work, an attacker framework was presented with false data injection and denial of service (DoS) attacks. The discussion was specific for power grid systems. In [18], a study on distinct attacker profiles and models for CPS was presented. In this work, distinct attacker profiles like terrorist, insiders, cybercriminals, hackers, and basic users were studied.

3 Design Challenges for Security Measures of CPS

Some of the design challenges arising due to the security threats and vulnerabilities include security by adaptivity, heterogeneity, design, real-timeliness, safe integration, and confidentiality [19, 20].

- Adaptivity: Conventional CPS systems are adaptable to only specific environments. This drawback limits the applicability of CPS systems in dynamic environments. Therefore, CPS systems should be designed such that they are flexible to changing environments.
- Heterogeneity: CPS comprises various components. These components require interfacing, distinct computation models, and interoperability over multiple platforms. Hence, heterogeneity is considered to be one of the crucial design challenges for CPS.
- Distributed: Present CPS components are generally networked. They can be isolated temporally and/or physically. Majority of applications prefer distributed CPS models. Therefore, CPS should be designed as per the application requirements.
- Security by Design: Malicious or natural disasters resulting in physical attacks lead to severe design concerns. The prime issue is to resolve the challenge of embedding the physical security measures in the CPS design phase.
- Real-timeliness: The CPS testing and realization for the vulnerabilities is performed at the design phase or at the post fabrication testing level. As various intentional and unforeseen vulnerabilities emerge during the runtime, it is essential to incorporate a runtime detection system in the CPS.
- Confidentiality: Majority of the measured CPSs and privacy information are sensed through the sensors, communication of this information to various devices through multiple transmission channels results in security challenges pertaining to confidentiality. The prime issue is to resolve the challenge of maintaining confidentiality during the communication in CPS.
- Safe Integration: Most often, the CPSs are highly complex and include various stakeholders, particularly, in the integration of inhomogeneous CPS components. This inhomogeneity complicates the integration of diversified CPS devices in a secure way. Hence, the significant task is to guarantee the safe integration of various heterogeneous CPS components in order to model the secure CPS.

4   Literature Review

In this section, the various attack detection models employed for CPS, application of ML in CPS security, design challenges of CPS have been surveyed.

4.1   Attack Detection Models

In [21], a unified system framework with uncertainties for CPS security control was formulated by analyzing the procedure of multi-sources cyber-attacks of information disclosure, DoS, replay attack, and stealthy attack. Powerful control theory was applied for designing the control scenarios in order to prevent cyber-attacks.

Utilizing the information technology, security requirements were derived for cyber-attack identification and detection. However, the effectiveness of the framework in detecting complex attacks was not discussed.

In [22], an approach was presented for detecting and evaluating cyber-attacks on control application namely automatic generation control (AGC). The impact of cyber-attacks on AGC and the effects of control-based and measurement-based attacks on system load and frequency were analyzed. Through this study, it was found that these attacks caused redundant load shedding. Although the robustness of approach for detecting stealthy attacks and other sophisticated attacks was not evaluated. In [23], a framework was developed for securing railway CPS. This framework implemented a smart reasoning procedure for configuring and controlling the system during runtime. The privacy, dependability, and security metrics were embodied with semantic information pertaining to the safety conditions of setting. At runtime, system was configured to counter attacks and control safety-related actions.

Physics-based intrusion detection framework was proposed by Agrawal et al. [24] for detecting internal threats. This work exploited the system dynamics and detected intrusions using physics of process. However, the effectiveness of this approach needs to be validated in complex CPS. A framework for detecting false data injection (FDI) attacks was proposed by Li et al. [25] in smart grid CPS. In this work, a rule dependent majority voting method was employed for detecting falsely measured values inserted by compromised measurement units. However, this work discussed detection of FDI attacks only.

In [26], a study on attack models and generalized attackers for CPS was presented. The several types of intrusions were discussed in detail. However, in the majority of the attack models, the complexity of designing and initiating an attack was not considered. In [27], an attack detecting scheme was proposed for detecting cyber-attacks on additive manufacturing CPS. In this work, the system behavior was modeled by statistically evaluating functions that determine the relation between cyber related data and corresponding analog emissions. However, this scheme failed to detect complex attacks.

In all the aforementioned works, intrusion detection frameworks were designed to handle specific attacks and were inefficient to deal with complex/multiple attacks in CPS. However, with the evolution of technology, advanced techniques like ML, DL are introduced. Use of ML techniques in attack detection frameworks has simplified the complex attack detection process in CPS to a broad extent.

4.2 ML in CPS

ML has been transformed to mainstream in various domains. ML can be used to address challenges such as networking, control, and computing in CPS. Various prerequisites are required for developing an efficient ML model. It is essential to balance the high detection abilities with resource consumption. Furthermore, it is crucial to deliver suitable and timely data to ML models so as to achieve better outcomes. Additionally, supply of input continuously enables real-time detection, however, timely verification is complicated. ML has the feasibility to be used in a variety of domains for an extensive range of applications such as smart transportation, wireless networks, smart grids, etc. Typically, ML techniques can enhance attack detection by training on intrusion datasets. Similarly in CPS, ML techniques not only enhance the system security but also automate the system [28].

An intrusion detection approach in aerospace CPS using ML techniques was proposed by Maleh [29]. In this study, IoT nodes were exposed to several intrusion threats. The detection model using different ML techniques was developed and trained for detecting routing attacks like sinkhole attack, wormhole attack, and hello-flood attack. Though this approach exhibited better performance with respect to attack detection, the need for extending this work for detecting more attacks in CPS was suggested.

A framework for detecting anomalies in CPS was proposed by Liu et al. [30] using an unsupervised graphical modeling scheme. In this work, for identifying and representing interactions between subsystems of CPS, a spatiotemporal feature extraction technique was employed. The spatiotemporal features thus extracted were utilized to study the system-wide behavior through a restricted Boltzmann machine. This framework detected anomalies by identifying low-probability events. However, the proposed method was inefficient to capture complicated anomaly patterns and identification of multiple faults simultaneously.

A CPS attack detection framework was proposed by Valdes et al. [31] for detecting attacks in the cyber plane and host audit logs using unsupervised ML methods. The effects of attacks in measurements obtained from distinct components, deployed in multiple locations were discussed. Furthermore, the physical constraints induced by current and voltages in electric grids, leveraging distributed algorithms in order to detect the abnormal conditions was accomplished by coding the physical constraints into a composite CPS attack detection framework. The proposed detection method predominantly improved the adversary potential to effectuate a successful, unrecognized injection attack. Additionally, an alternative method for identifying normal, attack, and fault positions in an intelligent distribution substation CPS using ML was presented. However, further research for identifying the label patterns for specific anomalies needs to be investigated.

A technique was proposed by Wang et al. [32] for detecting attacks pertinent to the time synchronization (TS) in CPS using ML techniques. The proposed technique was used to detect the unknown attacks. This technique exhibited better performance compared to existing schemes by effectively identifying TS attacks. Although further research related to the cost analysis and runtime behavior of ML detectors in real-time scenarios needs to be studied.

A falsification framework was proposed by Dreossi et al. [33] wherein a ML analyzer and temporal logic falsifier was employed to determine the falsifying executions. The efficiency of this framework was demonstrated on an intelligent emergency braking device with a perception element dependent on deep neural networks (DNN). The feature spaces were abstracted using ML classifier and misclassified sets of feature vectors required for the falsification process were provided by ML classifier.

However, to control industrial production systems, the proposed framework must be improved further through cloud computing. Moreover, techniques for applying this method in non-cyber-physical processes need to be explored.

In [34], ML approach was employed for performing safety assessment of water CPS. This approach was used for calculating the time to being unsafe (TTBU) of water CPS. The approach was fast, robust to disturbances, and scalable. Moreover, the proposed model could be updated easily to match the varying behavior of the environment and the system. However, further research needs to be carried out for enhancing the attack detection performance.

In [35], unsupervised ML approach was employed for anomaly detection in CPS. In this work, the performance of unsupervised ML techniques like DNN and SVM were evaluated under 36 distinct attack scenarios. It was found that DNN generated less false positives compared to SVM. Performance evaluation of the proposed approach indicated that DNN outperformed SVM in terms of precision and f-measure. However, DNN utilized more training time and attack detection time. Additionally, it involved higher computation cost. Moreover, the experimentation was performed using a single dataset, and hence, the results cannot be generalized to other CPSs.

In [36], supervised ML technique was employed for classifying abnormalities in network into attacks and faults in IoT-CPS. In this work, a differentiation operator was proposed for distinguishing attacks and faults in the system. The abnormalities analyzed in this work were generated by network attacks and fault components. The attack and faulty classes were classified with greater accuracy using ML technique. However, the need for further enhancing the classification accuracy and investigating ML techniques for detecting complex attacks in IoT-CPS environments was suggested.

An individual ML model alone will not be suitable for all the tasks in all situations as CPS are diverse, thus generalizing a single ML model for every situation is unacceptable. Furthermore, dynamically adjusting various parameters in NNs is complicated, as data in CPS is generated in real time that in turn need to be analyzed, stored, validated, and trained by learning models. ML models have definite requirements for sizes, input data types, and shapes. Data in CPS is collected continuously by a diverse range of sensors and thus handling the raw data is challenging. The rapid evolution and variety of malicious code and malware enhance the identification problems making the detection more challenging. Hence, through continuous updation and use of well-trained ML models can enhance the intrusion detection capability in CPS.

5   Analysis of Prior Works

In [37], an attack identification and detection approach based on smart sensor was proposed for automotive systems. In this work, deep neural networks (DNNs) were used for detecting attacks. This work addressed the issue of intrusion detection whenever multiple sensors are attacked in a smart CPS. The sequential data with real-time information was utilized for detection. However, the need for developing online sensor intrusion detection and resilient system for determining complex driving conditions and complex attacks were recommended. In [38], a DL-based Internet of Things (IoT) security framework was proposed for IoT-CPS. The proposed approach not only detected physical attacks but also cyber threats and attacks. The detection of anomalies was achieved by prediction errors capable of tracking both physical and cyber anomalous behaviors. However, this work used only energy consumption information for detecting attacks. In [39], various challenges and solutions for distinct system layers were discussed for enhancing the security of CPS-IoT systems that are exposed to several security and reliability threats. This work depicted how systems should be protected from such threats whenever ML subsystems are used in IoT/CPS. Furthermore, the need for developing advanced ML techniques for further security enhancement was suggested. In [40], a cyber-attack detection model was proposed for detecting attacks on water distribution systems. In this study, ML methods like extreme learning machine (ELM), KNN, and artificial neural network (ANN) were employed for attack detection. Comparison of ML methods in terms of various performance metrics pertaining to attack detection indicated that ELM exhibited better performance. Although, a high false positive detection ratio was achieved.

5.1 Analysis Based on Publication years

In this subsection, the analysis conducted depending on publication years of attack detection techniques considered in Table 1 is being presented. The number of technical papers published from 2016 to 2020 is being depicted in Fig. 1. Among the 25 papers reviewed, more number of studies, i.e., nine technical papers pertaining to attack detection based on ML for securing CPS were surveyed from papers published in 2017 and 2020. Among the papers published in 2016 and 2018, only one technical paper was considered for analysis and among the papers published 2019, only five technical papers were surveyed.

5.2 Analysis Based on Performance Measures

In this subsection, the analysis conducted depending on distinct performance measures considered in Table 1 is being presented. The various performance metrics used are being depicted in Fig. 2. From Fig. 2, it could be observed that accuracy metric is largely used, followed by precision, other metrics and then recall, f-measure, ROC, AUC, kappa value, specificity, TPR, sensitivity, FPR, PPV, and MSE.

Table 1 Analysis table based on attack detection techniques

Ref. No	Year	Performance measures	Techniques
[29]	2020	Accuracy, energy consumption, memory overhead, precision, recall	Random forest (RF), K-nearest neighbors (KNN), support vector machines (SVM), multilayer perceptron (MLP), Naive Bayes (NB)
[30]	2016	–	Restricted Boltzmann machine (RBM)
[32]	2017	Overall accuracy, secured precision, attacked precision, secured recall, attacked recall	Artificial neural network (ANN), SVM, KNN, NB, decision tree (DT)
[33]	2019	Region of uncertainty (ROU)	Deep neural network (DNN)
[35]	2017	Recall, precision, F-measure	DNN, SVM
[36]	2020	Accuracy, kappa value	DT, MLP, NB, ANN
[40]	2020	Accuracy, true positive rate (TPR), true negative rate (TNR), F-measure, positive predictor value (PPV), time to detection, score of time to detection, total performance score	Extreme learning machine (ELM), DNN, ANN
[41]	2018	AUC (area under curve) and ROC (receiver operating characteristic)	Extreme learning machine (ELM)
[42]	2017	AUC, accuracy, TPR, false positive rate (FPR)	SVM, NB, RF, J48
[43]	2019	MSE (mean squared error)	CNN
[43]	2017	Error rate and loss	ANN
[44]	2020	Accuracy, sensitivity, precision, specificity	SVM
[45]	2017	First rank, median rank, last rank, standard deviation	KNN
[46]	2017	Accuracy	SVM, ANN
[47]	2020	Recall, precision, F-measure	NB, RF
[48]	2020	Recall, precision, F-measure, AUC, ROC	DT, NB, RF, KNN, J48, MLP, long short-term memory (LSTM)
[49]	2017	Accuracy	Naïve Baye's (NB)
[50]	2020	Accuracy, recall, F-measure, precision	Stacked autoencoder (SAE), DT, DNN, RF, Adaboost (ADA)
[51]	2020	Recall, F-measure, precision	Decision tree (DT)
[52]	2019	Accuracy, kappa value	NB, J48, multi-layer perceptron (MLP), multinomial logistic regression (MLR)
[53]	2020	Recall, F-measure, precision	RF, SVM, NB
[54]	2019	Precision, accuracy	DNN

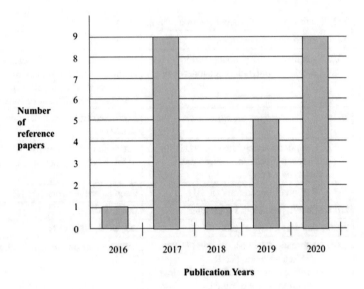

Fig. 1 Analysis on publication year basis

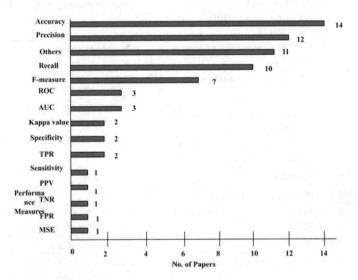

Fig. 2 Analysis of performance measures

5.3 Analysis Based on Attack Detection Techniques

In this subsection, the analysis conducted depending on distinct attack detection techniques considered in Table 1 is being presented. The various attack detection techniques are being depicted in Fig. 3. From Fig. 3, it could be observed that NB

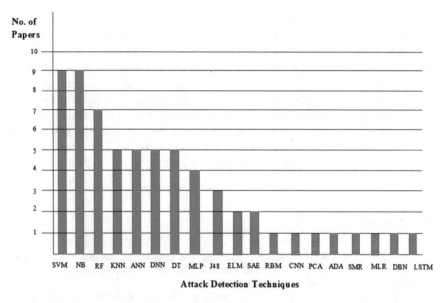

Fig. 3 Analysis of attack detection techniques

and SVM were used in nine papers, RF in seven papers, KNN, ANN, DT, and DNN in five papers, MLP in four papers, J48 in three papers, SAE and ELM in two papers, RBM, CNN, PCA, LSTM, ADA, MLR, DBN, and SMR in one paper.

6 Conclusion

CPSs are considered to be emerging technologies in the forthcoming generations. In CPS, control, communication, and computation are combined together for achieving high performance, efficiency, stability, and robustness, although security within CPS is highly ignored. Compromising security in CPS might result in catastrophic consequences. In ensuring CPS security against various threats, attack models and ML techniques play a vital role. In this review work, major design challenges for CPS security measures are discussed and reliability and security requirements are explored. The diverse attack detection models utilized in existing research works are investigated. The distinct ML approaches exploited in literary studies for detecting attacks on CPS are reviewed. Moreover, related studies are analyzed based on publication years, performance measures, and attack detection techniques. Though existing ML methods ensure efficacy they are not fully reliable. Moreover, exploitation of merely traditional ML techniques detected only specific attack patterns, showing inefficiency toward detecting complex attacks. Despite the employed attack detection methods in many works detected attacks on CPS, they lacked the potency and accuracy of identifying sophisticated CPS attacks. The vital findings discovered from reviewed studies suggest further research toward devising better and optimized ML techniques for superior attack detection, guaranteeing accuracy, and reducing the false attack detection rates. Future developments include designing efficient and powerful security frameworks using advanced ML techniques toward detecting diverse attacks more precisely or utilization of deep learning frameworks for further security optimization of CPS or hybrid framework comprising multiple ML techniques considering various security parameters and performance metrics for enhancing attack detection performance in CPS and for optimizing its security. The potential shown by ML and artificial intelligence, elucidates that a novel framework could be designed, which is more robust than existing attack detection frameworks.

References

1. Taylor JM, Sharif HR (2017) Security challenges and methods for protecting critical infrastructure cyber-physical systems. In: International conference on selected topics in mobile and wireless networking (MoWNeT), IEEE, pp 1–6
2. Mihalache SF, Pricop E, Fattahi J (2019) Resilience enhancement of cyber-physical systems: a review. In: Power systems resilience, Springer, Cham, pp 269–287
3. Sharma DM, Shandilya SK, Sharma AK (2021) A comprehensive review on cyber physical system and its applications in robotic process automation. In: Abraham A, Sasaki H, Rios R, Gandhi N, Singh U, Ma K (eds) Innovations in bio-inspired computing and applications. IBICA 2020. Advances in intelligent systems and computing, vol 1372. Springer, Cham
4. Konstantinou C, Maniatakos M, Saqib F, Hu S, Plusquellic J, Jin Y (2015) Cyber-physical systems: a security perspective. In: 2015 20th IEEE european test symposium (ETS), IEEE, pp 1–8
5. Sharma AK, Mehta IC, Sharma JR (2009) Development of fuzzy integrated quality function deployment software-a conceptual analysis. I-Manager's J Softw Eng 3(3):16
6. Purohit SK, Sharma AK (2017) Development of data mining driven software tool to forecast the customer requirement for quality function deployment. Int J Business Anal (IJBAN) 4(1):56–86
7. Sharma AK, Khandait S (2016) A novel software tool to generate customer needs for effective design of online shopping websites. Int J Inform Technol Comput Sci 83:85–92
8. Wurm J, Jin Y, Liu Y, Hu S, Heffner K, Rahman F, Tehranipoor M (2016) Introduction to cyber-physical system security: a cross-layer perspective. IEEE Trans Multi-Scale Comput Syst 3(3):215–227
9. Mishra S, Sagban R, Yakoob A, Gandhi N (2021) Swarm intelligence in anomaly detection systems: an overview. Int J Comput Appl 43(2):109–118
10. Keerthi CK, Jabbar MA, Seetharamulu B (2017) Cyber physical systems (CPS): security issues, challenges and solutions. In: 2017 IEEE international conference on computational intelligence and computing research (ICCIC), pp 1–4
11. Lu T, Zhao J, Zhao L, Li Y, Zhang X (2015) Towards a framework for assuring cyber physical system security. Int J Secur Its Appl 9(3):25–40
12. Dibaji SM, Pirani M, Flamholz DB, Annaswamy AM, Johansson KH, Chakrabortty A (2019) A systems and control perspective of CPS security. Annu Rev Control 47:394–411
13. Han S, Xie M, Chen HH, Ling Y (2014) Intrusion detection in cyber-physical systems: Techniques and challenges. IEEE Syst J 8(4):1052–1062
14. Gifty R, Bharathi R, Krishnakumar P (2019) Privacy and security of big data in cyber physical systems using Weibull distribution-based intrusion detection. Neural Comput Appl 31(1):23–34
15. Pandey SR, Hicks D, Goyal A, Gaurav D, Tiwari SM (2020) Mobile notification system for blood pressure and heartbeat anomaly detection. J Web Eng 19(5–6):747–773

16. Ding J, Atif Y, Andler SF, Lindström B, Jeusfeld M (2017) CPS-based threat modeling for critical infrastructure protection. ACM SIGMETRICS Perform Eval Rev 45(2):129–132
17. Krotofil M, Cárdenas AA, Manning B, Larsen J (2014) CPS: driving cyber-physical systems to unsafe operating conditions by timing DoS attacks on sensor signals. In: Proceedings of the 30th annual computer security applications conference, pp 146–155
18. Rocchetto M, Tippenhauer NO (2016) On attacker models and profiles for cyber-physical systems. In: European symposium on research in computer security. Springer, Cham, pp 427–449
19. Seshia SA, Hu S, Li W, Zhu Q (2016) Design automation of cyber-physical systems: challenges, advances, and opportunities. IEEE Trans Comput Aided Des Integr Circuits Syst 36(9):1421–1434
20. Lokesh M, Kumaraswamy Y, Tejaswini K (2016) Challenges and current solutions of cyber physical systems. IOSR J Comput Eng 18(2):104–110
21. Ge H, Yue D, Xie X, Deng S, Dou C (2019) A unified modeling of muti-sources cyber-attacks with uncertainties for CPS security control. J Franklin Institute
22. Ashok A, Wang P, Brown M, Govindarasu M (2015) Experimental evaluation of cyber-attacks on automatic generation control using a CPS security testbed. In: 2015 IEEE power and energy society general meeting, IEEE, pp 1–5
23. Hatzivasilis G, Papaefstathiou I, Manifavas C (2017) Real-time management of railway CPS secure administration of IoT and CPS infrastructure. In: 2017 6th Mediterranean conference on embedded computing (MECO), IEEE, pp 1–4
24. Agrawal A, Ahmed CM, Chang EC (2018) Poster: physics-based attack detection for an insider threat model in a cyber-physical system. In: Proceedings of the 2018 on Asia conference on computer and communications security, pp 821–823
25. Li B, Lu R, Wang W, Choo KKR (2017) Distributed host-based collaborative detection for false data injection attacks in smart grid cyber-physical system. J Parallel and Distrib Comput 103:32–41
26. Adepu S, Mathur A (2016) Generalized attacker and attack models for cyber physical systems. In: 2016 IEEE 40th annual computer software and applications conference (COMPSAC), vol 1. IEEE, pp 283–292
27. Chhetri SR, Canedo A, Al Faruque MA (2016) Kcad: kinetic cyber-attack detection method for cyber-physical additive manufacturing systems. In: 2016 IEEE/ACM international conference on computer-aided design (ICCAD), IEEE, pp 1–8
28. Liang F, Hatcher WG, Liao W, Gao W, Yu W (2019) Machine learning for security and the internet of things: the good, the bad, and the ugly. IEEE Access 7:158126–158147
29. Maleh Y (2020) Machine learning techniques for IoT intrusions detection in aerospace cyber-physical systems. In: Machine learning and data mining in aerospace technology, Springer, Cham, pp 205–232
30. Liu C, Ghosal S, Jiang Z, Sarkar S (2016) An unsupervised spatiotemporal graphical modeling approach to anomaly detection in distributed CPS. In: 2016 ACM/IEEE 7th international conference on cyber-physical systems (ICCPS), IEEE, pp 1–10
31. Valdes A, Macwan R, Backes M (2016) Anomaly detection in electrical substation circuits via unsupervised machine learning. In: 2016 IEEE 17th international conference on information reuse and integration (IRI), IEEE, pp 500–505
32. Wang J, Tu W, Hui LC, Yiu SM, Wang EK (2017) Detecting time synchronization attacks in cyber-physical systems with machine learning techniques. In: 2017 IEEE 37th international conference on distributed computing systems (ICDCS), IEEE, pp 2246–2251
33. Dreossi T, Donzé A, Seshia SA (2019) Compositional falsification of cyber-physical systems with machine learning components. J Autom Reason 63(4):1031–1053
34. Junejo KN (2020) Predictive safety assessment for storage tanks of water cyber physical systems using machine learning. Sādhanā 45(1):1–16
35. Inoue J, Yamagata Y, Chen Y, Poskitt CM, Sun J (2017) Anomaly detection for a water treatment system using unsupervised machine learning. In: 2017 IEEE international conference on data mining workshops (ICDMW), IEEE, pp 1058–1065

36. Tertytchny G, Nicolaou N, Michael MK (2020) Classifying network abnormalities into faults and attacks in IoT-based cyber physical systems using machine learning. Microprocessors and Microsyst 103121
37. Shin J, Baek Y, Lee J, Lee S (2019) Cyber-physical attack detection and recovery based on RNN in automotive brake systems. Appl Sci 9(1):82
38. Li F, Shi Y, Shinde A, Ye J, Song W (2019) Enhanced cyber-physical security in internet of things through energy auditing. IEEE Internet Things J 6(3):5224–5231
39. Kriebel F, Rehman S, Hanif MA, Khalid F, Shafique M (2018) Robustness for smart cyber physical systems and internet-of-things: from adaptive robustness methods to reliability and security for machine learning. In: 2018 IEEE computer society annual symposium on VLSI (ISVLSI), IEEE, pp 581–586
40. Choi YH, Sadollah A, Kim JH (2020) Improvement of cyber-attack detection accuracy from urban water systems using extreme learning machine. Appl Sci 10(22):8179
41. Yan W, Mestha L, John J, Holzhauer D, Abbaszadeh M, McKinley M (2018) Cyberattack detection for cyber physical systems security–a preliminary study. In: Annual conference of the PHM society, vol 10(1).
42. Huda S, Miah S, Hassan MM, Islam R, Yearwood J, Alrubaian M, Almogren A (2017) Defending unknown attacks on cyber-physical systems by semi-supervised approach and available unlabeled data. Inf Sci 379:211–228
43. Alpaño PVS, Pedrasa JRI, Atienza R (2017) Multilayer perceptron with binary weights and activations for intrusion detection of cyber-physical systems. In: TENCON 2017–2017 IEEE region 10 conference, IEEE, pp 2825–2829
44. Khalili A, Sami A (2017) SADCPS: Semi-supervised attack detection in cyber physical systems. In: 2017 international symposium on computer science and software engineering conference (CSSE), IEEE, pp 12–17
45. Wang Y, Amin MM, Fu J, Moussa HB (2017) A novel data analytical approach for false data injection cyber-physical attack mitigation in smart grids. IEEE Access 5:26022–26033
46. Panthi M (2020) Anomaly detection in smart grids using machine learning techniques. In: 2020 1st international conference on power, control and computing technologies (ICPC2T), IEEE, pp 220–222
47. Panthi M (2020) Anomaly detection in smart grids using machine learning techniques. In: 2020 first international conference on power, control and computing technologies (ICPC2T), January, IEEE, pp 220–222
48. Hossain MD, Ochiai H, Doudou F, Kadobayashi Y (2020) SSH and FTP brute-force attacks detection in computer networks: LSTM and machine learning approaches. In: 2020 5th international conference on computer and communication systems (ICCCS), IEEE, pp 491–497
49. Kreimel P, Eigner O, Tavolato P (2017) Anomaly-based detection and classification of attacks in cyber-physical systems. In: Proceedings of the 12th international conference on availability, reliability and security, pp 1–6
50. Al-Abassi A, Karimipour H, Dehghantanha A, Parizi RM (2020) An ystem. IEEE Access 8:83965–83973
51. Yeboah-Ofori A (2020) Classification of malware attacks using machine learning in decision tree. Int J Secur (IJS) 11(2):10
52. Tertytchny G, Nicolaou N, Michael MK (2019) Differentiating attacks and faults in energy aware smart home system using supervised machine learning. In: Proceedings of the international conference on omni-layer intelligent systems, pp 122–127
53. Gayatri R, Gayatri Y, Mitra CP, Mekala S, Priyatharishini M (2020) System level hardware trojan detection using side-channel power analysis and machine learning. In: 2020 5th international conference on communication and electronics systems (ICCES), IEEE, pp 650–654
54. Potluri S, Diedrich C (2019) Deep learning based efficient anomaly detection for securing process control systems against injection attacks. In: 2019 IEEE 15th international conference on automation science and engineering (CASE), IEEE, pp 854–860

Fake Account Detection in Social Networks with Supervised Machine Learning

Om Prakash and Rajeev Kumar

1 Introduction

Online Social Networks (OSNs) are an integral part of the modern Internet that came into existence with Web 2.0 in the early 2000s, and in a short period, they became popular worldwide due to their social aspects. Facebook and Twitter are the largest online social networking sites globally with more than 2.85 billion and 397 million monthly active users, respectively [1]. This popularity made OSNs a viably powerful tool for communication, especially during a national and an international crisis or events like sports, political events, natural calamities, epidemics, etc. This kind of popularity and public demand lead to the flourishing and mushrooming of many other social networking sites like YouTube, Instagram, WhatsApp, LinkedIn, Pinterest, Researchgate, Quora, Reddit, etc. To use these facilitating platforms, users need to create accounts by providing some personal pieces of information, whereas some users publicly store their personal and professional information on these networks. Hence, these OSNs prove to be a ground-breaking source of a massive amount of information, which attracts social analysts, researchers, and cybercriminals.

Fake accounts are anomalies on social media platforms and a baseline for cybercriminals to perform malicious or illegal activities. Cybercriminals create different types of fake accounts by feeding wrong credentials or bypassing social platform restrictions [2]. Verifying these false credentials is critical from security and privacy points of view. Our politicians commonly use fake accounts to gain political mileage by manipulating public opinions in recent trends. They use different bots that constantly post fake or manipulated content to control or manage these fake accounts. Common people like and share these adulterated beliefs and manipulated

O. Prakash (✉) · R. Kumar
School of Computer and Systems Sciences, Jawaharlal Nehru University, New Delhi, New Delhi 110 067, India
e-mail: omprak16_scs@jnu.ac.in

R. Agrawal et al. (eds.), *International Conference on IoT, Intelligent Computing and Security*, Lecture Notes in Electrical Engineering 982,
https://doi.org/10.1007/978-981-19-8136-4_24

facts without verifying with reliable sources. As a result, fake contents traverse more distance than the actual news and damage the image and credibility of a person or an organization.

Not everyone creates a fake account for misleading or malicious activities; sometimes, people create fake accounts to hide their details or for entertainment, but mostly cybercrime is done by fake accounts. Hence, it is critical to find all fake accounts on social media. Therefore, to maintain social networks integrity, we considered fake accounts as severe anomalies for social platforms in this paper. To deal with such hazards, we have proposed a supervised machine learning-based fake account detection model to identify fake accounts that can be useful in limiting the spread of online social crime.

The rest of the paper is organized as follows. Section 2 includes the literature survey. Section 3 demonstrates the proposed model and pre-processing. In Sect. 4, we exhibited results and analysis. Finally, we conclude the paper and include directions for future work in Sect. 5.

2   Related Work

A fake profile represents a person, organization, or enterprise that does not exist in reality. This section has explored the previous researcher's contribution dealing with fake accounts. Manuel Egele et al. [2] has built a statistical model to detect high-profile compromised accounts by characterizing users' social behavior. Authors created behavioral profile of users by observing historical information as daily time and duration spent by users on the social network, text language, source of the message (third party application), link (URL) in a message, direct interaction between users, and proximity of user in the network. They extract these features from past messages and build a statistical model. Similarly, Rahman et al. [3] have proposed an efficient hybrid system called DT-SVMNB by cascading several supervised machine-learning classifiers like Decision Tree (DT-C5.0), Support Vector Machine (SVM), and Naive Bayesian Classifier (NBC) for anomaly detection from social networks. They used two different types of datasets extracted from users' profiles and contents and detected the depression level of the user that is suicidal or not. At the first stage, they used DT to filter out all the anomalous entities, then SVM to filter out disappointed users, and finally, the NBC classifier was used to filter out suicidal users.

Torkey et al. [4] proposed a Fake Profile Recognizer (FPR) based on regular expression and Deterministic Finite Automata (DFA) with the assumption that all account is represented by DFA uniquely, where the fake account is unique kinds of cloned profile that are from the user friends list. The author tested and verified three datasets of Facebook, Twitter, and Google+. Jia et al. [5] have proposed a new random walk-based method called SybilWalk to detect Sybils (massive fake accounts) from online social networks. Ilias et al. [6] have emphasized the early detection of malicious activities from the social networking sites like Twitter. They have proposed two different models to identify fake accounts as (bots) from Twitter. In the first model, they have used five different supervised classifiers, whereas, in another model, they have proposed deep learning architecture with Bidirectional LSTM, including attention mechanism. Both proposed models are based on textual features, including account and tweets levels. They extracted 71 essential features from the existing dataset and applied them to oversample and undersampling due to imbalanced datasets. The dataset used in this paper contains 3474 genuine accounts and 8,377,522 tweets, and a set of 4912 social spambots and 3,457,344 tweets. To differentiate tweets between legitimate users and bots, they used another dataset containing 19,276 legitimate users and 22,223 fake accounts with 3,259,693 tweets and 2,353,473 tweets. The result is compared based on Precision, Recall, F-Measure, Accuracy, AUROC. In the first model, they got 99% accuracy as well as AURROC, whereas, in the second model, they had 82% Accuracy and 88% AUROC score.

Akyon et al. [7] have created an Instagram dataset and used various machine learning techniques such as naive Bayes, logistic regression, support vector machines, and neural networks to detect automated accounts. They manually created 1002 real and 201 fake accounts and used over-sampling and under-sampling to balance collected data. Finally, they generated 700 real and 700 automated accounts and selected features as the Total number of media, Follower, Following, Taggedphoto, highlight reel, and Profile picture of accounts. They used Precision, Recall, and F1-Score for performance comparison and found that SVM works better with 91% precision, 86% F1-score, whereas NB (Gaussian dist) has 98% highest score among all classifiers.

Breuer et al. [8] have introduced the SybilEdge algorithm, which aggregates over the choice of friend request targets and respective reaction as accept/reject to detect whether a new user is a fake account on Facebook social networks. SybilEdge is similarly robust to noise datasets and delivers excellent performance utilizing a graph-based method, according to the inventor. Kai Shu et al. [9] has described the importance of users' profile attributes to deal with fake accounts. The author includes implicit or explicit features of accounts. In another paper, Kai Shu et al. [10] used two real-world publically available benchmark datasets to claim that most of the fake content gets spread by fake accounts like bots or newly created accounts. The author found that fake users have low personalities than real users on social platforms. Van-Der et al. [11] have used three classifiers like SVM, RF, and AdaBoost to differentiate bots or human accounts from online social platforms.

Wanda et al. [12] have proposed a "Deep-Profile" CNN model to detect fake accounts from dynamic social networks. Instead of using a predefined standard max-pooling pooling layer in NN, they proposed their own pooling layer to improve performance. The author collected only textual features of high-profile accounts of OSNs such as user name, No of friends, followers, email-id, messages stream, location, URLs, and languages used by them. After collecting 1500 legitimate users and 1500 fake accounts, they manually labeled the dataset. They tested the dataset on different hyperparameter tunning and performed a comparison of the proposed model on Precision, Recall, F1-Score, and ROC-AUC. The output of ROC-AUC is 0.95%. Fazil et al. [13] have proposed a hybrid machine learning model to detect Social spam-

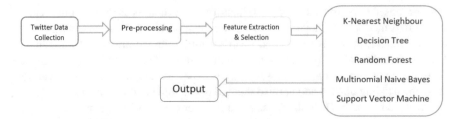

Fig. 1 Framework of the ML based Twitter fake account detector

mers from Twitter. They used community-based features, including metadata, contents, and interaction-based features. Nineteen different features are extracted, and three different classifiers are used as Random Forest, Decision Tree, and Bayesian networks to identify spammers. The author uses a benchmark dataset containing 11,000 labeled users, including 10,000 benign users with 12,09,522 tweets and 1000 spammers with 1,45,096 users. Newly legitimate users have no followers and followings initially; hence, only 128 benign users and 1000 spammers are taking, and 872 instances are generated using SMOTE oversampling techniques. The output of the proposed model is evaluated based on standard evaluation metrics as DR, FPR, and F-Score, and further 10-fold cross-validation is used to cross-examine the output.

3 The Proposed Method

A fake profile can be detected using various machine learning models from the social network platform. We have proposed a machine learning-based fake profile detector framework to detect a fake account. A pictorial representation of the framework is shown in Fig. 1. The process of fake account detectors has been divided into five blocks. The first block is the data collection part; we have collected the Twitter dataset using Twitter APIs. The second block is pre-processing, where we filter out raw data and make it suitable for machine learning, which increases the model's efficiency and accuracy. The third block is the features selection part; we select the important feature and drop the redundant features to speed the learning and generalization process. In the fourth block, we have selected five different machine learning algorithms as K-Nearest Neighbor (KNN), Decision Tree (DT), Random Forests (RF), Multinomial Naive Bayes (MNB), and Support Vector Machine (SVM), for classification. Then we have fine-tuned the hyperparameters for optimum results. Finally, the last block is the model's output, which produces a binary format, whether an account is fake or legitimate.

At first, we have considered the MNB classifier because it is straightforward to deal with textual data using Natural Language Processing (NLP), But accuracy is low compared to other algorithms. Second, we have selected Decision Tree because it is a non-parametric supervised learning algorithm, which does not require normalized data, and there is no effect on building a decision tree if there is some missing value. In parameter tuning, we used 'entropy' for criterion and set min_sample_leaf value 50. Third, We have selected kNN algorithm because it is faster than other algorithms as it stores datasets in the training phase and learns in the testing phase while making a real-time prediction. We set k_neighbors values 5 and random state 101 for the dataset. Fourth, We have taken SVM because It works more effectively when there is a clear margin of separation between classes or high-dimensional spaces. Here we used 'linear' kernel for the SVM and set random state value 42 for dataset. Finally, we have taken Random Forest classifier because It uses bagging and features randomness when building individual trees. It automates missing values in the dataset and is flexible for classification and regression problems. In parameter tuning, we set n_estimator 7, max_depth 7, and min_sample_split 5.

3.1 Data Collection

There are very few benchmark datasets available in the public domain due to data privacy and other social media norms. Therefore, we are using a self-collected Twitter dataset, and pre-processed ICC benchmark dataset [14] to validate the proposed model's efficiency. We collected 11,118 legitimate users and 5,394 fake accounts, including 21 attributes. We added some implicit features and removed redundant attributes. In another, ICC dataset contains 600 million tweets metadata; after optimizing, the author got 30 million labeled tweets, in which they found 6.5 million spam tweets. Thirteen different features are extracted and categorized into two subcategories as user profile-based and tweet content-based features. Profile-based features as the age of the account, the total number of followers, following, favorites received, total number of tweets done by users, whereas tweets-based contains information of particular tweets as number of a retweet or liked by others users, number of Hashtags, Characters, Digits, URLs, and Mentions in a particular tweet. Finally, the last column indicates the account as a fake (spammer) or legitimate (non-spammer). Both datasets contain user credentials and tweets-based features.

3.2 Preprocessing and Feature Selection

We have collected our own Twitter dataset using Twitter API. Firstly, we selected the verified account (with a blue tick) of a tweeter and collected their 20 friend lists and 30 tweets from each friend. Here we selected 20 different verified accounts of India's metropolitan or state police account and their friend to assume that they are not a spammer. After collecting metadata, we extract valuable attributes for classification, such as the number of URLs in tweets, number of Mentions, total number of hashtags, and many more other attributes. Similarly, we extracted different spammer metadata reported by some other users with the hypothesis that there are high chances of a fake account later analyzed.

After merging both legitimate and spammer user's data, we checked NaN values using heatmap and counted a specific word as vulgar words. In user's tweets with the hypothesis that legitimate users used very little as compared to a spammer. We removed all user's details having zero or less than 30 tweets, including repeated tweets, in the database because it will not contribute more to classification. To calculate average Hashtag, avg URL, avg Mention, and avg Favour for each user, we divide the total count of these attributes by 30. We also added a new column as SpammerOrNot by filling 0 (Zero) indicating Legitimate user and 1 (One) as Fake account, and finally Converted datatype from object to float. We also added an average reputation feature using Retweets and User Followers Counts to show the legitimacy of users. We also added other features such as AgeofAccount, TweetPerDay, TweetPerFollower, AgeByFollowing. Then, we divide the dataset into train and test by splitting 7:3 ratio and give input to the classification algorithms.

4 Results and Analysis

We have conducted a series of experiments using the proposed model on the Twitter dataset with various hyperparameter tuning and assessed these classification techniques on numerous metrics. This experiment aimed to classify legitimate and fake accounts accurately and efficiently. We have compared these classifier performances on Precision, Recall, F1 Score matrices of both fake and legitimate accounts demonstrated in Table 1. Training and testing AUC-ROC curve of these classifiers is also shown in Fig. 2 for comparative analysis. Finally, We have performed Ten-Fold Cross-validation and taken average accuracy to compare the overall performance of selected algorithms. The bold numerical values in Table 1 indicate the high performance achieved by classifiers.

We can observe from Table 1, that Multinomial Naive Bayes (MNB) classifier has only 58% accuracy. It has 97% Recall for the fake account and 97% precision

Table 1 Classifiers' performance on Twitter dataset

Method	Accuracy	Fake accounts (1)			Legitimate accounts (0)		
		Precision	Recall	F1 Score	Precision	Recall	F1 Score
MNB	0.58	0.40	**0.97**	0.56	**0.97**	0.44	0.60
DT	0.88	0.88	0.68	0.76	0.88	0.96	0.92
KNN	0.90	0.75	0.84	0.79	0.95	0.92	0.93
SVM	0.93	0.93	0.81	0.86	0.93	0.97	0.95
RF	**0.95**	**0.96**	0.88	**0.92**	0.95	**0.98**	**0.96**

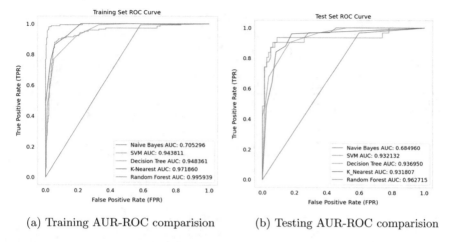

(a) Training AUR-ROC comparision   (b) Testing AUR-ROC comparision

Fig. 2 Training and testing AUC-ROC curve on Twitter dataset

for legitimate accounts detection, which is the highest among all classifiers. Training and testing AUR-ROC of MNB classifier is 70% and 68% respectively as shown in Fig. 2. After performing ten-fold cross-validation on the same data and getting 59% accuracy, which is more than 1% of actual output.

The performance of the Decision Tree (DT) classifier is much better than MNB, as we observe from Table 1. The accuracy of DT is 88%, which is 30% better than MNB classifiers. Figure 2 also indicates that training and testing AUC-ROC of DT is 94% and 93% respectively, which is much better than the MNB algorithm. When we performed ten-fold cross-validation, we got 88% accuracy, which is the same as the predicted value.

Similarly, the performance of the k-Nearest Neighbor(kNN) classifier increased by 2% than DT algorithm. For training data, the AUC-ROC is increased by 3%, but for testing the AUC-ROC remains constant at 93% as we can observe from Fig. 2. For this classifier, the performance of ten-fold cross-validation is 88% that is less than 2% of the actual prediction. In the same way, the SVM classifier has 93% accuracy, which is 3% more than the kNN. For training, the AUC is decreased by 3%.

Finally, for the Random Forest (RF), we got 95% accuracy that is the highest among all classifiers used in this model, as shown in Table 1. The F1 score of fake and legitimate accounts is higher among all classifiers with 92% and 96%, respectively. The precision of fake accounts is also highest among all classifiers with 96%, whereas Recall of legitimate accounts is 98% which is highest among all. The training and testing of the AUC-ROCs are also highest among all classifiers with 99% and 96% as shown in Fig. 2. In ten-fold cross-validation, the accuracy is 88%, which is 7% less than predicted values.

In another experiment, we used the same model on ICC benchmark dataset to check the model's accuracy. Here, we selected only three classifiers as k-Nearest Neighbor, Decision Tree, and Random Forest. We exclude the SVM because it takes

Table 2 Classifiers performance report on ICC dataset

Method	Accuracy	Fake accounts (1)			Legitimate accounts (0)		
		Precision	Recall	F1 Score	Precision	Recall	F1 Score
*k*NN	0.95	0.52	0.13	0.21	0.96	0.99	0.97
DT	1.00	1.00	1.00	1.00	1.00	1.00	1.00
RF	1.00	1.00	1.00	1.00	1.00	1.00	1.00

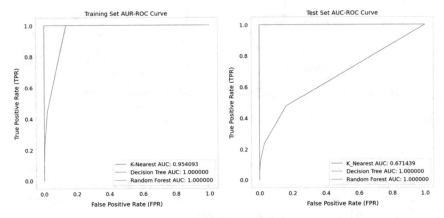

Fig. 3 Training and testing AUC-ROC curve on ICC dataset

much time, and the multinomial naive Bayes classifier had low accuracy. Table 2 demonstrates the performance of these three classifiers in terms of Accuracy, Precision, Recall, and F1 Score. By observing Table 2, We can say that DT and RF perform well with 100% accuracy, whereas *k*-Nearest Neighbor has 95% accuracy. *k*NN has a very low precision, Recall, and F1 score for fake accounts detection compared to Random Forest and Decision Tree. For training and testing, the AUC-ROCs of these classifiers are shown in Fig. 3, from where we can observe that for training, the AUC-ROC of *K*NN is 95%, But its testing is only 67%. Training and testing of DT and RF are outstanding with 100%, which is quite effective on the ICC dataset.

5 Conclusion

With the rise of digitization and social media, it is challenging to identify fake accounts because it has similar characteristics as legitimate accounts. Hence, we have proposed a fake accounts detector using supervised machine learning algorithms to deal with fake accounts. We used the five most popular supervised machine learning algorithms for classification. We used publicly available Twitter and ICC datasets to train and test the proposed model. We performed 10-fold cross-validation for validation, where we got similar results as classifiers did. Among these classifiers, Random Forest worked well with 95% accuracy on the Twitter dataset, whereas Random Forest and Decision Tree performed outstandingly with 100% accuracy on the ICC dataset. We also approximated AUC-ROC for the sensitivity of the results. As future work, we are exploring more effective attributes of social network accounts, including images features in tweets.

References

1. Elflein J, Puri-Mirza A, Sapun James P, Briony Key C, Kohl AK (2021) https://www.statista.com/statistics/272014/global-social-networks-ranked-by-number-of-users/, Aug 2021
2. Egele M, Stringhini G, Kruegel C, Vigna G (2015) Towards detecting compromised accounts on social networks. IEEE Trans Dependable Secure Comput 14(4):447–460
3. Rahman MS, Haldar S, Uddin MA, Acharjee UK (2021) An efficient hybrid system for anomaly detection in social networks. Cybersecurity 4:1–10
4. Meligy AM, Ibrahim HM, Torky MF (2017) Identity verification mechanism for detecting fake profiles in online social networks. Int J Comput Netw Inf Secur, 9(1):31–39
5. Jia J, Wang B, Zhenqiang Gong N (2017) Random walk based fake account detection in online social networks. In: Proceedings of 47th annual IEEE/IFIP international conference dependable systems and networks. IEEE, pp 273–284
6. Ilias L, Roussaki I (2021) Detecting malicious activity in twitter using deep learning techniques. Appl Soft Comput 107:107360
7. Cagatay Akyon F, Esat Kalfaoglu M (2019) Instagram fake and automated account detection. In: Proceedings of innovations in intelligent systems and applications conference (ASYU). IEEE, pp 1–7
8. Breuer A, Eilat R, Weinsberg U (2020) Friend or faux: graph-based early detection of fake accounts on social networks. In: Proceedings of the web conference, pp 1287–1297
9. Shu K, Zhou X, Wang S, Zafarani R, Liu H (2019) The role of user profiles for fake news detection. In: Proceedings of IEEE/ACM international conference advances in social networks analysis and mining, pp 436–439
10. Shu K, Wang S, Liu H (2018) Understanding user profiles on social media for fake news detection. In: Proceedings of IEEE conference on multimedia information processing and retrieval (MIPR). IEEE, pp 430–435
11. Van Der Walt E, Eloff J (2018) Using machine learning to detect fake identities: Bots vs. humans. IEEE Access 6:6540–6549
12. Wanda P, Jin Jie H (2020) DeepProfile: finding fake profile in online social network using dynamic CNN. J Secur Appl 52:102465
13. Fazil M, Abulaish M (2018) A hybrid approach for detecting automated spammers in twitter. IEEE Trans Info Forensics Secur, 13(11)
14. Chen C, Zhang J, Chen X, Xiang Y, Zhou W (2015) 6 million spam tweets: a large ground truth for timely twitter spam detection. In: Proceedings of IEEE international conference on communications (ICC). IEEE, pp 7065–7070

Peak Detector Circuits for Safeguarding Against Fault Injection Attacks

Shaminder Kaur, Sandhya Sharma⬤, Monika Parmar⬤,
and Lipika Gupta⬤

1 Introduction

Over the past 15 years, cybersecurity attacks have increased dramatically [1]. Hardware security is sub branch of cyber security which comprises 14% of attacks as shown in Fig. 1. These attacks are also known as physical attacks/hardware attacks. Hardware security is a field which deals with such types of attacks. It is a field, which challenges a hardware designer to build an efficient design and protect it against attacks known as physical attacks. It has been observed that security of cryptographic devices is threatened by these attacks, which further leaks secret information through leakages such as power consumption of embedded device, temperature, electromagnetic (EM) emanation, etc., which are known as side-channel leakages. Physical attacks/hardware attacks have become a worry since most recent couple of years and brought about serious examination exertion to create appropriate countermeasures that can at any rate make attacks more troublesome and tedious to perform [2–4]. Hardware attacks can be categorized as of two types: side-channel attacks and fault injection attacks. Side-channel attacks can be defined as attacks where the security of cryptographic devices is threatened by these attacks, which further leaks

S. Kaur · L. Gupta (✉)
Chitkara University Institute of Engineering and Technology, Chitkara University, Punjab, India
e-mail: lipikagupta21@gmail.com; gupta.lipika@chitkara.edu.in

S. Kaur
e-mail: shaminder.kaur@chitkara.edu.in

S. Sharma · M. Parmar
Chitkara University School of Engineering and Technology, Chitkara University, Solan, Himachal Pradesh, India
e-mail: sandhya.sharma@chitkarauniversity.edu.in

M. Parmar
e-mail: monika.parmar@chitkarauniversity.edu.in

© The Author(s), under exclusive license to Springer Nature Singapore Pte Ltd. 2023   297
R. Agrawal et al. (eds.), *International Conference on IoT, Intelligent Computing and Security*, Lecture Notes in Electrical Engineering 982,
https://doi.org/10.1007/978-981-19-8136-4_25

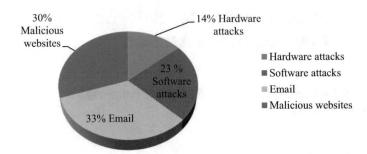

Fig. 1 Major sources of cyber-physical attacks

secret information through leakages known as side-channel leakage such as power consumption of device, temperature, electromagnetic emanation, etc. Second category of hardware attacks known as fault injection attacks proves to be more damaging than the first one. It is known as fault injection attacks. Fault injection attacks are extremely dangerous in breaking security of various cryptosystems [5]. It is a known fact that every embedded device is vulnerable to such attacks.

In contrast to the side-channel attacks, fault injection attacks actively attack a device and its operation by modifying physical/environmental parameters. Thus, it is possible to enforce faulty behavior or wrong conditions in the attacked device. As a consequence, induced faults typically result in erroneous calculations. This enables an attacker to gain details about the implementation or to even extract secret information [6]. Fault injection is a kind of active attack method which requires expertise to conduct it [7–12]. The main idea behind this attack is to produce faulty behavior and that behavior may lead to some erroneous changes in embedded device. Fault is produced either by tampering clock signal, power supply line, or optical injection as shown in the figure. These tampering of external signal may produce some glitch, and if the adversary is successful to produce glitch at the location he intends to, then he may be successful in obtaining the secret key [13, 14].

Categories of fault injection attacks: There are various fault injection techniques discussed in literature, viz: *voltage glitching, tampering with clock pin, EM disturbances, laser attacks* as depicted in Fig. 2. All these techniques aim at producing corrupted output. The approach we have used for generation of fault injection attacks is voltage glitching/supply glitching.

Variations in supply voltage: Voltage glitching/power supply glitching is one the most exploited factors to induce faults. The power is either underfed or overfed which further affects the gate delays. Gate delays create the setup and hold time violation, and faults are induced [15].

Hardware/Software/Simulation-based approaches for detection of attacks: Approach utilized by "hardware-controlled strategies" is it uses external equipment to inject fault deep within the hardware so that fault may propagate to the sufferer

Fig. 2 Injection attacks' categorization

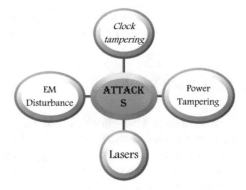

software program. Related mechanism is discovered via different platforms (software program platform) that is flowing at the hardware tool [16].

On the contrary, "software-managed" strategy utilizes a unsafe/adverse programming (i.e., shortcoming infusion and control programming), which runs on a comparable device stage because the goal programming does. It furthers hampers the external situations of the embedded hardware to supply defective output.

Another category is "simulation-based totally method." This paper makes use of simulation-based totally approach because of its several advantages: no need of prototype, less complicated, less expensive. Lastly, it lets in early analysis of detection of facet channel assaults. This trait of this method makes it one of the satisfactory options for doing early analysis of attacks in order to construct suitable countermeasures in a well-timed manner.

In this research work, we have presented simulation-based detection approach that successfully detects negative spike and positive spike fault injection attacks. Existing detection approaches either detect positive or negative fault injection attacks. Less literature is given on detection circuits that detect both types of attacks. Through this study, we tried to fill the existing research gaps. This experimental study will give upcoming engineers a platform to evaluate and analyze detection methods deeply so that suitable countermeasures can be built against them in an efficient and timely manner.

2 Contributions

- Analysis of the challenges associated with voltage glitch as recently emerged fault injection attack which further affects the safety of embedded devices.
- Implementation of detector circuit that detects positive as well as negative voltage glitch attacks which is less studied in existing literature.
- Simulation-based detection approach is used in this paper because of its several advantages: no need of prototype, much less complicated, less pricey. It further permits early analysis of detection of assaults. This trait of this approach makes

it one of the options for doing early evaluation of attacks in an effort to build suitable countermeasures in a well-timed manner.

3   Related Work

Various authors have either worked on software-based techniques or hardware-based techniques as discussed below. Limited authors worked on simulation-based approach which proves to be best approaches in doing prior evaluation. Through this study, we are the first one to investigate practical simulation-based approach on timing attacks against embedded devices. An effort has been made to distinguish between numerous strategies based on factors such as: price, overall performance, complexity, threat of harm, controllability, portability. Based on this information, hardware engineers can decide which technique to experiment with in order to get suitable countermeasures against side-channel attacks.

Gomina et al. in [17] stated that supply voltage variation poses a serious threat to embedded devices. Attacker uses this approach since it requires less expertise to perform it. Authors in this paper have analyzed various mechanisms associated with such kinds of attacks. They also characterized various detection techniques such as critical path replica (CPR), tunable replica circuit (TRC), path delay detection techniques. It concluded that the properties related to timing of gates are very sensitive to these attacks. Among the three detection techniques, CRC approach was the best in terms of area and complexity.

Raychowdhury et al. in [16] discussed about error detection and correction methods in microprocessor core. Authors focused on effects of voltage droop, i.e., sudden variation in power supply since it is very difficult to mitigate its effects. Techniques and algorithms are designed to study effects of voltage droop. Tunable replica circuit (TRC) is designed to study effects of fast and sudden variation in power supply line. The methods used in this paper require calibration which is quite complex process, and it furthers increases the processing time.

Bowman et al. in [18] presented both error detection and correction techniques against voltage drop. Voltage droops are fast/sudden change in voltage signal which are difficult to mitigate. To do analysis, two circuits are designed: TRC and error detection sequential (EDS) circuits. Detection circuits designed detect both kind of variations, i.e., fast and slow variations. Drawback of these detection circuits is that they require post-silicon calibration which further increases its time. Comparison between two techniques is given in terms of area overhead, power overhead, minimum buffer delay insertion overhead, error detection, and accumulation area overhead.

4 Glitch Detector Circuit

Glitch detector circuit: Glitch detector is a device that detects glitch in a signal. Glitch consumes lot of power consumption; hence, they are necessary to detect and then removed from the signal in order to get uninterrupted output [19]. We have designed a circuit that detects glitch successfully.

4.1 Simulated Response of Glitch Detector Circuit

Figures 3 and 5 represent circuit diagrams of glitch detector circuit. This detector circuit comprises two parts. The first part is defined as glitch generation circuit which is responsible for generation of glitches so that they can further be detected. Next stage/part comprises the circuit known as glitch detection circuit which plays a very important role in designing of embedded circuits. This part is responsible for detection of glitches so that these glitches further may not harm the circuit or produce erroneous outputs. Here in our circuit, we have shown that first part generates glitch in stable power rail. As shown in Fig. 4, the first line shows some spikes. It represents that these are positive as well as negative glitches in stable power rail which should be removed once detected. After this, detection part is responsible for detecting them. Simulated results shown in Figs. 4 and 6 represent detection of glitch in power rail.

In this research article, we have designed op-amp-based glitch detector circuits which detect glitches in the stable power rail as already represented in the figures below. Second line shown in Figs. 4 and 6 shows detection of glitch. Designed circuits detect both types of glitches, viz: positive spike glitch and negative spike glitch. Existing detector circuit detects negative glitch attack successfully. Less literature was given on detector circuits which detects both types of attacks. It is desirable to

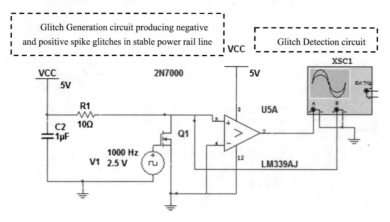

Fig. 3 Fault injection detector circuit 1

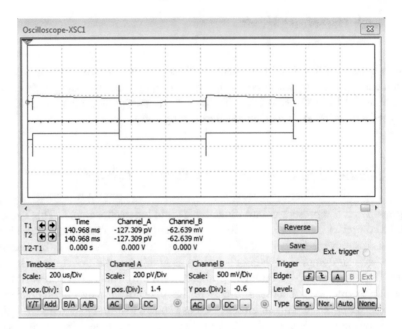

Fig. 4 Simulated response of detector circuit 1

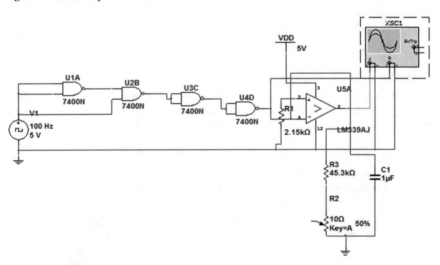

Fig. 5 Fault injection detector circuit 2

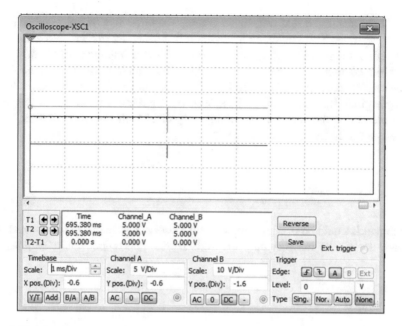

Fig. 6 Simulated response of detector circuit 2

detect negative as well as positive glitch attacks. Through this research article, we have tried to fill the existing gap that detects either positive or negative glitches. Our circuit detects both types of glitches: positive as well as negative.

Proposed detector is designed using op-amp which works as peak detector. The values of capacitor $C = 1 \, \mu F$ and resistor $R = 10 \, \Omega$. Glitch generator circuit generates glitches/spikes using N-channel metal–oxide–semiconductor field-effect transistor (MOSFET) depicted as Q_1 in Fig. 3. We need to generate narrow glitches so this experimental study used low-power MOSFET with the specifications as: N-channel enhancement-mode vertical CMOS FET 0.2 A; 60 V; 5 Ω. If broader glitches are required, then high-power MOSFET may be required. Since our experimental analysis required generation of narrow glitches/spikes, so we used low-power MOSFET. Second part of the proposed circuit is glitch detection circuit. Op-amp as peak detector acting as glitch detector circuit has been used. Whenever there is a peak in input power rail line, it will be detected by peak detecting circuit.

5 Conclusion

Simulation-based approach of generation of glitches to analyze performance of proposed glitch detector circuit is presented. Simulation-based methods prove to be quit economical, and they prove to be one of the best strategies for doing early analysis of attacks so that well-timed countermeasures may be built in the future. In this paper, we have built combined circuit that generates glitches first and then it further detects them. Detection of glitches plays a very important role in designing of embedded circuits. Once detected, only then they can be removed successfully or some suitable countermeasures can be built against them. Op-amp as peak detector acting as glitch detector circuit has been used. Also, strive has been made in providing deep perception into the new and rising subject of fault injection assaults which impacts the security of embedded devices.

6 Future Work

Future work would be to build suitable countermeasures against glitches which are already detected so that they may not produce erroneous outputs/faulty outputs.

References

1. Bar-El H, Choukri H, Naccache D, Tunstall M, Whelan C (2006) The sorcerer's apprentice guide to fault attacks. Proc IEEE 94(2):370–382
2. Karaklajić D, Schmidt JM, Verbauwhede I (2013) Hardware designer's guide to fault attacks. IEEE Trans Very Large Scale Integr (VLSI) Syst 21(12):2295–306
3. Skorobogatov S (2018) Hardware security implications of reliability, remanence, and recovery in embedded memory. J Hardware Syst Secur 2(4):314–321
4. Lyu Y, Mishra P (2018) A survey of side-channel attacks on caches and countermeasures. J Hardware Syst Secur 2(1):33–50
5. Kaur S, Singh B, Kaur H (2021) Silicon based security for protection against hardware vulnerabilities. Silicon. https://doi.org/10.1007/s12633-021-00989-6
6. Oswald D (2013) Implementation attacks: from theory to practice. Bochum
7. Mangard S, Oswald E, Popp T (2008) Power analysis attacks: revealing the secrets of smart cards. Springer Science & Business Media
8. Dubeuf J, Hely D, Beroulle V (2016) ECDSA passive attacks, leakage sources, and common design mistakes. ACM Trans Des Autom Electron Syst (TODAES) 21(2):1–24
9. Yen SM, Joye M (2000) Checking before output may not be enough against fault-based cryptanalysis. IEEE Trans Comput 49(9):967–970
10. Amiel F, Villegas K, Feix B, Marcel L (2008) Passive and active combined attacks: combining fault attacks and side channel analysis. In: Workshop on fault diagnosis and tolerance in cryptography (FDTC 2008), 10 Sept 2008. IEEE, pp 92–102
11. Clavier C, Feix B, Gagnerot G, Roussellet M (2010) Passive and active combined attacks on AES combining fault attacks and side channel analysis. In: Workshop on fault diagnosis and tolerance in cryptography, 21 Aug 2010. IEEE, pp 10–19
12. Roche T, Lomné V, Khalfallah K (2011) Combined fault and side-channel attack on protected implementations of AES. In: International conference on smart card research and advanced applications, 14 Sept 2011. Springer, Leuven, pp 65–83
13. Schmidt JM, Tunstall M, Avanzi R, Kizhvatov I, Kasper T, Oswald D (2010) Combined implementation attack resistant exponentiation. In: International conference on cryptology and information security in Latin America, 8 Aug 2010. Springer, Puebla, pp 305–322

14. Guilley S, Sauvage L, Danger JL, Selmane N, Pacalet R (2008) Silicon-level solutions to counteract passive and active attacks. In: 5th Workshop on fault diagnosis and tolerance in cryptography, 10 Aug 2008. IEEE, pp 3–17
15. Korak T, Hoefler M (2014) On the effects of clock and power supply tampering on two microcontroller platforms. In: Workshop on fault diagnosis and tolerance in cryptography, 23 Sept 2014. IEEE, pp 8–17
16. Raychowdhury A, Tschanz J, Bowman K, Lu SL, Aseron P, Khellah M, Geuskens B, Tokunaga C, Wilkerson C, Karnik T, De V (2011) Error detection and correction in microprocessor core and memory due to fast dynamic voltage droops. IEEE J Emerg Sel Top Circ Syst 1(3):208–217
17. Gomina K, Rigaud JB, Gendrier P, Candelier P, Tria A (2014) Power supply glitch attacks: design and evaluation of detection circuits. In: International symposium on hardware-oriented security and trust (HOST), 6 May 2014. IEEE, pp 136–141
18. Bowman KA, Tschanz JW, Lu SL, Aseron PA, Khellah MM, Raychowdhury A, Geuskens BM, Tokunaga C, Wilkerson CB, Karnik T, De VK (2010) A 45 nm resilient microprocessor core for dynamic variation tolerance. IEEE J Solid-State Circ 46(1):194–208
19. Kaur S, Singh B, Kaur H, Gupta L (2021) Injecting power attacks with voltage glitching and generation of clock attacks for testing fault injection attacks. In: Singh PK, Noor A, Kolekar MH, Tanwar S, Bhatnagar RK, Khanna S (eds) Evolving technologies for computing, communication and smart world. Lecture notes in electrical engineering, vol 694. Springer, Singapore. https://doi.org/10.1007/978-981-15-7804-5_3

An Intuitionistic Fuzzy Approach to Analysis Financial Risk Tolerance with MATLAB in Business

Vinesh Kumar, Sandeep Kumar Gupta⬭, Rohit Kaushik,
Subhask Kumar Verma, and Olena Sakovska

1 Introduction

The concept of fuzzy sets was introduced by Zadeh [1] in 1965. Later in 1983, Atanassov [2] introduced the concept of intuitionistic fuzzy sets as a generalization of fuzzy sets. There are so many applications of intuitionistic fuzzy sets in different fields. In this paper, we consider the risk tolerance capacity in business [3]. Risk tolerance relates to the amount of market risk such as volatility, market ups and downs which can be tolerated by an investor. The financial service institutions aim to help the financial planner build a portfolio of investment that the investor will be comfortable with over a long period. Every investor needs to measure their risk tolerance (RT) before choosing their investment.

In this research paper, we present a model of an investor's risk tolerance capacity which depends on his/her current income (CI) and total net worth (TNW).

V. Kumar
Phonics Group of Institutions, Roorkee, India

S. K. Gupta (✉) · S. K. Verma
AMET University, Chennai, India
e-mail: skguptabhu@gmail.com

R. Kaushik
G. L. Bajaj Institute of Technology and Management, Greater Noida, India

S. K. Verma
School of Business Management, Noida International University, Greater Noida, India

O. Sakovska
Uman National University of Horticulture, Uman, Ukraine

© The Author(s), under exclusive license to Springer Nature Singapore Pte Ltd. 2023   307
R. Agrawal et al. (eds.), *International Conference on IoT, Intelligent Computing
and Security*, Lecture Notes in Electrical Engineering 982,
https://doi.org/10.1007/978-981-19-8136-4_26

2 Preliminaries

In this section, we review some elementary concepts related to this paper.

Definition 2.1 [1] Let X be a non-empty set. A fuzzy set $\widetilde{A}$ drawn from X is defined as

$$\widetilde{A} = \{\langle x, \mu_{\widetilde{A}}(x)\rangle : x \in X\}, \tag{2.1}$$

where $\mu_{\widetilde{A}}(x) : X \to [0, 1]$ is a membership function of the fuzzy set $\widetilde{A}$.

Definition 2.2 [2] An Intuitionistic fuzzy set $\widetilde{A}^i$ on X is given by

$$\widetilde{A}^i = \{\langle x, \mu_{\widetilde{A}^i}(x), \nu_{\widetilde{A}^i}(x)\rangle : x \in X\}, \tag{2.2}$$

where $\mu_{\widetilde{A}^i}(x) : X \to [0, 1]$ and $\nu_{\widetilde{A}^i}(x) : X \to [0, 1]$ such that $0 \leqslant \mu_{\widetilde{A}^i}(x) + \nu_{\widetilde{A}^i}(x) \leqslant 1, \forall x \in X$.

The value of $\mu_{\widetilde{A}^i}(x)$ is a lover bond on the degree of membership of x derived from the evidence for x, and $\nu_{\widetilde{A}^i}(x)$ is a lower bond on the negation of x derived from the evidence x.

We call $\pi_{\widetilde{A}^i}(x) = 1 - \mu_{\widetilde{A}^i}(x) - \nu_{\widetilde{A}^i}(x), x \in X$ to be hesitation or the intuitionistic index of x in $\widetilde{A}^i$. This index indicates the lack of knowledge to fact whether the element belongs to the set or not.

For two intuitionistic fuzzy sets, $\widetilde{A}^i$ and $\widetilde{A}^i$ in X it hold that

- $\widetilde{A}^i \cap \widetilde{B}^i = \{\langle x, \min(\mu_{\widetilde{A}^i}(x), \mu_{\widetilde{B}^i}(x)), \max(\nu_{\widetilde{A}^i}(x), \nu_{\widetilde{B}^i}(x))\rangle : x \in X\}$
- $\widetilde{A}^i \cup \widetilde{B}^i = \{\langle x, \max(\mu_{\widetilde{A}^i}(x), \mu_{\widetilde{B}^i}(x)), \min(\nu_{\widetilde{A}^i}(x), \nu_{\widetilde{B}^i}(x))\rangle : x \in X\}$
- $\widetilde{A}^i \subset \widetilde{B}^i$ iff $\forall x \in X, (\mu_{\widetilde{A}^i}(x)) \leq (\mu_{\widetilde{B}^i}(x))$ and $(\nu_{\widetilde{A}^i}(x)) \geq (\nu_{\widetilde{B}^i}(x))$

Definition 2.3 [4] An intuitionistic fuzzy set is said to be an intuitionistic fuzzy number if it has the following properties:

1. It is an intuitionistic fuzzy subset of the real line.
2. It is normal that is, there is some $x_0 \in R$ such that $\mu_{\widetilde{A}^i}(x_0) = 1$ and $\nu_{\widetilde{A}^i}(x_0) = 0$.
3. It is convex for the membership function $\mu_{\widetilde{A}^i}(x)$ that is

$$\mu_{\widetilde{A}^i}(\lambda x_1 + (1 - \lambda x_2)) \geq \min(\mu_{\widetilde{A}^i}(x_1), \mu_{\widetilde{A}^i}(x_2)) \forall x_1, x_2 \in R, \lambda \in [0, 1]$$

4. It is concave for non-membership function $\nu_{\widetilde{A}^i}(x)$ that is

$$\nu_{\widetilde{A}^i}(\lambda x_1 + (1 - \lambda x_2)) \leq \max(\nu_{\widetilde{A}^i}(x_1), \nu_{\widetilde{A}^i}(x_2)) \forall x_1, x_2 \in R, \lambda \in [0, 1]$$

Definition 2.4 [5] Triangular intuitionistic fuzzy number (TIFN) $\widetilde{A}^i$ is denoted by

$$\widetilde{A}^i = \langle(a_1, a_2, a_3)(a_1', a_2, a_3')\rangle$$

such that $a_1' \leq a_1 \leq a_2 \leq a_3 \leq a_3'$, where the membership function

$$\mu_{\tilde{A}^i}(x) = \begin{cases} \frac{x-a_1}{a_2-a_1} & \text{for } a_1 \leq x \leq a_2 \\ \frac{a_3-x}{a_3-a_2} & \text{for } a_2 \leq x \leq a_3 \\ 0 & \text{otherwise} \end{cases} \tag{2.3}$$

and non-membership function

$$\nu_{\tilde{A}^i}(x) = \begin{cases} \frac{a_2-x}{a_2-a_1'} & \text{for } a_1' \leq x \leq a_2 \\ \frac{x-a_2}{a_3'-a_2} & \text{for } a_2 \leq x \leq a_3' \\ 1 & \text{otherwise} \end{cases} \tag{2.4}$$

3 Components of the Proposed Intuitionistic Fuzzy Inference System (IFIS)

The basic components of the proposed intuitionistic fuzzy inference system are shown in Fig. 1.

In this intuitionistic fuzzy inference system (IFIS), we consider two input variables as current income (CI), total net worth (TNW) and one output variable as risk tolerance (RT) level. In this problem, we consider five linguistic variables as VL: very low, L: low, M: medium, H: high, and VH: very high. All linguistic variables are taken as triangular intuitionistic fuzzy numbers.

Current income (CI) = {VL, L, M, H, VH}
Total net worth (TNW) = {VL, L, M, H, VH}
Risk tolerance (RT) = {VL, L, M, H, VH}

The range of input and output variables taken as [3]

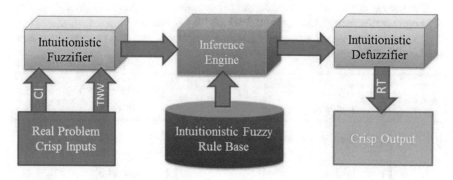

Fig. 1 Intuitionistic fuzzy inference system

$$CI = \{x * 10^3 : 0 \leqslant x \leqslant 100\}, \quad TNW = \{y * 10^5 : 0 \leqslant y \leqslant 100\},$$
$$RT = \{z : 0 \leqslant z \leqslant 100\},$$

where x, y are real numbers in \$ and z represent the risk % between 0 and 100.

4 Fuzzification of Input and Output Variables

The membership and non-membership functions of input variable current income (CI) are represented as

$\mu_{VL}(x) = \begin{cases} \frac{12-x}{12-0} & \text{for } 0 \leq x \leq 12 \\ 0 & \text{otherwise,} \end{cases}$	$\nu_{VL}(x) = \begin{cases} \frac{x-0}{15-0} & \text{for } 0 \leq x \leq 15 \\ 1 & \text{otherwise,} \end{cases}$
$\mu_{L}(x) = \begin{cases} \frac{x-10}{20-10} & \text{for } 10 \leq x \leq 20 \\ \frac{40-x}{40-20} & \text{for } 20 \leq x \leq 40 \\ 0 & \text{otherwise,} \end{cases}$	$\nu_{L}(x) = \begin{cases} \frac{20-x}{20-8} & \text{for } 8 \leq x \leq 20 \\ \frac{x-20}{42-20} & \text{for } 20 \leq x \leq 42 \\ 1 & \text{otherwise,} \end{cases}$
$\mu_{M}(x) = \begin{cases} \frac{x-30}{50-30} & \text{for } 30 \leq x \leq 50 \\ \frac{70-x}{70-50} & \text{for } 50 \leq x \leq 70 \\ 0 & \text{otherwise,} \end{cases}$	$\nu_{M}(x) = \begin{cases} \frac{50-x}{50-28} & \text{for } 28 \leq x \leq 50 \\ \frac{x-50}{73-50} & \text{for } 50 \leq x \leq 73 \\ 1 & \text{otherwise,} \end{cases}$
$\mu_{H}(x) = \begin{cases} \frac{x-60}{75-60} & \text{for } 60 \leq x \leq 75 \\ \frac{85-x}{85-75} & \text{for } 75 \leq x \leq 85 \\ 0 & \text{otherwise,} \end{cases}$	$\nu_{H}(x) = \begin{cases} \frac{75-x}{75-58} & \text{for } 58 \leq x \leq 75 \\ \frac{x-75}{87-75} & \text{for } 75 \leq x \leq 87 \\ 1 & \text{otherwise,} \end{cases}$
$\mu_{VH}(x) = \begin{cases} \frac{x-80}{100-80} & \text{for } 80 \leq x \leq 100 \\ 0 & \text{otherwise,} \end{cases}$	$\nu_{VH}(x) = \begin{cases} \frac{100-x}{100-78} & \text{for } 78 \leq x \leq 100 \\ 1 & \text{otherwise.} \end{cases}$

The membership and non-membership functions of input variable current income (CI) represented using MATLAB are such as Figs. 2 and 3.

The membership and non-membership functions of the another input variable total net worth (TNW) are represented as

$\mu_{VL}(x) = \begin{cases} \frac{15-x}{15-0} & \text{for } 0 \leq x \leq 15 \\ 0 & \text{otherwise,} \end{cases}$	$\nu_{VL}(x) = \begin{cases} \frac{x-0}{17-0} & \text{for } 0 \leq x \leq 17 \\ 1 & \text{otherwise,} \end{cases}$
$\mu_{L}(x) = \begin{cases} \frac{x-8}{25-8} & \text{for } 8 \leq x \leq 25 \\ \frac{35-x}{35-25} & \text{for } 25 \leq x \leq 35 \\ 0 & \text{otherwise,} \end{cases}$	$\nu_{L}(x) = \begin{cases} \frac{25-x}{25-8} & \text{for } 8 \leq x \leq 25 \\ \frac{x-25}{36-25} & \text{for } 25 \leq x \leq 36 \\ 1 & \text{otherwise,} \end{cases}$

(continued)

(continued)

$$\mu_M(x) = \begin{cases} \frac{x-30}{45-30} & \text{for } 30 \leq x \leq 45 \\ \frac{65-x}{65-45} & \text{for } 45 \leq x \leq 65 \\ 0 & \text{otherwise,} \end{cases} \qquad \nu_M(x) = \begin{cases} \frac{45-x}{45-28} & \text{for } 28 \leq x \leq 45 \\ \frac{x-45}{67-45} & \text{for } 45 \leq x \leq 67 \\ 1 & \text{otherwise,} \end{cases}$$

$$\mu_H(x) = \begin{cases} \frac{x-50}{70-50} & \text{for } 50 \leq x \leq 70 \\ \frac{80-x}{80-70} & \text{for } 70 \leq x \leq 80 \\ 0 & \text{otherwise,} \end{cases} \qquad \nu_H(x) = \begin{cases} \frac{70-x}{70-48} & \text{for } 48 \leq x \leq 70 \\ \frac{x-70}{82-70} & \text{for } 70 \leq x \leq 82 \\ 1 & \text{otherwise,} \end{cases}$$

$$\mu_{VH}(x) = \begin{cases} \frac{x-75}{100-75} & \text{for } 75 \leq x \leq 100 \\ 0 & \text{otherwise,} \end{cases} \qquad \nu_{VH}(x) = \begin{cases} \frac{100-x}{100-73} & \text{for } 73 \leq x \leq 100 \\ 1 & \text{otherwise.} \end{cases}$$

The membership and non-membership functions of the output variable risk tolerance (RT) are represented as

$$\mu_{VL}(x) = \begin{cases} \frac{20-x}{20-0} & \text{for } 0 \leq x \leq 20 \\ 0 & \text{otherwise,} \end{cases} \qquad \nu_{VL}(x) = \begin{cases} \frac{x-0}{22-0} & \text{for } 0 \leq x \leq 22 \\ 1 & \text{otherwise,} \end{cases}$$

$$\mu_L(x) = \begin{cases} \frac{x-15}{30-15} & \text{for } 15 \leq x \leq 30 \\ \frac{40-x}{40-30} & \text{for } 30 \leq x \leq 40 \\ 0 & \text{otherwise,} \end{cases} \qquad \nu_L(x) = \begin{cases} \frac{30-x}{30-13} & \text{for } 13 \leq x \leq 30 \\ \frac{x-30}{42-30} & \text{for } 30 \leq x \leq 42 \\ 1 & \text{otherwise,} \end{cases}$$

$$\mu_M(x) = \begin{cases} \frac{x-35}{45-35} & \text{for } 35 \leq x \leq 45 \\ \frac{60-x}{60-45} & \text{for } 45 \leq x \leq 60 \\ 0 & \text{otherwise,} \end{cases} \qquad \nu_M(x) = \begin{cases} \frac{45-x}{45-32} & \text{for } 32 \leq x \leq 45 \\ \frac{x-45}{63-45} & \text{for } 45 \leq x \leq 63 \\ 1 & \text{otherwise,} \end{cases}$$

$$\mu_H(x) = \begin{cases} \frac{x-50}{70-50} & \text{for } 50 \leq x \leq 70 \\ \frac{80-x}{80-70} & \text{for } 70 \leq x \leq 80 \\ 0 & \text{otherwise,} \end{cases} \qquad \nu_H(x) = \begin{cases} \frac{70-x}{70-48} & \text{for } 48 \leq x \leq 70 \\ \frac{x-70}{85-70} & \text{for } 70 \leq x \leq 85 \\ 1 & \text{otherwise,} \end{cases}$$

$$\mu_{VH}(x) = \begin{cases} \frac{x-75}{100-75} & \text{for } 75 \leq x \leq 100 \\ 0 & \text{otherwise,} \end{cases} \qquad \nu_{VH}(x) = \begin{cases} \frac{100-x}{100-72} & \text{for } 72 \leq x \leq 100 \\ 1 & \text{otherwise.} \end{cases}$$

5 Intuitionistic Fuzzy Inference Rules

In this model, we consider two inputs heaving five linguistic variables each. There are total twenty five, if and then rules that are used for intuitionistic fuzzy inference shown in the table below

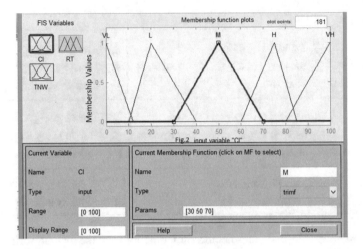

Fig. 2 Input variable "CI"

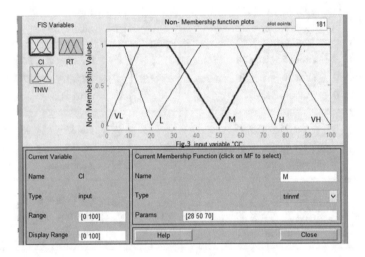

Fig. 3 Input variable "CI"

CI	TNW				
	VL	L	M	H	VH
VL	VL	VL	L	L	M
L	VL	L	L	M	M
M	L	L	M	M	H
H	L	M	M	H	VH
VH	M	M	H	VH	VH

Rule (1): If (CI is VL) and (TNW is VL), then (RT is VL)
Rule (2): If (CI is VL) and (TNW is L), then (RT is VL)
...
Rule (25): (CI is VH) and (TNW is VH), then (RT is VH).

6 Defuzzification Using Centroid Method (COA)

To execute our intuitionistic fuzzy inference model, we are choosing the value of input variables randomly. Let current income (CI) = 40 and total net worth (TNW) = 70 (Fig. 4).

Using if and then rules we get the result risk tolerance (RT) is 46.9 for membership function and 50.4 for non-membership function (Fig. 5).

7 Conclusion

An intuitionistic fuzzy control system provides a flexible model to elaborate the uncertainty and vagueness involved in real-world problems. In this paper, we proposed to develop the non-membership function and defuzzification using MATLAB and applied it in a business model to calculate the risk tolerance based on current income and total net worth. In this paper, we choose a random value of current income (CI) as 40 and total net worth (TNW) as70 of a businessman. We find that acceptance of risk tolerance (RT) is 46.9% and non-acceptance of risk tolerance is 50.4%.The remaining 2.7% is doubtful it may be not maybe acceptance of risk tolerance. We observed that the non-membership functions may improve the performance of an intuitionistic fuzzy control system.

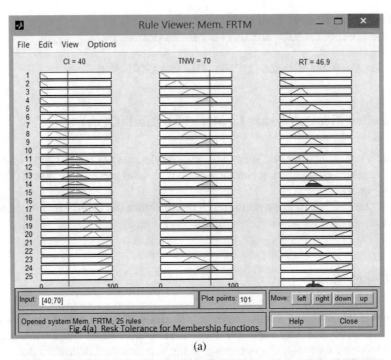

(a)

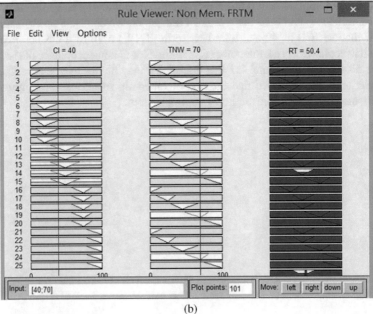

(b)

Fig. 4 **a** Risk tolerance for membership function. **b** Risk tolerance for non-membership function

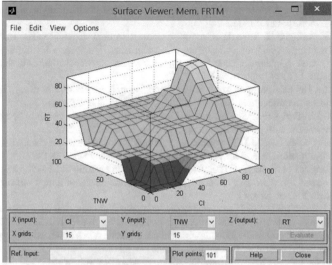

(a) Risk Tolarance for Membership Functions

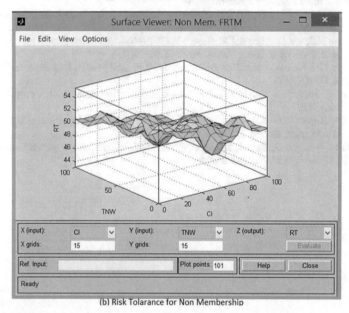

(b) Risk Tolarance for Non Membership

Fig. 5 **a** Risk tolerance for membership function. **b** Risk tolerance for non-membership function

References

1. Zadeh LA (1965) Fuzzy sets. Inf Control 8:338–353
2. Atanassov KT (1983) Intuitionistic fuzzy sets. VII ITKR's session, Sofia. Deposed in Central Science—Technology Library of Bulgaria Academy of Science, 1697/84 (in Bulgarian)

3. Bojadziev G, Bojadziev M (2007) Fuzzy logic for business, finance and management. World Scientific Publishing Co Pte Ltd.
4. Burillo P, Bustince H, Mohedano V (1994) Some definition of intuitionistic fuzzy number. Fuzzy based expert systems, Fuzzy Bulgarian enthusiasts, 28–30 Sept, Sofia, Bulgaria
5. Rajarajeswari P, Menaka G (2017) A new approach for ranking of octagonal intuitionistic fuzzy numbers. Int J Fuzzy Logic Syst (IJFLS) 7(2)
6. Hamdy M, Helmy S, Magdy M (2020) Design of adaptive intuitionistic fuzzy controller for synchronisation of uncertain chaotic systems. CAAI Trans Intell Technol 1–10
7. Yu C-M, Lin K-P, Liu G-S, Chang C-H (2020) A parameterized intuitionistic type-2 fuzzy inference system with particle swarm optimization. Symmetry 12:562
8. Zarzycki H, Dobrosielski WT, Apiecionek Ł, Vince T (2020) Center of circles intersection, a new defuzzification method for fuzzy numbers. Bull Pol Acad Sci Tech Sci 68(2)
9. Lu Y, Wang J, Bai X, Wang H (2020) Design and implementation of LED lighting intelligent control system for expressway tunnel entrance based on Internet of things and fuzzy control. Int J Distrib Sens Netw 16(5)
10. Rahilaa J, Santhib M (2020) Integrated fuzzy and phase shift controller for output step voltage control in multilevel inverter with reduced switch count. Automatika 61(2):238–249
11. Bas E, Yolcu U, Egrioglu E (2019) Intuitionistic fuzzy time series functions approach for time series forecasting. Granular Computing, 17 Mar
12. Kuoa RJ, Cheng WC (2019) An intuitionistic fuzzy neural network with Gaussian membership function. J Intell Fuzzy Syst 36:6731–6741
13. Azimi SM, Miar-Naimi H (2019) Designing an analogue CMOS fuzzy logic controller for the inverted pendulum with a novel triangular membership function. Sci Iranica D 26(3):1736–1748
14. Mabalane MD, van Schalkwyk CH, Reyers M (2019) National differences in the financial risk tolerance of financial planning clients. Manag Dyn 28(2)
15. Dhinaiya GM (2016) A study on determinants of financial risk tolerance: a review of the evidence. GJRIM 8(1)
16. Radhika C, Parvathi R (2016) Intuitionistic fuzzification functions. Glob J Pure Appl Math 12(2):1211–1227
17. Butt MA, Akram M (2016) A new intuitionistic fuzzy rule based decision making system for an operating system process scheduler. SpringerPlus 5:1547
18. Akram M, Shahzad S, Butt A, Khaliq A (2013) Intuitionistic fuzzy logic control for heater fans. In: Mathematics in computer science

Contemporary Computing Applications

Deep Learning for Self-learning in Yoga and Fitness: A Literature Review

Dhananjay Sharma, Harshil Panwar, Harshit Goel, and Rahul Katarya

1 Introduction

Yoga, calisthenics, sports and minor gym exercises have slowly started finding their way into everyone's daily routines due to their numerous benefits and minimal prereq-uisites. One major factor that determines the effectiveness of the aforementioned exercises is posture correctness. The correct posture will not only allow the user to reap maximum benefits but it will also prove to be instrumental in preventing injuries. Unfortunately, professional instructors and institutes providing such guid-ance to beginners are often very expensive and have a very tight schedule due to their popular nature. This paper aims to summarize the role and scope of deep learning in building assistive technology to provide efficient and inexpensive alternatives for the same.

Pose estimation [1] in deep learning is the subdomain associated with analyzing and approximating various key points on a human being in order to estimate the pose. It broadly follows two approaches: the bottom-up approach where each joint is estimated and then connected to form a pose and the top-down approach where a human's bounding box is estimated and then joints are approximated within the region. Pose estimation today forms the base for many other research fields such as human activity recognition, pedestrian analysis for self-driving cars. This paper

D. Sharma (✉) · H. Panwar · H. Goel · R. Katarya
Department of Computer Science and Engineering, Delhi Technological University, Delhi, India
e-mail: dhanajaysharma_2k18co119@dtu.ac.in

H. Panwar
e-mail: harshilpanwar_2k18co142@dtu.ac.in

H. Goel
e-mail: harshitgoel_2k18co143@dtu.ac.in

R. Katarya
e-mail: rahulkatarya@dtu.ac.in

analyzes the use of various pose estimation techniques in building posture correction technology for the various forms of exercise mentioned above.

Nowadays, most end-to-end models in the domain use libraries/hardware such as OpenPose [2], Kinect, along with convolutional neural networks (CNNs) [3] to detect key points on a human being. These pose analysis libraries are often paired with LSTMs [4] and/or CNNs to classify the exact pose being performed ("asana" in the case of yoga). In the case of video input, the combination of both proves to work the best as LSTMs help record the temporal relation between the frames, while CNNs record the spatial data in every frame, thus combining to produce fast and accurate real time results. Once the pose has been classified and the joint locations have been estimated using key points, various mathematical and trigonometric models are employed to perform a comparison between the obtained joint locations and ideal joint locations for a particular pose. These data are then used to generate feedback reports allowing users to fix their posture in real time.

Thus, by utilizing deep learning techniques, many assistive solutions have been and can be developed to tackle the problem of ensuring posture correctness during exercises. These solutions are not only accurate but also they provide real-time feedback and are affordable enough to provide a better alternative to posh institutes and personal trainers.

2 Methodology

The general approach that has been adopted by most researchers in the domain comprises two major steps. Firstly, a pose detection algorithm is employed to detect key points on a human being. A pose detection network first localizes human body joints and then groups them into valid pose configuration (Figs. 1 and 2).

2.1 Backbone Architecture

The backbone architecture used in these networks primarily ranges from AlexNet [5], to the recently developed ones such as fast R-CNN [6], mask R-CNN [7] and feature pyramid networks (FPN) [8]. However, VGG [9] and ResNet [10] remain the popular choice.

Fig. 1 General yoga self-learning pipeline

Fig. 2 Detected joint coordinates

2.2 Loss Functions

Loss functions are at the core of any machine–learning or deep learning model to learn from the dataset. In case of human pose estimation models, cross-entropy loss, mean absolute error (MAE) and mean-squared error (MSE) are mostly used loss functions.

MAE or L_1 loss function is measured as mean of absolute error between prediction and true values. Being outlier insensitive, this loss function is more robust.

$$L_1 = 1/n \sum_{i=1}^{n} |y_i - f(x_i)| \tag{1}$$

MSE or L_2 loss function is mean of squared sum of errors between true and predicted values. This loss function penalizes outliers in a dataset.

$$L_2 = 1/n \sum_{i=1}^{n} (y_i - f(x_i))^2 \tag{2}$$

For a classification model with probability of output between 0 and 1, cross-entropy loss is used for measuring performance. Similar to other loss function, cross-entropy loss increases for increasing deviation from true value.

$$\mathrm{Log}_{\mathrm{loss}} = -(y_i \log(f(x_i)) + (1 - y_i) \log(1 - f(x_i))) \tag{3}$$

2.3 Evaluation Metrics

For performance comparison of human pose estimation models, several custom evaluation metrics have been developed.

Percentage of Correct Parts (PCPs)
It focuses on correct prediction of limbs. A body limb is identified, if distance between two joint positions is less than half in comparison to true value between these key points. Hence due to scaling issues, it penalizes for shorter limbs.

Percentage of Detected Joints (PDJs)
This metric is based on distance between true and predicted joints within a threshold of torso diameter, thus achieving localization precision over former metric.

Percentage of Correct Key Points (PCKs)
It measures if true joint and predicted key point are within a certain threshold. Due to smaller head bone connections and torsos, it overcomes the shorter body limb problem.

The estimated body joint key point data or the frame itself is then passed through a classifier network that identifies the yoga pose or exercise being performed.

In the second step, the joint body key points are then used to compute the joint angles using various trigonometric functions. These angle values are then compared to the angle values obtained from the instructor (the ideal values, in this scenario). This comparison helps generate a feedback score and detect flaws in the user's posture, if any (Fig. 3).

$$m = (y_2 - y_1)/(x_2 - x_1) \tag{4}$$

$$\text{angle} = \left| \tan^{-1} m \right| \tag{5}$$

This feedback score is then utilized to generate a report to help users learn by themselves. These approaches are summarized in Table 1.

Fig. 3 Estimated angle differences

```
In [47]:  diff_matrix[1]

Out[47]:  [2.3509990606347797,
          9.782407031807287,
          6.314564091110093,
          12.329753344666454,
          33.92254327774141,
          14.272830843813622,
          20.16585144175515,
          3.465795525230117,
          3.783385236533917]
```

Table 1 Scoring algorithm based on angle differences observed

Angle difference	Score
< 10	10
< 15	9
< 20	8
< 25	7
≥ 25	Make a better attempt

Further, we have also analyzed recent advancements in the field of pose detection. DeepPose [30] approaches the problem as a CNN-based regression problem. It refines roughly estimated pose using stream of regressors, and images are cropped around predicted joints and fed to next stage, thus allowing feature learning for finer scales. Convolutional pose machines (CPMs) consist of series of convolutional networks which yield belief maps for each predicted key point. Hence, feature representation and implicit spatial information are learned at the same time, with supervision at each stage to solve the problem of vanishing gradients. Unlike conventional technique, i.e., down sampling high-resolution feature maps and then recovering it later, high-resolution Net (HRNet) [31] maintains a high-resolution representation throughout, thus comprising parallel high low-resolution networks with information exchange across multiple resolution networks.

These cutting-edge algorithms, while exceptional in analyzing the human pose, have not yet been utilized for self-learning and posture correction in the field of exercise. The new state of the art pipelines would be able to detect humans and poses more accurately and efficiently, thus generating a much more relevant feedback report than their predecessors. These new algorithms and networks are summarized in Table 3.

3 Result and Discussion

In Table 2, the paper employing a novel IOT method in order to reduce privacy related issues associated with the use of Kinect and other camera devices makes stellar progress in the self-learning and posture correction domain. It averages an F_1 score of 0.93 along all axes and an accuracy of 99.45%. The paper conducting a comparison study between OpenPose and mask-R-CNN as pose analysis modules also achieves excellent results. OpenPose as the pose module gives an accuracy of 99.91%, whereas the mask-R-CNN algorithm surpasses that result and achieves a 99.96% accuracy on the test dataset. The paper employing the convolutional pose machine and faster R-CNN also achieved a mean accuracy precision value of 99.9%. Lastly, the paper using regional-based networks also improves upon its predecessor on the same dataset by achieving an F_1 score of 0.71. As other papers have not proposed a classification algorithm, accuracy metrics are not available for them.

For each recent human pose estimation model, parameters such as backbone architecture, loss function have been weighed as shown in Table 3. The majority of models opt for ResNet as the backbone architecture due to its ability to negate the vanishing gradient problem, hence providing a deeper model with greater accuracy. Apart from average precision (AP) and mean average precision (mAP), many of the models prefer PCKh@0.5 as a good evaluation metric since it directly deals with human images. For training purposes, apart from the standard cross-entropy loss, most models are shifting toward least square errors (L_2) even though it is sensitive

Table 2 Analysis of existing posture correction pipelines

Authors	Pose estimation techniques	Correction techniques	Pose classification accuracy
J. Palanimeera, K. Ponmozhi [11]	KNNs [12], SVMs [13], logistic regression [14] and Naive Bayes [15] are used on dataset of 12 joints obtained from 17 detected key points	Only pose classification approaches mentioned	0.9902 (KNN), 0.9817 (SVM), 0.7347 (NB), 0.8461 (LR)
Munkhjargal Gochoo, Tan-Hsu Tan, Shih-Chia Huang, Tsedevdorj Batjargal, Jun-Wei Hsieh, Fady S. Alnajjar, Yung-Fu Chen [16]	Introduces an IOT-based approach using a very deep CNN along with a low-resolution infrared sensor. This infrared sensor is based a wireless sensor network (WSN). WSN is used on the client side to scan the user's movements and then employ a deep CNN to detect the yoga poses	Only pose classification approaches mentioned	0.9945 (mean all axes)
Maybel Chan Thar, Khine Zar Ne Winn, Nobuo Funabiki [17]	OpenPose is used along with a CNN to detect yoga poses accurately	Tan-inverse formula is used to estimate the joint angles using joint key points. These angles are compared to ideal joint angle values of a particular pose. The output pose is displayed using a greed–red gradient to display degree of correctness in posture	0.9777

(continued)

Table 2 (continued)

Authors	Pose estimation techniques	Correction techniques	Pose classification accuracy
Fazil Rishan, Binali De Silva, Sasmini Alawathugoda, Shakeel Nijabdeen, Lakmal Rupasinghe, Chethana Liyanapathirana [18]	OpenPose and mask-R-CNN are used as key point estimators in video frames. Time-distributed CNNs are used to capture spatial features. An LSTM is then used to capture temporal and spatial changes based on the features. A Softmax layer then uses these changes to display final pose probabilities	Only pose classification approaches mentioned	0.9991 (OpenPose), 0.9996 (mask-R-CNN)
Manisha Verma, Sudhakar Kumawat, Yuta Nakashima, Shanmuganathan Raman [19]	A novel fine-grained hierarchical classification method based on visual similarity of poses is used. Many famous CNN architectures along with a modified dense net 201 with hierarchical connections are proposed. Increased classification accuracy and exemplar performance are observed on an 82-pose dataset	Only pose classification approaches mentioned	0.9347 (Var 1 on L_3)
Rehnhao Huang, Jiqing Wang, Haowei Lou, Haodang Lu, Bofei Wang [20]	Kinect combined with OpenPose is used to improve upon previous results	Joint angles are calculated by finding the tan inverse of the value of the vector difference between the two vectors joining to make a joint. These angles are used for comparison with ideal values	None

(continued)

Table 2 (continued)

Authors	Pose estimation techniques	Correction techniques	Pose classification accuracy
Ian Gregory, Samuel Mahatmaputra Tedjojuwono [21]	OpenPose is used to detect 18 key points from multiple angles	Cosine-distance formula is used to obtain joint angles. The difference between the user's joint angles and instructor's joint angles is then used to provide a feedback score	None
Edwin W. Trejo, Peijiang Yuan [22]	Kinect v2 is paired with the AdaBoost algorithm to implement a yoga pose classifier for six poses for up to six people. AdaBoost selects the final dataset for training in order to obtain maximum accuracy	Uses joint angle comparisons based on joint coordinates obtained with the help of Kinect v2	0.9478 (DB-3 mean)
Grandel Dsouza, Deepak Maurya, Anoop Patel [23]	CNN trained on different body part images is used	Comparison of shoulder and joint angles with that of an athlete's is displayed on a graph over time	None
Yuxin Hou, Hongxun Yao, Haoran Li, Xiaoshuai Sun [24]	Convolutional pose machine [25] and faster R-CNN [26] are used to detect a person and map their position	An action-guided network is used to rate how accurately a pose is being performed	None
Amit Nagarkoti, Revant Teotia, Amith K. Mahale, Pankaj K. Das [27]	Real-time two-dimensional multiperson pose estimation is performed. The proposed approach is based on Part Affinity Fields	For comparison, dynamic time warping (DTW) between each frame of the user and trainer is done. For output, feedback used affine transformations to solve camera problems	None

(continued)

Table 2 (continued)

Authors	Pose estimation techniques	Correction techniques	Pose classification accuracy
Jianbo Wang, Kai Qiu, Houwen Peng, Jianlong Fu and Jianke Zhu [28]	Region-based networks are used to detect humans in an image and isolate them. A custom single person tracking algorithm based off of Siamese tracking [29] is used. A custom pose estimation algorithm is built to analyze both spatial and temporal key points	Based on the features obtained, a classifier identifies all bad poses so athletes can work on their posture	None

Table 3 Comparison of pose estimation algorithms

Model	Year	Backbone architecture	Approach	Loss function	Evaluation metrics	Training datasets
DeepPose	2014	AlexNet	Top-down	L_2	PDJ, PCP	LSP, FLIC
DeeperCut [32]	2016	ResNet	Bottom-up	L_1, cross-entropy	AP, mAP, AUC, PCKh@0.5	MPII, COCO, LSP
Convolutional pose machines	2016	VGG	Top-down	L_2	PCKh@0.1, PCKh@0.2, PCKh@0.5	FLIC, LSP, MPII
RMPE: Regional multiperson pose estimation [33]	2017	VGG, ResNet	Top-down	L_2	mAP	MPII, COCO
DensePose [34]	2018	FCN, mask-R-CNN	Bottom-up	Cross-entropy	AP, PCP, GPS	COCO
High-resolution network (HRNet)	2019	ResNet	Bottom-up	L_2	AP, mAP, PCKh@0.5	MPII, COCO
Human pose estimation for real-world crowded scenarios [35]	2019	ResNet	Top-down	L_2	AP, OKS	CrowdPose, JTA

toward outliers. Last but not the least, in these recent works, COCO and MPII have been, by far, default choices of dataset with richer training data.

4 Future Work

There is potential for the introduction of numerous improvements and improvisations along every step of the pipeline. Firstly, a novel IOT method was proposed to capture user input in the place of a traditional camera or a Kinect to tackle privacy issues that their predecessors were not even aware of. Further improving this approach or finding better and safer methods is one track future authors could research along.

Pose detection networks still have to overcome many challenges. To this day, multiperson pose detection and dynamic activity tracking have not yet reached their saturation points. Thus, improving upon this part of the pipeline is also a track researchers could entertain.

Moreover, the aforementioned pose analysis networks have successfully surpassed their predecessors in terms of accuracy, efficiency and raw speed. Still, these networks have not yet been tested in the domain of self-learning and posture correction for exercise and fitness. Thus, by utilizing the potential of these models, future researchers could help make accurate and truly real-time posture classification and analysis possible.

Lastly, by using natural language processing (NLP), rather than tree-based hard-coded correction output mechanisms, the pipeline could be perfected in a more end-to-end manner. The use of NLP could allow the model to give a personalized and more relevant feedback instead of relying on a few pre-coded statements. This will help create impactful products that the end-users could truly benefit from.

5 Conclusion

Thus, this study manages to showcase the increasing importance of yoga, calisthenics, gyms and general fitness in the daily life of an average worker. Furthermore, the paper neatly summarizes the relevant works and benchmarks that have been set by past researchers aiming to develop an end-to-end pipeline capable of identifying the pose being performed by a user and suggesting corrective measures accordingly. The paper also mentions the algorithms, networks and devices required for the same.

Moreover, the paper also manages to summarize all the cutting-edge pose detection networks that have been created in the past few years. These networks, while much better than their predecessors, have not yet been applied in the field of posture correction and self-learning and posture correction in exercise and fitness.

Lastly, the paper also breaks down the pipeline in a systematic format and manages to point out the scope for improvement and improvisation along each step of the pipeline for future researchers to pick up on.

References

1. Munea TL, Jembre YZ, Weldegebriel HT, Chen L, Huang C, Yang C (2020) The progress of human pose estimation: a survey and taxonomy of models applied in 2D human pose estimation. IEEE Access 8:133330–133348. https://doi.org/10.1109/ACCESS.2020.3010248
2. Cao Z, Hidalgo G, Simon T, Wei S-E, Sheikh Y (2018) OpenPose: realtime multi-person 2D pose estimation using part affinity fields
3. Yamashita R, Nishio M, Do RKG et al (2018) Convolutional neural networks: an overview and application in radiology. Insights Imaging 9:611–629
4. Hochreiter S, Schmidhuber J (1997) Long short-term memory. Neural Comput 9:1735–1780. https://doi.org/10.1162/neco.1997.9.8.1735
5. Krizhevsky A, Sutskever I, Hinton G (2012) ImageNet classification with deep convolutional neural networks. In: Advances in neural information processing systems. Curran Associates, Inc.
6. Girshick R (2015) Fast r-cnn. In: Proceedings of the IEEE international conference on computer vision
7. He K, Gkioxari G, Dollár P, Girshick R (2017) Mask R-CNN
8. Lin T-Y et al (2017) Feature pyramid networks for object detection. In: Proceedings of the IEEE conference on computer vision and pattern recognition
9. Simonyan K, Zisserman A (2014) Very deep convolutional networks for large-scale image recognition. arXiv preprint arXiv:1409.1556
10. He K et al (2016) Deep residual learning for image recognition. In: Proceedings of the IEEE conference on computer vision and pattern recognition
11. Palanimeera J, Ponmozhi K (2021) Classification of yoga pose using machine learning techniques. Mater Today Proc 37(Part 2):2930–2933. ISSN 2214-7853
12. Guo G, Wang H, Bell D, Bi Y (2004) KNN model-based approach in classification
13. Evgeniou T, Pontil M (2001) Support vector machines: theory and applications, vol 2049, pp 249–257. https://doi.org/10.1007/3-540-44673-7_12
14. Peng J, Lee K, Ingersoll G (2002) An introduction to logistic regression analysis and reporting. J Educ Res 96:3–14. https://doi.org/10.1080/00220670209598786
15. Rish I (2001) An empirical study of the Naïve Bayes classifier. IJCAI 2001 Work Empirica Methods Artif Intell 3
16. Gochoo M et al (2019) Novel IoT-based privacy-preserving yoga posture recognition system using low-resolution infrared sensors and deep learning. IEEE Internet Things J 6(4):7192–7200. https://doi.org/10.1109/JIOT.2019.2915095
17. Thar MC, Winn KZN, Funabiki N (2019) A proposal of yoga pose assessment method using pose detection for self-learning. In: 2019 International conference on advanced information technologies (ICAIT), pp 137–142. https://doi.org/10.1109/AITC.2019.8920892
18. Rishan F, De Silva B, Alawathugoda S, Nijabdeen S, Rupasinghe L, Liyanapathirana C (2020) Infinity yoga tutor: yoga posture detection and correction system. In: 2020 5th International conference on information technology research (ICITR), pp 1–6. https://doi.org/10.1109/ICITR51448.2020.9310832
19. Verma M, Kumawat S, Nakashima Y, Raman S (2020) Yoga-82: a new dataset for fine-grained classification of human poses. In: 2020 IEEE/CVF conference on computer vision and pattern recognition workshops (CVPRW), pp 4472–4479. https://doi.org/10.1109/CVPRW50498.2020.00527
20. Huang R, Wang J, Lou H, Lu H, Wang B (2020) Miss Yoga: a yoga assistant mobile application based on keypoint detection. In: 2020 Digital image computing: techniques and applications (DICTA), pp 1–3. https://doi.org/10.1109/DICTA51227.2020.9363384
21. Gregory I, Tedjojuwono SM (2020) Implementation of computer vision in detecting human poses. In: 2020 International conference on information management and technology (ICIMTech), pp 271–276. https://doi.org/10.1109/ICIMTech50083.2020.9211145

22. Trejo EW, Yuan P (2018) Recognition of yoga poses through an interactive system with Kinect device. In: 2018 2nd International conference on robotics and automation sciences (ICRAS), pp 1–5. https://doi.org/10.1109/ICRAS.2018.8443267

23. Dsouza G, Maurya D, Patel A (2020) Smart gym trainer using Human pose estimation. In: 2020 IEEE International conference for innovation in technology (INOCON), pp 1–4. https://doi.org/10.1109/INOCON50539.2020.9298212

24. Hou Y, Yao H, Li H, Sun X (2017) Dancing like a superstar: action guidance based on pose estimation and conditional pose alignment. In: 2017 IEEE International conference on image processing (ICIP), pp 1312–1316. https://doi.org/10.1109/ICIP.2017.8296494

25. Wei S-E, Ramakrishna V, Kanade T, Sheikh Y (2016) Convolutional pose machines

26. Ren S, He K, Girshick R, Sun J (2015) Faster R-CNN: towards real-time object detection with region proposal networks. IEEE Trans. Pattern Anal Mach Intell 39. https://doi.org/10.1109/TPAMI.2016.2577031

27. Nagarkoti A, Teotia R, Mahale AK, Das PK (2019) Realtime indoor workout analysis using machine learning & computer vision. In: 2019 41st Annual international conference of the IEEE engineering in medicine and biology society (EMBC), pp 1440–1443. https://doi.org/10.1109/EMBC.2019.8856547

28. Wang J, Peng KOH, Fu J, Zhu J (2019) AI coach: deep human pose estimation and analysis for personalized athletic training assistance. In: Proceedings of the 27th ACM international conference on multimedia (MM'19)

29. Bertinetto L, Valmadre J, Henriques J, Vedaldi A, Torr P (2016) Fully-convolutional siamese networks for object tracking, vol 9914, pp 850–865. https://doi.org/10.1007/978-3-319-48881-3_56

30. Toshey A, Szegedy C (2014) DeepPose: human pose estimation via deep neural networks. In: 2014 IEEE Conference on computer vision and pattern recognition, pp 1653–1660. https://doi.org/10.1109/CVPR.2014.214

31. Wang J, Sun K, Cheng T, Jiang B, Deng C, Zhao Y, Liu D, Mu Y, Tan M, Wang X, Liu W, Xiao B (2020) Deep high-resolution representation learning for visual recognition

32. Insafutdinov E, Pishchulin L, Andres B, Andriluka M, Schiele B (2016) DeeperCut: a deeper, stronger, and faster multi-person pose estimation model, vol 9910, pp 34–50. https://doi.org/10.1007/978-3-319-46466-4_3

33. Fang H-S, Xie S, Tai Y-W, Lu C (2017) RMPE: regional multi-person pose estimation, pp 2353–2362. https://doi.org/10.1109/ICCV.2017.256

34. Güler RA, Neverova N, Kokkinos I (2018) DensePose: dense human pose estimation in the wild. In: 2018 IEEE/CVF Conference on computer vision and pattern recognition, pp 7297–7306. https://doi.org/10.1109/CVPR.2018.00762

35. Golda T, Kalb T, Schumann A, Beyerer J (2019) Human pose estimation for real-world crowded scenarios. In: 2019 16th IEEE International conference on advanced video and signal based surveillance (AVSS), pp 1–8. https://doi.org/10.1109/AVSS.2019.8909823

Cardio Vascular Disease Prediction Using Ensemble Machine Learning Techniques

Shivangi Diwan and Mridu Sahu

1 Introduction and Background

Cardiovascular diseases are major cause of death among the Asian population and worldwide. The annual number of deaths from cardiovascular disease (CVD) in India has increased to 2 times in the past 30 years. In 2020, the number was found to be 4.77 million [1]. A timed diagnosis is extremely important to save lives. As medical data interpretation is quite tough and human mistakes make it more challenging because of their complex nature. The efficiency and accuracy of diagnosis can be improved effectively using artificial intelligence. Machine learning techniques are deployed in this study for the prediction of cardiac arrest. The comprehensive database repository with 11 features is used [2]. ECG is the electrical signal generated due to contraction (depolarization) and relaxation (repolarization) of the heart. The ECG wave pattern is as depicted in Fig. 1. It comprises P, QRS, T, and U waves [3].

ECG waveform conveys important information regarding the heart condition and health of the patient in terms of the interval, amplitude, and shape of the particular wave, where the QRS complex shows depolarization of ventricles, the P wave reflects atrioventricular depolarization, and the T wave denotes repolarization of ventricles and U wave because of smaller size may not be observed and represents repolarization of the Purkinje fibers. Every depolarization is followed by repolarization, so arterial repolarization lies within the QRS complex. As the QRS complex is tall and strong it masks arterial repolarization, it exists but can't be seen [4]. For heart disease prediction, Ghosh et al. used feature selection techniques like least absolute shrinkage and selection operator (LASSO) and relief and opted ensem-

S. Diwan (✉) · M. Sahu
National Institute of Technology, Raipur, Chattisgarh, India
e-mail: shivangi.diwan10@gmail.com

M. Sahu
e-mail: mrisahu.it@nitrr.ac.in

© The Author(s), under exclusive license to Springer Nature Singapore Pte Ltd. 2023
R. Agrawal et al. (eds.), *International Conference on IoT, Intelligent Computing and Security*, Lecture Notes in Electrical Engineering 982,
https://doi.org/10.1007/978-981-19-8136-4_28

Fig. 1 ECG signal and its
components

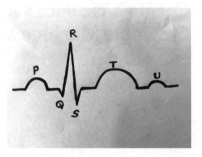

ble method of machine learning to get hybrid classifiers like decision tree bagging
method (DTBM), random forest bagging method (RFBM), K-nearest neighbors bagging method (KNNBM), AdaBoost boosting method (ABBM), and GradientBoost-
ing boosting method (GBBM) [5] for training and testing process, and the comparison
resulted in the highest increase in accuracy for AdaBoost and GBBM. Mohan et al.
proposed a novel method for finding the most appropriate features and classification
using machine algorithms and hence improved the prediction accuracy to 88.7%
using hybrid random forest with a linear model (HRFLM) [6]. Li et al. suggested
four standard feature selection algorithms including relief, MRMR, LASSO, LLBFS,
and proposed a novel feature selection algorithm fast conditional mutual information (FCMIM). For the best hyperparameters selection, the LOSO cross-validation
method is used in the system. Dataset used was the Cleveland heart disease dataset.
The result showed ANN with relief is the best predictive system for detection. And
the most suitable features are chest pain and exercise-induced angina [7]. Bharti et
al. deployed logistic regression, K-neighbors classifier, random forest classifier for
the prediction of heart diseases [8, 9]. Data from patients with cardiovascular illness
has been subjected to a variety of methodologies. Simple data mining techniques
are used for cleaning and preparation of data by removing the outliers [10], features
are normalized, and cross-validation is performed to find the better base models.
Optimal feature selection is one of the targets of this paper. The ensemble method is
used to combine various models [11], therefore, giving better prediction results.

2 Methodology

Models predicting the most appropriate and concise features without compromising
the accuracy is the main target of the paper which is accomplished by ML algorithms
and data mining techniques. The baseline models built are-

1. **Soft Voting**-A voting classifier is effective for leveling out the flaws of a group
   of models that are all doing well. The concept is to combine conceptually diverse
   machine learning classifiers and predict class labels using a majority vote (hard
   vote) or the average predicted probability (soft vote). Sherazi et al suggested

"Based on probability values of individual decisions of the combination of classifiers is termed as soft voting ensemble classifier. Depending on importance of a particular classifier, predictions are being weighted and then combined to find the weighted probabilities". The highest sum of these weighted probabilities target is finally chosen. To give more importance to any particular model to find the weighted average, customized weights can also be used [11, 12].

2. **Random Forest Classifiers**-An ensemble algorithm random forest (RF) classifier is made up of multiple algorithms. It usually consists of numerous decision trees, and thus, RF creates a full forest out of a collection of trees. Forests with infused randomness produce decision trees with dissociated prediction errors; upon averaging, the estimates error is minimized. By merging distinct trees, random forests reduce volatility, sometimes at the expense of a modest increase in bias. Individual decision trees usually have a lot of diversity and are prone to overfitting [10, 12, 13].

3. **Extra Tree classifier**-It fits several randomized decision trees to various subsamples of the dataset, and to prevent and control overfitting and boost the accuracy, averaging is used [12, 14].

4. **XGBoost**-It is the implementation of gradient boosting. Because of a block structure in its system design, XGBoost may utilize several cores on the CPU, allowing for faster learning through parallel and distributed computing, as well as effective memory utilization [12, 14].

The flow of the work is -

Import dataset and perform data cleaning and preprocessing—Build different baseline models and perform tenfold cross-validation—Test models and evaluate them—Feature selection (minimization)—Apply soft voting—Evaluate the model—Compare individual models metrics with soft voting model—Evaluate the results.

2.1 Dataset

This dataset is the largest openly available dataset with 1190 records and 11 features which is being obtained from the 5 popular heart disease datasets, which are-

1. Cleveland dataset
2. Statlog dataset
3. Hungarian dataset
4. Switzerland
5. Long Beach VA.

Alizadehsani et al. [2] have presented this dataset, and the description is briefly shown in Table 1.

Target variable: 1-heart disease, 0-normal.

Table 1 Dataset: features with datatype

Feature	Description	Datatype
Age	Patients age in years	Numeric
Sex	Gender, Male-1, Female-0	Nominal
Chest pain	Type 1-typical, 2-typical angina, 3-non-anginal pain, 4-asymptomatic	Nominal
Resting bps	Blood pressure at resting mode in mm/HG	Numeric
Cholesterol	Serum cholestrol in mg/dl	Numeric
Fasting blood sugar	1: True, 0: False, Condition: greater than 120 mg/dl	Nominal
ECG at rest	Left ventricular hypertrophy-2, Abnormality in ST-T wave-1, Normal-0	Nominal
Max heart rate	highest heart rate	Numeric
Exercise angina	0 depicting No, 1 depicting Yes	Nominal
Oldpeak	In comparison to, while you're at rest, ST depression because of exercise	Numeric
ST slope	0: Normal, 1: Upsloping, 2: Flat, 3: Downsloping	Nominal

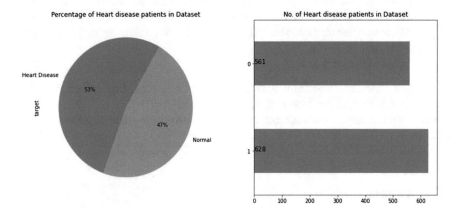

Fig. 2 From the dataset, the figure shows the percentage of CVD patients and normal

2.2 Preprocessing

In this comprehensive dataset first, the name of columns is changed; then, the features are encoded into categorical variables, and null values are dropped. There was no missing value found in the dataset. The distribution of heart disease(target variable) is shown in Fig. 2; the dataset is balanced with normal patients 561 and heart disease patients 629 [13]. The various distributions of features are depicted in Figs. 2, 3, 4, 5, 6, 7, 8, 9, and 10.

The gender and age-wise distribution clearly shows that the male percentage is more than female and the patients average age is 55.

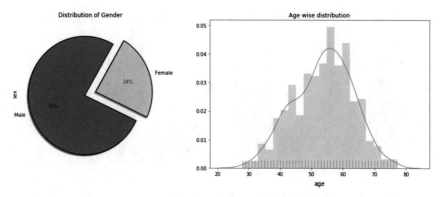

Fig. 3 Gender and age-wise distribution

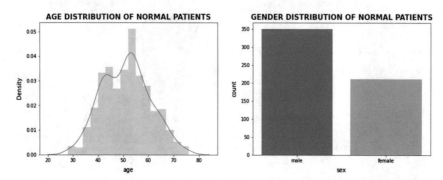

Fig. 4 Age and gender distribution of normal patient

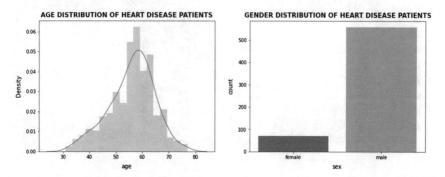

Fig. 5 Age and gender distribution of heart patient

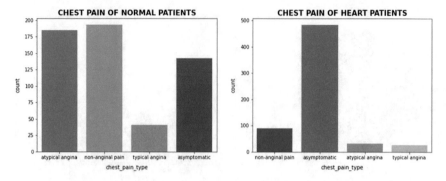

Fig. 6 Type of chest pain seen in normal and heart patients

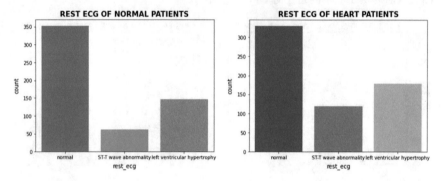

Fig. 7 Distribution of rest ECG of normal and heart patients

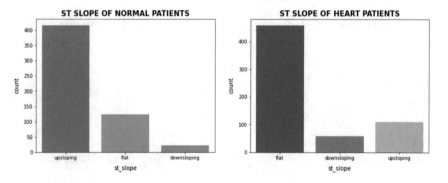

Fig. 8 ST-S lope of normal and heart patients

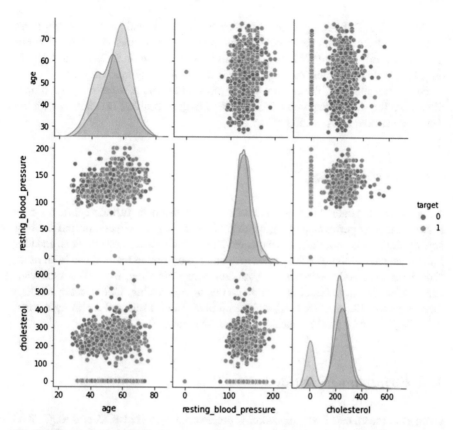

Fig. 9 Numeric features distribution

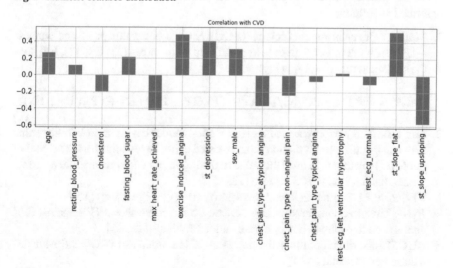

Fig. 10 Correlation of features with response variable class

The mean age of heart patients is approx 58–60 years. Male patients suffer more from heart disease than females. Asymptomatic heart attacks also termed **silent myocardial infarction (SMI)** [4] results in 45–50% of morbidities due to CVD in our country. And the chest pain of CVD patients is 76% asymptomatic.

For outlier removal, the z-score with a threshold value is used to ensure proper filtering. After the segregation of features and target variables, correlation with the response variable class is computed.

2.3 Model Building

Train and test the data with 80% train data and 20% test data, further feature normalization has been performed using MinMax Scalar [12]. To select and find the best performing baseline models, this paper developed multiple baseline models, and tenfold cross-validation has been performed, and the ensemble method has been used. This paper uses XGBoost classifier, random forest classifier, extra trees classifier, and further ensemble learning is deployed using soft voting. Chi-selector, random forest selector [12], etc., are used, and the optimal features selected are brought down to 7. The 7 features gave an acceptable accuracy value.

2.4 Performance Metrics

Geometric mean, sensitivity, specificity, precision, F1-score, MCC, and ROC AUC curve are the most relevant assessment criteria for this problem area. These are described as follows:

1. **Mathew Correlation coefficient (MCC)**: MCC's best value is +1 and worst is −1. Only, if the forecast gives better results in all the areas (TP, FN, TN, FP) does the statistical rate result in a high score [15].

$$MCC = (TP.TN - FP.FN)/\sqrt{((TP + FP).(TP + FN).(TN + FP).(TN + FN))}$$
(1)

2. **Log loss**: The performance of a classification model having a probability value between 0 and 1 as a prediction input is tested using logarithmic loss. The goal of our ML algorithm is to drop the value to the minimum possible level. In a perfect model, the log loss would be zero [12, 15].
3. **F1 Score**: F1-Score = 2(Recall Precision) / (Recall + Precision) [13]
4. **ROC**: The receiver operator characteristic (ROC) curve shows TPR against FPR and is used to evaluate binary classification problems [12, 15]
5. **AUC**: To discriminate among the classes, AUC is a summary of ROC and evaluates a classifier's capacity. [14]
6. **Precision**: The fraction of true positives among all the recovered examples is called precision (positive predictive value) [13]. Precision= TP/(TP + FP)

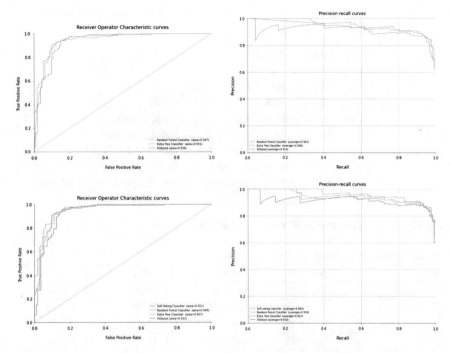

Fig. 11 ROC curve for random forest, extra tree, XGBoost classifiers before and after soft voting and precision-recall curve for random forest, extra tree, XGBoost classifiers before and after soft voting

7. **Recall**: Recall or sensitivity can be mathematically expressed as TP/(TP+FN) [13] .
8. **Accuracy** = (TP + TN)/(TP + FN + TN + FP)
9. **Specificity**= TN/(TN + FP) [13]
10. **Confusion Matrix**: It is an NxN matrix (say A) that aids in the evaluation of a machine learning model's performance in a classification problem. A11 = TP, A12 = FP, A21 = FN, 22 = TN [13, 14].

3 Results and Discussions

The anticipated probabilities for class labels are added together after soft voting as in a soft voting ensemble, and prediction is of the maximum sum class label. The performance of various models and the metrics are all depicted in the results as shown in Fig. 11, and the ROC and precision-recall values verify that the model with soft voting out performed all individual models. Figure 12 shows important features for disease prediction.

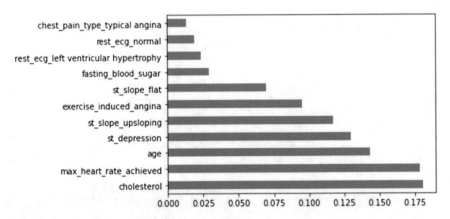

Fig. 12 Feature importance for heart disease prediction

Table 2 Performance metrics of different models

Model	Accuracy	Precision	Sensitivity	Specificity
Soft voting	0.906383	0.878788	0.943089	0.785714
RF entropy	0.902128	0.878788	0.943089	0.857143
Extra tree classifier	0.902128	0.884615	0.934959	0.866071
XGB	0.897872	0.872180	0.943089	0.848214

Table 3 Performance metrics of different models

Model	F1-score	ROC	Log loss	MCC
Soft voting	0.909804	0.899717	3.380449	0.806549
RF entropy	0.909804	0.823751	3.380445	0.805512
Extra tree classifier	0.909091	0.900515	3.380442	0.804719
XGB	0.906250	0.895652	3.527422	0.797405

Table 4 Comparative analysis

Parameters	Katarya et al. [16]	This paper
Best performing algorithm	Random forest	Soft voting and random forest
Accuracy	95.60%	RF = 90.63%, Soft voting = 90.21%
Precision	0.5528	RF = 0.878788, Soft voting = 0.878788
Recall	0.9768	RF = 0.943089, Soft voting = 0.943089

Tables 2 and 3 shows the performance of the model accuracy, precision, sensitivity, specificity, F1-score, ROC, log loss, and MCC values. The results from Katarya et al. [16] are compared with this paper, and the results are shown in Table 4.

4   Conclusion and Future Work

This paper presents the ensemble machine learning methods for heart disease prediction and achieves better performance through the stacked ensemble (after applying soft voting) in comparison to other ML models. Chi-selector, random forest selector are used for optimal feature selection, and the best-contributed features found in this dataset for prediction are as follows: (1) Max heart Rate (2) Cholesterol (3) ST depression (4) Age (5) Exercise-induced angina. The future plan is to implement deep learning algorithm on the dataset to further improve its performance in disease prediction and classification.

References

1. Huffman MD, Prabhakaran D, Osmond C, Ramji S, Khalil A, Poornima Prabhakaran GT, Biswas SKD, Srinath Reddy K (2011) Incidence of cardiovascular risk factors in an Indian Urban Cohort
2. Alizadehsani R, Roshanzamir M, Abdar M, Beykikhoshk A, Khosravi A, Panahiazar M, Koohestani A, Khozeimeh F, Nahavandi S, Sarrafzadegan N et al (2019) A database for using machine learning and data mining techniques for coronary artery disease diagnosis. Sci Data 6(1)
3. Tripathi RP, Tiwari A, Mishra GR, Bhatia D (2018) Study of ecg waveform and development of an algorithm for removal of power line interference. In: 2018 international conference on computational and characterization techniques in engineering and sciences (CCTES). IEEE, pp 40–48
4. Clifford GD, Azuaje F, McSharry P et al (2006) Advanced methods and tools for ECG data analysis. Artech house Boston
5. Ghosh P, Azam S, Jonkman M, Karim A, Javed Mehedi Shamrat FM, Ignatious E, Shultana S, Beeravolu AR, De Boer F (2021) Efficient prediction of cardiovascular disease using machine learning algorithms with relief and lasso feature selection techniques. IEEE Access 9:19304–19326
6. Senthilkumar M, Chandrasegar T, Gautam S (2019) Effective heart disease prediction using hybrid machine learning techniques. IEEE Access 7:81542–81554
7. Li JP, Haq AU, Din SU, Khan J, Khan A, Saboor A (2020) Heart disease identification method using machine learning classification in e-healthcare. IEEE Access 8:107562–107582
8. Bharti R, Khamparia A, Shabaz M, Dhiman G, Pande S, Singh P (2021) Prediction of heart disease using a combination of machine learning and deep learning. Comput Intell Neurosc
9. Obasi T, Omair Shafiq M (2019) Towards comparing and using machine learning techniques for detecting and predicting heart attack and diseases. In: 2019 IEEE international conference on big data (big data), pp 2393–2402
10. Witten IH, Frank E, Hall MA, Pal CJ (2005) Mining data. Practical machine learning tools and techniques. In: Data Mining, vol 2, pp 4
11. Sherazi SWA, Bae J-W, Lee JY (2021) A soft voting ensemble classifier for early prediction and diagnosis of occurrences of major adverse cardiovascular events for stemi and nstemi during 2-year follow-up in patients with acute coronary syndrome. Plos one 16(6):e0249338
12. Zhang C, Ma Y (2012) Ensemble machine learning: methods and applications. Springer

13. Tan P-N (2016) Michael Steinbach, and Vipin Kumar. Introduction to data mining, Pearson Education India
14. Layton R (2015) Learning data mining with python. Packt Publishing Ltd
15. Harrington P (2012) Machine learning in action. Simon and Schuster
16. Katarya R, Meena SK (2020) Machine learning techniques for heart disease prediction: a comparative study and analysis. Health Technol 11:87–97

Deep Learning Approach for Breast Cancer Detection

Prashant Ahlawat, Manoj Kumar Sharma, Hitesh Kumar Sharma, and Mukul Gupta

1 Introduction

Any type of cancer arises when cells grow abnormally in specific part of body. When abnormally grown cells comes in contact with the healthy ones to make them infectious. One of the most frequently occurring cancers is the breast cancer. Different parts of the breast can be affected by the cancer. Women breast has three main parts in breast: ducts, lobules, and connective tissue. The glands which are responsible for producing milk are lobules. The ducts are tubes that are responsible for carrying milk to nipple. And the connective tissue (mainly consists of fat and fibrous) neighbored and clench everything all together. It (cancer) can be present in any of these parts but mostly breast cancers starts in the ducts and (or) lobules. There are different approach which are used for the classification of breast image for cancer detection and segmentation but the most relying among them is machine learning and artificial intelligence. It has been observed and studied in literature that for achieving high grade of accuracy for the study of histopathology images of breast cancer concepts of machine learning and deep learning are used. During research, these concepts are implemented for detection, segmentation, and classification of cell nuclei with machine learning concepts (especially deep learning concepts).

Human body comprises different types of nuclei; out of them only two types are of more interest in experimental studies: lymphocyte nuclei and epithelial nuclei.

P. Ahlawat (✉) · M. K. Sharma
School of Information Technology, Manipal University, Jaipur, India
e-mail: ahlawatprashant2@gmail.com

H. K. Sharma
School of Computer Science, University of Petroleum and Energy Studies, EnergyAcres, Bidholi, Dehradun, India

M. Gupta
G.L. Bajaj Institute of Management, Greater Noida, India

© The Author(s), under exclusive license to Springer Nature Singapore Pte Ltd. 2023
R. Agrawal et al. (eds.), *International Conference on IoT, Intelligent Computing and Security*, Lecture Notes in Electrical Engineering 982,
https://doi.org/10.1007/978-981-19-8136-4_29

Each of the nuclei can look different because of factors like nuclei type, metastatic tumor of disease, and life period of nuclei. First, lymphocyte is a class from white blood cells which plays an important part in the defense mechanism of human body. Lymphocyte nuclei (LN) are inflaming nucleus which has structured appearance and small in structure as compared to the (EN) epithelial nuclei (Fig. 1a). No pathological epithelial nucleus has almost consistent chromosome spread location with even borderlines (in Fig. 1b). In lofty class cancerous tissue, epithelial nuclei are bigger in shape as compared to others, certainly possess some diverse chromatin arrangement, non-uniform borderline, introduce to as nuclear polymorphism, and distinctly seen nuclei in contrast to usual EN (in Fig. 1c).

The automation in nuclei segmentation is nowadays a favorite topic of interest in recent research, for that an enormous number of techniques and methods are highlighted in the previous study and many new approach pursue to look over.

The DSC, i.e., detect, segment, and classify cell nuclei in normally dyed histopathological images give rise to a complex computerized vision issue because of major irregularity in pictures due to many factors comprising of different ways in preparing slide for microscopic analysis (dye absorption, smoothness of the tissue cut, having some unknown antiquity or some destruction in tissue specimen, and so on) and picture acquisition (antiquity found due to compaction of the picture, specific features of the slide scanner, presence of digital noise, etc.). Besides this,

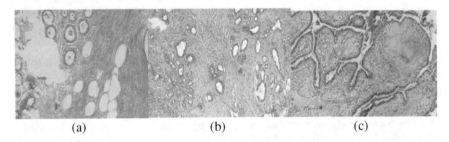

(a)　　　　　　　　(b)　　　　　　　　(c)

Fig. 1 Structure of cell nucleus. **a** Lymphocyte nucleus. **b** Epithelial nucleus. **c** Epithelial nucleus (Cancer)

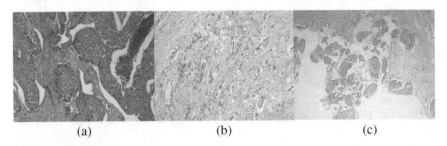

(a)　　　　　　　　(b)　　　　　　　　(c)

Fig. 2 Examples of nuclei which are hard to segment and detect. **a** Blur nuclei. **b** Over-lapped nuclei. **c** Heterogeneous nuclei

nuclei are sometimes arranged in overlay (Fig. 2). Diagrammatic examples of testing nuclei for detection and segmentation: (a) blur, (b) overlaps, and (c) heterogeneity. Above problems (shown in Fig. 2) force the nuclei classification, segmentation, and detection of a difficult task. A victorious picture preparing plan will have to get the better of these problems in strong way in such to keep a high standard (quality) and precision in every circumstances [9].

2 Literature Review

Many researchers have proposed various models and algorithms for detection and diagnosis of breast cancer using medical images or featured data from medical images. In [1] author, after forming, the PMaps post processing is done on PMaps which was discovered to have big role to identify clarity of a nuclei. And two networks architectonic was acknowledged which enhances detection standard or the run time of the system. In [2], authors developed a computerized approach for computable image investigation method of BCH pictures is presented. Top Bottom Transform is used for segmentation, decomposition wavelet technique used for ROI, and space scale curvature for splitting cells. This paper [3] proposes use of Hematoxylin and Eosin (H&E) dyed histopathology pictures of breast for detecting individual nuclei using novel segmentation algorithm. This detection framework estimates boundary using a Propagation Loopy Back algorithm. This technique is executed and applied on all whole-slide pictures and framework of histopathology pictures of breast cancer for accuracy. In this paper [4], automatic system is suggested and proposed for identifying or segment nucleus with cancer indication. In distinction, this system identifies diseased nucleus having asymmetric dimensions or size, borderline, and compactness in greater scores. In this paper [5], the pathological research is conducted on many cancer detection and grading applications and systems, comprising of prostate cancer grading, brain, cervix, breast, and lungs. It also explains highlights, reviews, and gets the main drift from the deep survey of nuclei identification, disjuncture, feature complexity, and classify methods used in histopathology image process, especially in H&E and chemical immune histo dyes technique. In this paper [6], study is done to gather understanding and the acknowledgement involving the threat factors, risk involvements, healing factors, and the safeguard guidelines to early stage identification of cancer in breast in the females of the pure city of Varanasi in Uttar Pradesh. This paper [7] studies and reviews early age cancer in breast which has different clinic and biology indications which is never present in old age victims. The biology of female chest cancer is very prominent and also connected with the critical prediction in young females. This paper [8] briefly overviews of how surprising breast tumor is, ways it develops inside different parts of breast, and how common this cancer is. This paper [9] addresses and explains the kinds, causes, effects, clinic indications, or many approaches including without medicine-like operation and rays therapy and medicine diagnosis which consist of chemo, gene transplant therapy, etc., of breast tumor. In this paper [10], author summarizes specific research of pathogenic, same genetic structure, lots of risky parameters and preventative, and precautions techniques on breast tumor over the past few decades [11]. That studies help during the exhaustive diagnosis for the breast tumor.

3   Machine Learning Algorithms Used for Classification

At present many unsupervised and supervised learning algorithm [12, 13] are used for guessing (prediction) and classification but dataset which is used here, two preferred popular algorithms are random forest and decision tree. Highlighted features of these two algorithms are mentioned below.

Decision Tree

For classification of cell nucleus under supervised machine learning algorithm, we use decision tree. It this tree like structure is there, decision has to be taken at each node to find the new or next path to ingress. Each leaf node acts as the solution to reach [14]. At every node, a splitting basis needs to set for jumping to low level. Splitting basis are:

Information Gain

By transforming a dataset, one can calculate the reduction in the entropy which is known as Information Gain. In case of decision trees, we calculate information gain for the each variable and after that we choose a variable that increases the information gain, which in turn decreases the entropy and effectively splits the dataset.

$$\text{Entropy} = E(S) = \Sigma(-\text{PI} \log 2\text{PI})$$

Pi is the probability of class i.

$$\text{Information Gain} = \text{IG}(Y, X) = E(Y) - E(Y|X)$$

Gini impurity

To split a decision tree when the target variable is categorical, we use a method known as Gini Impurity.

$$\text{Gini Impurity} = (1) - (\text{Gini})$$

That's the likelihood of detecting a correct label for a chosen element aimlessly if it was labeled randomly.

$$\text{Gini} = \sum \text{PI2}$$

Chi-Square

It was used for calculating the analytical important differences between parent and child node.

$$\text{Chi-Square} = \left((\text{Actual} - \text{Expected})^2 / (\text{Expected})\right)^{1/2}$$

Heterogeneous type of data is effectively and efficiently handled by the decision tree algorithms.

Random Forest

The combination of many decision trees forms a random forest. Random forest is a group approach. Prediction of model is represented by each tree in random forest, and best prediction is taken as the prediction for whole model. The main advantage of using this supervised algorithm in machine learning is that it can be implemented on regression and classification issues both.

4 Dataset Description

For this study, the dataset used is from Breast Cancer Wisconsin (Diagnostic) Dataset. This dataset contain total 569 records. It contains 32 columns which represents 32 features of cancerous/non-cancerous cell image. Fine Needle Aspirate (FNA) of a breast mass is used for computing image features. Given 569 image records contain 357 benign and 212 malignant classification of breast tissue image.

The summary of the dataset generated by python code is displayed in Table 1. We have displayed some limited columns out of 32 columns. Some brief statistics about the dataset is given for a snapshot.

5 Exploratory Analysis of Dataset

The exploratory analysis of dataset helps to find insights and correlation among the various features of the dataset. We have plotted various graphs for visualizing the data with different prospective.

In Fig. 3, we have shown the bar plot for malignant and benign cells data. It shows that in given dataset 357 are benign and 212 are malignant records.

In Fig. 4, we have shown the correlation graph. It shows that how much attributes are correlated with each other. As we can see from the figure, radius_mean is more correlated to parameter_mean.

In Fig. 5, we have shown the heat map to show more accurate correlation in number between various attributes of the dataset, as we can see that again radius_mean is fully correlated to parameter_mean. This is more accurate visualization technique for exploratory analysis dataset.

Table 1 Summary statistics of Boston Crime Dataset

	1. radius_mean	2. texture_mean	3. perimeter_mean	4. area_mean	5. smoothness_mean	6. compactness_mean
1. Count	569.000001	569.000001	569.000001	569.000001	569.000001	569.000001
2. Mean	14.12729	19.28964	91.96903	654.88910	0.09636	0.104340
3. Std	3.524040	4.301030	24.298980	351.914120	0.014060	0.052810
4. Min	6.981001	9.710001	43.790001	143.500001	0.052631	0.019381
5. 25.5%	11.700001	16.170001	75.170001	420.300001	0.086371	0.064921
6. 50.5%	13.370001	18.840001	86.240001	551.100001	0.095871	0.092631
7. 75.5%	15.780001	21.800001	104.100001	782.700001	0.105301	0.130401
Maximum	28.110001	39.280001	188.500001	2501.000001	0.163401	0.345401

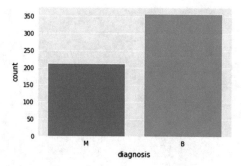

Fig. 3 Bar plot for malignant and benign cell

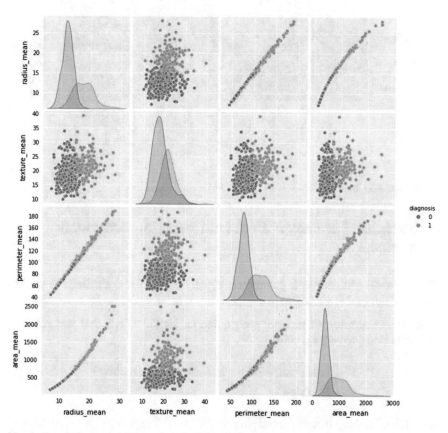

Fig. 4 Correlation map within various attribute of dataset

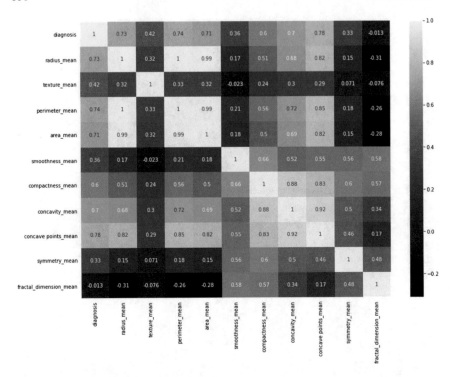

Fig. 5 Heat map for correlation between parameters

6 Statistical Parameters Used for Testing the Performance of Model

The statistical and analytical parameters which are used for evaluating the performance of our model are as follows.

Precision: Proportion of positive recognition or identifications that was actually correct is evaluated with precision.

$$P = \text{True Positive}/(\text{True Positive} + \text{False Positive})$$
$$P = \text{Precision}$$

Recall (R): The proportion of actual positives that was identified correctly is evaluated by Recall.

$$R = \text{True Positive}/(\text{True Positive} + \text{False Negative})$$
$$R = \text{Recall}$$

F1-score: The F1-score is interpreted as a weighted average of the precision and recall, where F1-score of value 1 is best and score of value 0 is worst.

$$F1 = 2 * [(R * P)/(R + P)]$$

Accuracy (A): Accuracy is a metric for evaluating classification models. Accuracy gives the information about the fraction of predictions that our model got right. In percentage,

$$A = 100 * [(\text{True Positive} + \text{True Negative})/(\text{True Positive} + \text{True Negative}$$
$$+\text{False Positive} + \text{False Negative})]$$

A = Accuracy

Here, True Positive = TP, True Negative = TN, False Positive = FP and False Negative = FN.TP, TN, FP, and FN are decided based on following matrix. PP = Predicted positive, PN = Predicted Negative

	AP (1)	AN (0)
PP (1)	TP	TN
PN (0)	FN	FP

7 Results and Comparison

The results of the study is given in this section. The validation accuracy of two proposed model decision tree and random forest on given dataset is given below.

(1) Decision tree training accuracy: 1.0.
(2) Random forest training accuracy: 0.9949748743718593.

The transpose matrix generated by the system for both the model is also given below. Model 1 (decision tree)

	AP (1)	AN (0)
PP (1)	100	5
PN (0)	7	59

Testing accuracy = 0.9298245614035000

Model 2 (random forest)

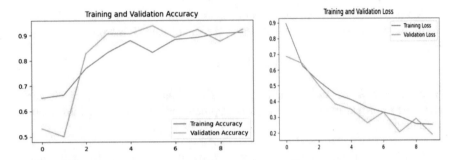

Fig. 6 Trainingand validation accuracy and loss

	Actual positive (1)	Actual negative (0)
PP (1)	103	2
PN (0)	7	59

Testing Accuracy = 0.9473684210526300

The testing accuracy for both the machine learning model is more than 90%. The model classified the cancerous and non-cancerous record in accurate manner.

In Fig. 6, we have also shown training and validation accuracy and loss after applying deep learning model on given dataset. The accuracy enhanced in each iteration and loss decreased in each iteration. The graphs shows the satisfactory results.

8 Conclusion

Implementing above mentioned methods and frameworks, we can expect much high percentage increased accuracy in detection, segments of cell nucleus in histopathology images of breast cancer. Number of researchers is currently working on deep learning concepts to achieve higher accuracy in detection and segmentation of cell nuclei. But complete accuracy is still not achieved. By the end of this research, a better and improved method of identification and segments of cell nucleus in histopathology pictures of breast cancer will be implemented. Using the concepts and parameters of machine learning, a better method will be evolved in image processing for breast cancer.

In this work, we have implemented decision tree algorithm and random forest algorithms of machine learning for classification of benign and malignant patient from given dataset.

References

1. Ma WW, Adjei AA (2009) Novel agents on the horizon for cancer therapy, CA: a cancer. J Clinic 59(2):111–137
2. May M (2010) A better lens on disease: computerized pathology slides may help doctors make faster and more accurate diagnoses. Sci Amer 302:74–77
3. Rubin R, Strayer DS (2004) In: Rubin EM, Gorstein F, Schwarting R, Strayer DS (eds) Rubin's pathology: clinicopathologic foundations of medicine, 4th ed. Lippincott Williams & Wilkins, Philadelphia, PA, USA, April2004
4. Zhou X, Wong S (2006) Informatics challenges of high-throughput microscopy. IEEE Sig Process Mag 23(3):63–72
5. Gurcan MN, Boucheron LE, Can A, Madabhushi A, Rajpoot NM, Yener B (2009) Histopathological image analysis: a review. IEEE Rev Biomed Eng 2:147–171
6. Dalle J-R, Li H, Huang C-H, Leow WK, Racoceanu D, Putti TC (2009) Nuclear pleomorphism scoring by selective cell nuclei detection. In: Proceedings of IEEE workshop applications of computer vision, p 6
7. Kulikova M, Veillard A, Roux L, Racoceanu D (2021) Nuclei extraction from histopathological images using a marked point process approach. Presented at the proceedings of SPIE medical imaging, San Diego, CA, USA
8. Mosaliganti K, Cooper L, Sharp R, Machiraju R, Leone G, Huang K, Saltz J (2008) Reconstruction of cellular biological structures from optical microscopy data. IEEE Trans Vis Comput Graph 14(4):863–876
9. Irshad H, Veillard A, Roux L, Racoceanu D Methods for nuclei detection, segmentation, and classification in digital histopathology: a review—current status and future potential
10. Wang P, Hu X, Li Y, Liu Q, Zhu X (2015) Automatic cell nuclei segmentation and classification of breast cancer histopathology images. Sig Process. ISSN 0165-1684
11. Paramanandam M, O'Byrne M, Ghosh B, Mammen JJ, Manipadam MT, Thamburaj R, Pakrashi V (2016) Automated segmentation of nuclei in breast cancer histopathology images. Public Libr Sci (PLOS). Academic Jounals
12. Faridi P, Danyali H, Helfroush MS, Jahromi MA (2016) An automatic system for cell nuclei pleomorphism segmentation in histopathological images of breast cancer. In: IEEE conference: signal processing in medicine and biology symposium (SPMB)
13. Höfener H, Homeyer A, Weiss N, Molin J, Lundström CF, Hahn HK (2018) Deep learning nuclei detection: a simple approach can deliver state-of-the-art results. J Comput Med Imaging Graph
14. He XC, Yung NHC (2004) Curvature scale space corner detector with adaptive threshold and dynamic region of support. In: IEEE international conference on pattern recognition, computer society,pp 791–794

Iterated Shape-Bias Graph Cut-Based Segmentation for Detecting Cervical Cancer from Pap Smear Cells

Sengathir Janakiraman, M. Deva Priya, A. Christy Jeba Malar, S. Padmavathi, and T. Raghunathan

1 Introduction

Pap smear test is considered to be the most potential test for diagnosing cervical cancer. Women in the age group of 30–40 must be screened at least once with pap smear test for reducing the risk by 28–38% [1]. Pap smear test is the most vital screening test for diagnosing cervical cancer in women in order to detect the infection of non-visible Human Papilloma Virus (HPV) [2]. This test refers to the sequential procedure conducted by clinicians, during which cell samples from the cervix region are taken using a swab or brush in order to examine them for identifying the existence of abnormal microscopic appearances caused due to the infections of HPV [3].

The plethora of computer-assisted cervical cancer detection techniques available in literature suffer from challenges in segmenting cervical pap smear cells [4].

S. Janakiraman
Department of Information Technology, CVR College of Engineering, Mangalpally, Vastunagar, Hyderabad, Telangana, India

M. Deva Priya (✉)
Department of Computer Science and Engineering, Sri Eshwar College of Engineering, Coimbatore, Tamilnadu, India
e-mail: devapriya.m@sece.ac.in

A. Christy Jeba Malar
Department of Information Technology, Sri Krishna College of Technology, Coimbatore, Tamilnadu, India
e-mail: a.christyjebamalar@skct.edu.in

S. Padmavathi · T. Raghunathan
Department of Computer Science and Engineering, Sri Krishna College of Technology, Coimbatore, Tamilnadu, India
e-mail: padmavathi.s@skct.edu.in

T. Raghunathan
e-mail: raghunathan.t@skct.edu.in

© The Author(s), under exclusive license to Springer Nature Singapore Pte Ltd. 2023
R. Agrawal et al. (eds.), *International Conference on IoT, Intelligent Computing and Security*, Lecture Notes in Electrical Engineering 982,
https://doi.org/10.1007/978-981-19-8136-4_30

355

The process of segmentation is considered to be more complex during the detection of abnormalities in cervical region, since the presence of irregular boundaries makes separation of cytoplasm and nucleus more challenging [3]. The process of segmentation is even more complex when the nucleus of the cell overlaps with one another making it difficult to identify the overlapping boundaries, and estimate the nucleus and cytoplasm ratio. Traditionally, the segmentation process involved in cervical cancer detection segments the nucleus and cytoplasm, and finally detects the nucleus and cytoplasm ratio. The segmentation schemes propounded for cervical cancer detection initially considers pap smear slides as the input image [5]. Then, images are pre-processed and morphological operations are performed, thus eliminating unwanted noise from input image. The segmentation methods used for cervical cancer detection are detailed below [6]. The incorporation of advanced graph cut techniques such as Neutrosophic Graph cut, Semantic Super pixel inspired Conditional Random Field-based improved Boykov Graph Cut and Hybrid linear clustering, and Bayesian Classification-based Grab cut have wide opened the way for detecting abnormalities in a cervical cell derived through pap smear test. Thus, the main scope of this research concentrates on enhancement of segmentation process by incorporating graph cuts for detecting accurate boundaries of abnormal nuclei and cytoplasm derived from cervical cell used for examination in pap smear test. This research focuses on improvement in the automated reading of cervical cytology for facilitating superior sensitivity during segmentation of abnormal nuclei and cytoplasm of smear cells.

In this paper, Iterated Shape-Bias Graph Cut-based Segmentation (ISBGCS) method is proposed for initially focusing on the development of nucleus presegmenting process that defines the size and coarse of the nucleus in order to construct a graph by the process of image unfolding. It adopts an image unfolding process in the graph cut-segmentation scheme that aids in mapping the ellipse-like border presented in Cartesian coordinate system to the lines in the polar coordinate system. The cost function estimated through the inclusion of this ISBGCS-based graph cut-segmentation scheme combines the characteristics of nucleus region and border. In addition, the proposed ISBGCS concentrates on the feasibility of estimating global optimal path from the constructed graph through the inclusion of iterative dynamic programming, which facilitates formation of optimal closed contour. This aids in estimating the precise boundaries of cytoplasm and nucleus for detecting abnormalities in the cervical cells taken for examination.

2 Related Works

The graph cut segmentation-based cervical cancer detection was first introduced by Boykov and Jolly in the year 2001 [7]. This detection scheme supports interactive segmentation by using white and black stokes for representing the white and black brushes. This scheme uses max-flow algorithm for segmenting the object, considers stokes as seeds and identifies background using seeds. However, this scheme is only a semi-automatic segmentation method as it includes intervention of human experts for segmenting regions in a potential manner. Zhang et al. [8] have adaptively and locally used graph cuts for automated segmentation of healthy and abnormal cervical cells. This scheme combines intensity, boundary, texture and region information. Concave point-based schemes are combined to divide the touching nuclei. Further, a graph cut segmentation-based cervical cancer detection approach based on nuclei and cytoplast contour detector is contributed for automated segmentation of cytoplasm and nuclei from cervical smear image considered for examination [9]. This graph cut scheme uses a thresholding approach for differentiating the cytoplast from the background, since they can be potentially distinguished based on their gray level intensities. Unfortunately, Otsu's method of thresholding is not able to achieve potential threshold when the number of data and standard deviations in different classes exhibit variation. Another, graph cut segmentation-based cervical cancer detection approach is contributed using Adaptable Threshold Detection (ATD) for automated segmentation of cytoplasm and nuclei from the cervical smear images [10]. This ATD-based graph cut segmentation scheme considers quantity of data, standard deviation and group interval as decision parameters for determining optimal thresholds that are adopted in cervical cell segmentation process: This graph cut technique is a suitable method for resolving the issues of Otsu's method of thresholding, since they are highly suitable for deriving the cytoplast contour in a predominant way. Then, an integrated local and global graph cut-based segmentation scheme is proposed for effective segmentation of cervical cells in order to effectively differentiate normal cervical cells from cancerous cervical cells [11]. This graph cut scheme significantly extracts cytoplasm boundaries by the application of A* channel algorithm and global multi-way graph cut technique when the image histograms of cervical cell images are distributed in non-bimodal nature. This graph cut scheme integrates texture, intensity, region and boundary information for segmenting abnormal nucleus. This graph cut scheme also integrates two concave approaches for splitting the touching nuclei in a predominant manner. This graph cut scheme offers 88.4% of F-measure and 93% of accuracy for segmentation of cytoplasm and nuclei boundaries derived from the abnormal nuclei. Furthermore, a graph cut segmentation scheme using local graph cuts based on Poisson distribution is proposed for modeling the background of cytoplasm and nuclei for cervical cancer detection [12]. This local graph cut-based segmentation scheme is used for improving the segmentation degree for the task of detecting cancerous cell from normal healthy cervical cells. It improves sensitivity and specificity involved in detecting cervical cancer, since it explores the structure of cytoplasm and nuclei based on the multi-dimensional constraints that helps in potential detection of pre-cancerous and cancerous cells in an effective manner.

3   Proposed Iterated Shape-Bias Graph Cut-Based Segmentation (ISBGCS) Method

In this section, the basics of graph cut and the process of establishing graphs from input image which undergoes graph cut segmentation process are detailed.

Let an undirected graph be represented as a collection of vertices and edges denoted by $G = \langle V, E \rangle$ with 'V' and 'E' representing the series of vertices and edges which connect the neighboring vertices. The set of vertices consists of two categories of nodes, in which first class of nodes pertains to the neighborhood nodes associated with each pixel image and the second class named terminal nodes represent the source (object) and sink (background). Thus, this type of graph is termed as s-t graph with 's' and 't' highlighting the object and its associated background. Further, there are two types of edges labeled as n-link and t-link in these kinds of graphs. The n-links refer to the first category of edges which connects each neighboring pixel within the image and t-link relates to the second category of edges which connects neighborhood nodes with terminal nodes. In this type of graphs, every individual edge is assigned a cost (non-negative weight). At this juncture, a cut 'C' is defined as the edges subset (E) with $C \in E$. In this context, the cost of the graph cut $|G_C|$ is determined based on the sum of the cost associated with each edge of the cut as presented in Eq. (1).

$$|G_C| = \sum_{c \in e} W_{\text{edges}} \tag{1}$$

Further, minimum cut termed as min-cut is the cut that possess minimum cost, and it is estimated by determining the maximum flow. The min-cut is considered to be equivalent to the max-flow, since the min-cut or max flow algorithm contributed by Boykov and Kolmogorov can be utilized for estimating the minimum cut for the s-t graph. Thus, the graph is partitioned by this cut, and the vertices are divided into two mutually exclusive subsets.

In general, the process of graph cut-based segmentation [13] is a pixel labeling problem. This pixel labeling process includes source node (objects) and terminal nodes (background set to 1 and 0 respectively). Further, the process of graph cut-based segmentation is facilitated by reducing the energy function by employing minimum graph cut. This minimum graph cut originates at the boundary between the object and background in order to make segmentation reasonable. However, the energy or cut must be minimized at the boundaries of the object. At this juncture, the energy function defined in Eq. (2) can be minimized by incorporating min cut over the s-t graph.

$$E_f(I) = \alpha R_t(I) + B_t(I) \tag{2}$$

where $R_t(I)$ and $B_t(I)$ correspond to the regional term and boundary term that represent the regional term and boundary term incorporating regional information and boundary constraint in the process of segmentation. 'α' is the factor of relative importance. If the value of 'α' is assigned '0', then the boundary alone is considered in energy minimization of the pixel under investigation. Further, the regional term related to energy function in Eq. (2) is defined in Eq. (3).

$$R_t(I) = \sum_{p=0}^{n} R_{t(p)}(I_P) \tag{3}$$

where 'a' and 'b' are the neighborhood pixels, and $\delta(I_a, I_b)$ is defined as

$$\delta(I_a, I_b) = \begin{cases} 1, \text{if } I_a = I_b \\ 0, \text{if } I_a \neq I_b \end{cases} \tag{4}$$

It is proved that the labels need to be assigned for each neighboring pixel corresponding to the regional constraint. In addition, if the value of penalty is '0', then the regional term is the sum of penalty used at the boundary of segmentation. It is defined as a non-decreasing function of $|I_a - I_b|$ as shown in Eq. (5)

$$B_t(a, b) = \exp\left(\frac{(I_a - 1_b)^2}{2\sigma^2}\right) \tag{5}$$

where 'σ' is the degree of noise in each pixel taken for investigation. The penalty is considered to be very high when the intensity of the neighboring pixels is similar in characteristics. In contrast, in case the intensity of neighboring pixels are non-similar, then penalty is not considered. Hence, it is evident that the energy function is capable of achieving minimum value only at the object boundaries. Moreover, Boykov and Jolly have proved that min-cut through max-flow can impactfully minimize the energy of pixels. Hence, the problem to minimise energy is handled using graph cut strategy for attaining good segmentation output. Moreover, cost assignment in s-t graph is essential. Hence, the weight of the s-t graph is defined as,

$$\text{cost} = \begin{cases} B_t(a, b)[p, q] \in \text{Neigborhood\_pixel} \\ \alpha R_{t(p)}(0) \text{for Edge } I, S \\ \alpha R_{t(p)}(1) \text{for Edge } I, S \end{cases} \tag{6}$$

From Eq. (6), it is evident that the cut is present more likely at the edges with less degree of weight, when pixel intensity is related to the object [14].

Algorithm Proposed Iterated Shape-Bias Graph Cut [15]-based Segmentation (ISBGCS) Method

Input: Cervical Pap Smear Cell Image (I_M), Prior shape model (PS_M) comprising of silhouettes of interested objects to be derived during segmentation process defined through $PS_M = \{F_{M(i)}\}_{i=1}^{|PS_M|}$.

Output: Optimal segmented cytoplasm and nuclei portion (OSCN$_{\text{Part}}$) of input Cervical Pap Smear Cell Image (I_M).

Initialization: Assign relative weights of pixel derived from input image as $H_{\text{PS}(M)}(x)$ and $H_{T0(M)}(x)$ respectively.

Process of Iterative Segmentation

Step 1: Partition the Image Gradient edge related to the Cervical Pap Smear Cell Image (I_M) into smooth fragments $\text{SF}_M = \left\{E_{M(i)}\right\}_{i=1}^{|\text{SF}_M|}$.
Step 2: Utilize shape model PS$_M$ for evaluating the edge fragment and determining its associated confidence score based on Eq. (7)

$$(\text{SF}_M)^C = \left\{V_{M(i)}, E_{M(i)}\right\}_{i=1}^{|\text{SF}_M|} \tag{7}$$

Step 3: Use the edge confidence score to determine the edge item of the specific shape bias based on Eq. (8)

$$E_W(x_i, x_j) = \delta \cdot \exp\left(\frac{-\left(V_p|x_i \in E_p - V_q|x_j \in E_p\right)^2}{2\beta^2}\right) \tag{8}$$

Step 4: Determine the global shape ($G_{\text{SH}(\tau)}$) of the object using the boundary ($\varepsilon_{(\tau)}$) estimated through the current process of segmentation ($C_{\text{SP}(\tau)}$).
Step 5: Use the global shape ($G_{\text{SH}(\tau)}$) to refine the region item determined from the specific shape bias as presented in Eqs. (9) and (10)

$$R_L(x) = \text{Max}\left\{(G_{T(\tau)}(x) - \varphi\omega_\tau(x)) * E\left(G_{\text{SH}(\tau)}, \varepsilon_{(\tau)}\right), 0\right\} \tag{9}$$

$$R_U(x) = \text{Max}\left\{(G_{T(\tau)}(x)\varphi + \omega_\tau(x)) * E\left(G_{\text{SH}(\tau)}, \varepsilon_{(\tau)}\right), 0\right\} \tag{10}$$

Step 6: Use the process of min-cut and estimate the segmentation using Eq. (11)

$$C_{\text{SP}(\tau+1)} = \underset{C_{\text{SP}(\tau)}}{\text{Arg min}}\, E\left(C_{\text{SP}(\tau)}, R_L(x), E_W(x_i, x_j)\right) \tag{11}$$

and pixels' t-link weights.
Step 7: Until convergence, repeat the steps from 4 to 6.
Step 8: Return Optimal segmented cytoplasm and nuclei portion (OSCN$_{\text{Part}}$) based on $G_{\text{SH}(\tau)}$ and $C_{\text{SP}(\tau)}$.

4   Experimental Results and Discussion

The experiments of the proposed ISBGCS method are conducted using MATLAB R2016a with respect to the Herlev Dataset for considering different cervical cancer images for investigation. Figure 1 presents the results of the systematic steps included during the implementation of the proposed ISBGCS method.

Comparative investigations of the proposed ISBGCS-based graph cut segmentation-based cervical cancer detection schemes are conducted and presented for identifying the predominant among them in terms of precision, recall, F-measure and overlap. In this comparative analysis, the proposed ISBGCS-based graph cut segmentation scheme and the baseline schemes are evaluated and studied based on percentage in precision, recall, F-Measure, and overlap for increasing number of mild dysplasia cancer cells derived from the Herlev dataset.

Figures 2 and 3 highlight the results of comparative investigation conducted based on percentage increase in precision and recall of the proposed ISBGCS and the benchmarked schemes for increasing number of mild dysplasia cancer cells. The percentage increase in the precision and recall of the proposed ISBGCS for increasing number of mild dysplasia cancer cells is determined to be superior in contrast to LGCSS and ATDGCSS schemes, since it incorporates the potential of pixel-level forecasting method that uses conditional random fields for improving the degree of semantic-based segmentation accuracy. The percentage increase in the precision

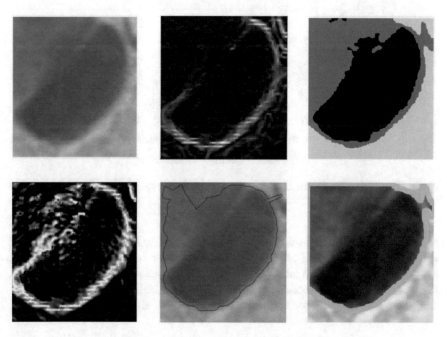

Fig. 1 Output derived from individual steps of the proposed ISBGCS method

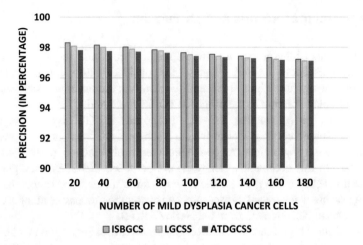

Fig. 2 Comparative investigation of proposed graph cut segmentation schemes–Precision for increasing number of mild dysplasia cancer cells

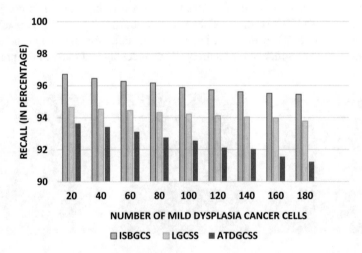

Fig. 3 Comparative investigation of proposed graph cut segmentation schemes–Recall for increasing number of mild dysplasia cancer cells

of the proposed ISBGCS on an average is confirmed to be 6% and 9% excellent when compared to LGCSS and ATDGCSS schemes for increasing number of mild dysplasia cancer cells. Similarly, the percentage increase in the recall of the proposed ISBGCS on an average is confirmed to be 5% and 7% excellent in contrast to LGCSS and ATDGCSS schemes for increasing number of mild dysplasia cancer cells.

Figures 4 and 5 demonstrate the results of the comparative investigation conducted using percentage increase in F-measure and overlap of the proposed ISBGCS and the benchmarked LGCSS and ATDGCSS schemes for increasing number of mild dysplasia cancer cells. The percentage increase in the F-measure and overlap of the proposed ISBGCS for increasing number of mild dysplasia cancer cells is determined to be superior in contrast to the compared LGCSS and ATDGCSS schemes, since it inherits the benefits of boundary adherence and superior feature extraction potential during the detection of cervical cancer. The percentage increase in F-measure precision of the proposed ISBGCS on an average is confirmed to be 8% and 10% excellent to the compared LGCSS and ATDGCSS schemes for increasing number of mild dysplasia cancer cells. Similarly, the percentage increase in the overlap of the proposed ISBGCS on an average is confirmed to be 6% and 9% excellent when compared to LGCSS and ATDGCSS schemes for increasing number of mild dysplasia cancer cells.

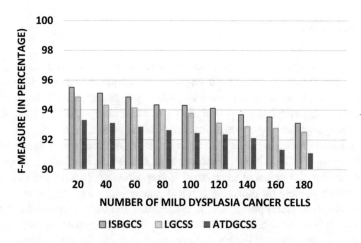

Fig. 4 Comparative investigation of proposed graph cut segmentation schemes–F-measure for increasing number of mild dysplasia cancer cells

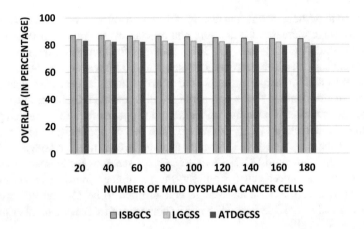

Fig. 5 Comparative investigation of proposed graph cut segmentation schemes–Overlap for increasing number of mild dysplasia cancer cells

5 Conclusion

In this paper, ISBGCS scheme is proposed with the merits of Iterated Shape-Bias Graph Cut-based Segmentation for accurate identification of nuclei and cytoplasmic boundaries in the detectio of cervical cancer from pap smear cells. It also focuses on the process of quantifying the potential of the proposed graph cut-based segmentation techniques based on evaluation metrics of classification accuracy, precision, recall and mean processing time for varying number of rounds involved in segmentation. It is proposed as a rapid graph cut approach which is efficient in appropriately segmenting the object. The interactive graph cuts are determined to be highly suitable and applicable for detecting cervical cancer that completely relies on a pure automatic segmentation approach. It is also considered to be suitable for detecting abnormalities in the cervical images, as they require maximum degree of cytoplasm and nuclei segmentation. It further focuses on identifying the predominant scheme among the three proposed schemes by investigating them for different kinds of cervical pap smear cells during the detection of cervical cancer.

References

1. Zhang L, Kong H, Chin CT, Wang T, Chen S (2014) Cytoplasm segmentation on cervical cell images using graph cut-based approach. Bio-Med Mater Eng 24(1):1125–1131
2. Zhang L, Kong H, Ting Chin C, Liu S, Fan X, Wang T, Chen S (2013) Automation-assisted cervical cancer screening in manual liquid-based cytology with hematoxylin and eosin staining. Cytometry A 85(3):214–230
3. Liu J, Wang H (2011) A graph cuts based interactive image segmentation method. J Electron Inf Technol 30(8):1973–1976

4. Kong D, Wang G (2010) Localized graph-cuts based multiphase active contour model for image segmentation. J Electron Inf Technol 32(9):2126–2132
5. Balaji G, Suryanarayana S, Sengathir J (2018) Enhanced Boykov's graph cuts based segmentation for cervical cancer detection. EAI Endorsed Trans Pervasive Health Technol 2(1):170284
6. Rajarao C, Singh RP (2019) Improved normalized graph cut with generalized data for enhanced segmentation in cervical cancer detection. Evol Intel 13(1):3–8
7. Boykov YY, Jolly MP (2001, July) Interactive graph cuts for optimal boundary & region segmentation of objects in ND images. In: Proceedings eighth IEEE international conference on computer vision. ICCV 2001, vol 1, pp 105–112. IEEE
8. Zhang L, Kong H, Chin CT, Liu S, Chen Z, Wang T, Chen S (2014) Segmentation of cytoplasm and nuclei of abnormal cells in cervical cytology using global and local graph cuts. Comput Med Imaging Graph 38(5):369–380
9. Kolmogorov V, Rother C (2007) Minimizing nonsubmodular functions with graph cuts—a review. IEEE Trans Pattern Anal Mach Intell 29(7):1274–1279
10. Egger J, Colen RR, Freisleben B, Nimsky C (2012) Manual refinement system for graph-based segmentation results in the medical domain. J Med Syst 36(5):2829–2839
11. Zheng Q, Warner S, Tasian G, Fan Y (2018) A dynamic graph cuts method with integrated multiple feature maps for segmenting kidneys in 2D ultrasound images. Acad Radiol 25(9):1136–1145
12. Ijaz MF, Attique M, Son Y (2020) Data-driven cervical cancer prediction model with outlier detection and over-sampling methods. Sensors (Basel) 20(10):2809
13. Huang J, Wang T, Zheng D, He Y (2020) Nucleus segmentation of cervical cytology images based on multi-scale fuzzy clustering algorithm. Bioengineered 11(1):484–501
14. Lee HK, Kim CH, Bhattacharjee S, Park HG, Prakash D, Choi HK (2021) A paradigm shift in nuclear chromatin interpretation: from qualitative intuitive recognition to quantitative texture analysis of breast cancer cell nuclei. Cytometry A 99(7):698–706
15. Sun X, Li D, Wang W, Yao H, Xu D, Du Z, Sun M (2021) Iterated shape-bias graph cut with application to ellipse segmentation. J Intell Fuzzy Syst 40(1):53–63

Evaluation of Deep Learning Approaches for Lung Pneumonia Classification

S. Asha⊕, Shola Usharani, and Sarvottam Ola

1 Introduction

The degree of danger of pneumonia is massive all through the world, particularly in the developing nations where still billions of individuals face vitality neediness and at last depend on the dirtying types of the vitality. As per the WHO, more than 4 million [1, 2] unexpected losses everywhere throughout the world happen yearly from the illnesses identified with family unit air contamination including pneumonia. As per estimation, on a yearly basis, more than 150 million of individuals [3] get contaminated with pneumonia which particularly incorporates youngsters younger than 5 years. In these sorts of locales, the issue can additionally increment because of the absence of clinical assets and clinical staff. For instance, as indicated by information, there is a whole of 2.4 million [4, 5] specialists and attendants in Africa's 57 countries. Here, exact and quick finding of this sickness means the world for this tremendous populace. This can offer access to treatment on opportune time to those individuals who are now encountering destitution and can spare their truly necessary time and cash. The deep neural system models have been planned traditionally, and human specialists performed probes in a standard experimentation technique. To overcome from this issue, a novel [6, 7] yet basic model is set up to consequently utilize the deep neural system design and play out the ideal characterization

S. Asha (✉)
School of Computer Science and Engineering and Centre for Cyber Physical Systems, VIT University, Chennai, Tamil Nadu, India
e-mail: asha.s@vit.ac.in

S. Usharani
School of Computer Science and Engineering, VIT University, Chennai, Tamil Nadu, India
e-mail: sholausha.rani@vit.ac.in

S. Ola
VIT University, Chennai, Tamil Nadu, India

errands. So as to pneumonia picture grouping task, the neural system engineering was uncommonly planned. The proposed method here depends on the convolutional neural system calculation, using the arrangement of neurons so as to convolve on a given picture and afterward remove the important highlights from the picture. In this investigation, characterization of chest X-beam pictures into a specific class of typical or pneumonia case will be engaged. One of the important things in this research is the classification of lung images into one of the categories among normal and pneumonia cases.

- To collect a large chest X-ray image dataset [8] containing both normal and pneumonia cases.
- To perform preprocessing, in order to extract the features from the images.
- To design and build a deep neural network model for performing the classification task successfully.
- To train the proposed model, using the available training dataset and to learn from them.
- To test the proposed model on the available test dataset in order to achieve the classification of images into normal or pneumonia cases.
- To work on the model architecture to achieve high validation accuracy for more accurate results.

In the ongoing occasions, calculations inspired by CNN profound learning [9] have become the most wanted decision with the end goal of picture order in the clinical field in spite of the fact that the cutting-edge CNN-based arrangement methods have comparatively focused neural system engineering of experimentation-based framework that have been their structuring standard. Here, SegNet, U-Net, and ResNet are a portion of the norm and unmistakable model designs with the end goal of clinical picture characterization.

To plan these kinds of models, the applicable experts have an enormous number of choices for settling on the structure choices, and the instinct here aids the manual inquiry process. The models like reinforcement learning (RL) and transformative-based models have been proposed to find or identify the ideal system hyperparameters during the way toward preparing. Nonetheless, these kinds of strategies are computationally over the top expensive, which requires a tremendous measure of preparing power. For alternate option, this research proposes a network model which is simple and efficient for handling pneumonia classification. CNNs are better than DNNs as they pose a visual scheme for processing which is similar to that of the human beings and their structure is also highly optimized for the purpose of handling the images in 2D and 3D shapes; also, they have the power to detect the two-dimensional features by process of learning. Here, in CNN, gradient-based learning algorithm is used for training the complete network which directly decreases the error criterion, and ultimately highly accurate and optimized weights can be produced by CNNs. Thus, CNN model architectures can be used for the task of image examination purpose in the medical field. The motivation behind this attempt was to explore the efficiency of existing medical image examination models and to build a deep neural network from scratch for the purpose of classification of X-ray pictures of lungs into normal or pneumonia category. In this paper, an attempt has been made to propose a model which takes chest X-ray images as input and gives output by classifying that whether the image belongs to normal or pneumonia-infected person.

The rest of the paper is organized as follows: In Sect. 2, we present the related literature to pneumonia research. Section 3 describes the proposed methodology, Sect. 4 gives the materials and methods, Sect. 5 explains the results and discussion and finally the Sect. 6 gives the conclusion and future scope of the research work.

2 Literature Survey

The previous studies have significantly contributed to the research [10] but have been unable to present a highly accurate result. It is thus essential to improve or come up with better deep learning model architectures for medical image examination which ultimately improves the disease diagnosing process in medical field.

Pneumonia is included among the top [11] diseases all over the world which causes most of deaths of people. It can be caused by bacteria, virus, fungi, etc. However, it is very difficult to identify it just by having look at the chest X-ray images. In order to save lives from pneumonia, its early detection is a much-needed condition which can be carried out through the use of deep learning-based neural network models.

Previous research has demonstrated that deep neural network models have been designed conventionally, and human specialists performed experiments on them in a regular trial and error method, but these models were not automatic in operational mode; and thus, they used to take a lot of time for the image examination process.

Afterward, calculations persuaded by CNN [12] profound learning have become the most standard decision with the end goal of picture assessment in the clinical field despite the fact that the cutting-edge CNN-based grouping procedures have comparable focused neural system engineering of experimentation framework which we can say have been their planning guideline. Here, U-Net, SegNet, and CardiacNet are a portion of the conspicuous model structures for the errand of clinical picture assessment. To plan these kinds of models, the expert's workforce has countless choices for settling on the structure choices, and the instinct here aids the manual inquiry process.

Most recent changes and upgrades in the [13] profound learning models and furthermore the accessibility of enormous datasets have helped models beat the workforce in clinical field if there should arise an occurrence of various clinical imaging undertakings, for example, discharge distinguishing proof, arrangement of skin malignant growth, diabetic retinopathy identification, and arrhythmia location. With time, the intrigue is developed in mechanized conclusion which is empowered by chest radiographs. The presentation of numerous convolutional models on assortment of variations from the norm depending on accessible open dataset found that equivalent convolutional organization does not perform respectable over all irregularities, the outfit models to a great extent improved the exactness for order when compared to the single model, and lastly, profound learning models or strategies improved the precision when compared with rule-based models.

The models like transformative-based calculations and reinforcement learning (RL) have been proposed to identify ideal system hyperparameters during the way toward preparing. Notwithstanding, these methods are computationally over the top expensive and requires a lot of computational force.

CNNs have an edge over DNNs as they contain a visual plan for preparing which is undoubtedly identical to that of the people, and their structure is additionally exceptionally advanced for the undertaking of dealing with the pictures in two-dimensional (2D) and three-dimensional (3D) shapes, just as they can remove the two-dimensional highlights through learning.

In these profound neural systems, the maximum pooling layer in the convolutional neural system is viable fit as a fiddle ingestion, and it additionally represents the inadequate associations related to the related tied loads. At the point when comparison is finished with completely associated layer systems of the identical size, CNNs have more modest number of parameters. One of the most significant elements is that CNNs utilize the inclination-based learning calculations during the preparation on the grounds as they are less inclined to the diminishing angle issue.

Since, in CNNs, the inclination-based calculation is for the most part liable to prepare the entire system with the goal that it straightforwardly diminishes a mistake measure, lastly profoundly upgraded loads can be accomplished by CCNs. Subsequently, profound learning models perform well in precision in comparison to manage based models.

Deep learning model in today's time can reach up to accuracy of human level in segmenting and analyzing an image. As the medical field is one of the most important industries, deep learning concepts can play a prominent role, especially when it is about imaging. All new techniques and advancement in deep learning are important part of the medical industry. Deep learning can be used in large variety of areas like the detection of tumors in medical images, analysis of electronic data related to health, computer-aided treatment, and the planning of drug intake and treatment, which aims at coming with the decision support for the examination of person's health. Here, key factor of success of deep learning is because of the capability of the neural networks in order to learn high-level abstraction from raw input data and images through a general learning procedure. TB diagnosis from patient's X-ray is implemented through a computer-sided detection [14]. In this paper, the authors improves the diagnosis of TB via CNN and deep learning models. Transfer learning model is used for classifying TB normal and abnormal cases [15] through pre-trained weights. Similarly, CNN-based diagnostic tool is used [16] to detect the COVID-19 patients by analyzing the features of scanned images. In [17], assist the radiologists in detecting the disease. It even evaluates and validates the generalization of errors.

In this approach, several deep learning models were introduced for the purpose of analyzing and examination of medical images. Exploration of these existing deep learning models is done here in order to know about their working methodology and architecture as follows:

2.1 LeNet-5

LeNet-5 was utilized for huge scope to consequently [18] classify manually written digits on bank checks in the USA. Then, neural system architecture for written by hand and machine-printed character acknowledgment in the 1990's was proposed which they called LeNet-5. This system is a convolutional neural system (CNN). CNNs are the establishment of current best in class profound learning-based [19] computer vision. These systems are based upon three principle thoughts: receptive fields, spatial subsampling, and shared weights. Receptive fields with shared weights are the substance of the convolutional layer, and most structures portrayed underneath use convolutional layers in different forms.

Another motivation behind why LeNet is a significant [3] design is that before it was imagined, character acknowledgment had been done generally by hand engineering, trailed by an AI model to figure out how to arrange hand-built features as shown in Fig. 1. LeNet made hand designing features excess, on the grounds that the system learns best inside representation from raw pictures naturally.

The LeNet-5 architecture comprises of two arrangements of convolutional and normal pooling layers, trailed by 3 layers of convolutional and two layers of pooling followed by a Softmax classifier.

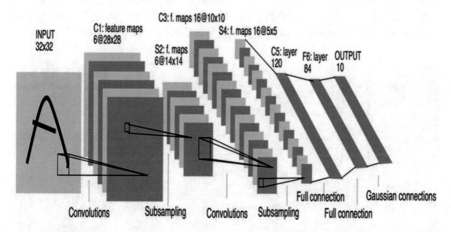

Fig. 1 LeNet architecture (taken from [20])

2.1.1 AlexNet

A couple of years back, we despite everything utilized little [3] datasets like CIFAR and NORB comprising a ten of thousands of numbers of pictures. The ongoing accessibility of enormous datasets like ImageNet, which comprises many thousands to a million number of labeled pictures, has pushed the requirement for a very competent deep learning model. At that point came AlexNet. AlexNet is a staggeringly ground-breaking model equipped for accomplishing high accuracy on exceptionally challenging datasets.

AlexNet shown in Fig. 2 is a main design [18] for any object identification task and may have gigantic applications in the computer vision segment of artificial intelligence issues. The design comprises eight layers: five convolutional layers and three fully associated layers. In any case, this is not what makes AlexNet exceptional; these are a portion of features utilized that are new ways to deal with convolutional neural networks:

- ReLU nonlinearity: AlexNet utilizes rectified linear units (ReLUs) rather than the tanh work, which was standard at that point. ReLU's preferred position is in training time; a CNN utilizing ReLU had the option to arrive at a 25% error on the image dataset multiple times quicker than a CNN utilizing tanh.
- Multiple GPUs: This was particularly awful on the grounds that the training set had 1.2 million pictures. AlexNet considers multi-GPU preparing by putting half of the model's neurons on one GPU and the other half on another GPU.

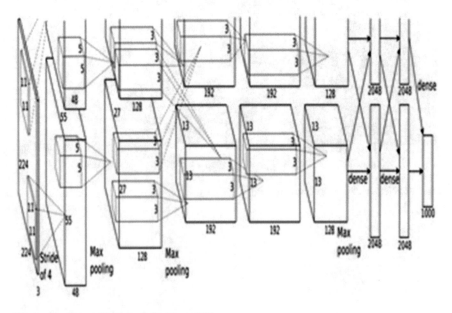

Fig. 2 AlexNet architecture (taken from [21])

- Overlapping pooling: CNNs customarily "pool" outputs of neighboring gatherings of neurons with no covering or overlapping. Nonetheless, when the creators presented overlap, they saw a decrease in error by about 0.5% and found that models with covering pooling for the most part think that it is harder to overfit.
- The overfitting problem is reduced by use of two factors as follows:

  (a) Data augmentation
  (b) Dropout.

2.1.2 VGGNet

Karen Simonyan and Andrew Zisserman introduced the VGG model architecture as shown in Fig. 3. They investigated about the effect on accuracy of the convolutional network depth in case of large-scale image examination and recognition task. They enhanced the depth of their network [18] architecture to 16 and 19 layers with the use of small 3×3 convolution filters, whereas previous networks of AlexNet emphasized on smaller strides and window sizes in first convolutional layer. The other important aspect that VGG addresses is the depth of CNNs.

Differences that separate VGG from the other existing models are:

- In place of employing huge receptive fields such as of AlexNet (11×11 including stride of 4), VGG employs small receptive fields (3×3 with just a stride of 1) as shown in Table 3. Because there present only 3 ReLUs in place of just 1, the

ConvNet Configuration					
A	A-LRN	B	C	D	E
11 weight layers	11 weight layers	13 weight layers	16 weight layers	16 weight layers	19 weight layers
input (224 × 224 RGB image)					
conv3-64	conv3-64 LRN	conv3-64 conv3-64	conv3-64 conv3-64	conv3-64 conv3-64	conv3-64 conv3-64
maxpool					
conv3-128	conv3-128	conv3-128 conv3-128	conv3-128 conv3-128	conv3-128 conv3-128	conv3-128 conv3-128
maxpool					
conv3-256 conv3-256	conv3-256 conv3-256	conv3-256 conv3-256	conv3-256 conv3-256 conv1-256	conv3-256 conv3-256 conv3-256	conv3-256 conv3-256 conv3-256 conv3-256
maxpool					
conv3-512 conv3-512	conv3-512 conv3-512	conv3-512 conv3-512	conv3-512 conv3-512 conv1-512	conv3-512 conv3-512 conv3-512	conv3-512 conv3-512 conv3-512 conv3-512
maxpool					
conv3-512 conv3-512	conv3-512 conv3-512	conv3-512 conv3-512	conv3-512 conv3-512 conv1-512	conv3-512 conv3-512 conv3-512	conv3-512 conv3-512 conv3-512 conv3-512
maxpool					
FC-4096					
FC-4096					
FC-1000					
soft-max					

Fig. 3 VGG configuration (taken from [22])

decision function nature is more discriminative. There exist fewer parameters in VGG (27 times the number of channels in place of 49 channels of AlexNet).

- VGG employs (1×1) convolutional layers in order to make decision function highly nonlinear without making change in the network's receptive fields. The smaller size of convolution filters enhances VGG to employ a large number of weight layers that ultimately would lead to enhanced or improved performance.

Two major drawbacks with VGGNet are:

(a) It is very slow to train.
(b) The weights in the network architecture are very large (in terms of bandwidth).

2.1.3 ResNet

After the triumphant presentation of AlexNet at the LSVRC2012 picture order challenge, profound learning residual network was the most earth-shattering model in the field of computer vision or in profound learning community in most recent couple of years. ResNet architecture made it conceivable [23] to prepare up to hundreds or thousands of layers which still accomplishes high compelling execution [18] as shown in Fig. 4. The presentation of numerous computer vision applications other than picture grouping has been supported as a result of its solid illustrative capacity like face acknowledgment and item identification.

The authors brought up a hyperparameter known as cardinality which represents the number of independent paths, to provide a fresh method of adjusting the network

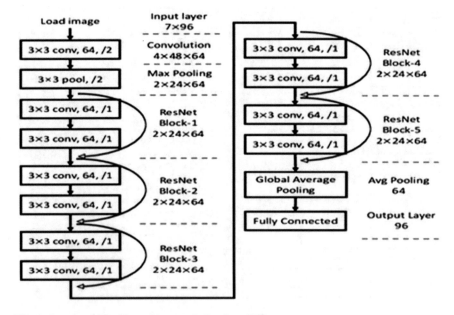

Fig. 4 Structure of ResNet architecture (taken from [23])

capacity. Experiments' result shows that the accuracy can be enhanced more efficiently by raising the cardinality than by going wider or deeper. The authors estimate that in comparison with inception, this unique architecture is more easier to adapt to fresh datasets or tasks, as it has a simple and straight forward paradigm; and, only one of its hyperparameters needs to be adjusted, whereas in case of inception, many hyperparameters (like for each path, the kernel size of convolutional layer) needs to be tuned.

The most noteworthy adjustment to know is the "Skip Connection" and the character mapping. Here, the character mapping is just to include the yield of the past layer to the following layer, and it does not have any parameters moreover. The skip connection present between layers includes the yields of past layers to yields of straightaway or stacked layers. These outcomes are eventually in the capacity to prepare more profound systems than what was before conceivable.

Advantages of ResNet:

(a) To increase the speed of the training process of the deep networks.
(b) In place of widening the network, increasing or enhancing network depth results in minimum extra parameters.
(c) Vanishing gradient problem effect is also reduced.
(d) Higher accuracy in network working performance can be achieved, especially in task of image classification.

2.1.4 Inception Network

The inception network is a very heavy engineered and complex deep learning model. It employs a lot of tricks to increase the performance in both terms, i.e., accuracy and [18] speed. Its regular evolution leads to the introduction of various versions of the network. Each version of the network is an improved and enhanced form of the previous version. Its architecture is shown in Fig. 5.

Inception V1

This version was the starting point of the inception network. It is 22 layers deep (27, when pooling layers are included). It uses global average pooling in its last inception module. This is indeed a very deep classifier. It also aims at vanishing the gradient problem.

Inception V2

It included the upgrades which decreased the computational complexity and increased the accuracy. Other upgrades to this version are:

(a) It employed the factorization of 5×5 convolution into two (3×3) convolution operations to increase the computational speed.
(b) To remove the representational congestion, the filter banks in the network module were expanded. There would be large reduction in dimensions in case

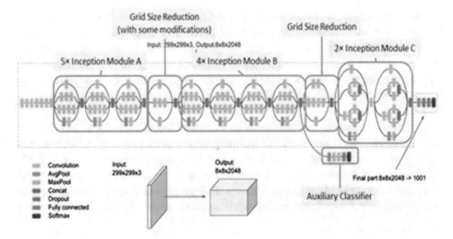

Fig. 5 Inception network architecture (taken from [24])

the module was made deeper and ultimately loss of information would have taken place.

Inception V3

Inception v3 included all above upgrades and changes stated for inception v2, and in addition to those upgrades, used the following:

(a) RMSProp optimizer.
(b) Factorized 7×7 convolutions.
(c) Label smoothing: It prevents the network from becoming too much confident about a particular class.

Advantages of inception network:

(a) In employs multilevel extraction of features. For example, it extracts the common (5×5) and local (1×1) features simultaneously.
(b) Use of multiple features' extraction from multiple filters improves the overall performance of network.
(c) The inception block is employing cross-channel correlations through performing the 1×1 convolution. Here, the spatial dimensions are ignored.

3   Proposed Model and Implementation

One of the most important requirements needed by transfer learning is the existence of models which perform quiet well on different source tasks. Luckily, the deep learning world believes in sharing. Many of the deep learning models are openly

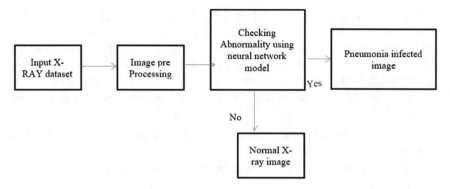

Fig. 6 Proposed model for disease identification

shared by their respective developers publicly, and these also include the two most important fields, i.e., computer vision and NLP of deep learning applications.

While sharing the pre-trained models, they also contain millions of weights and parameters gained while training them to a stable state. Existing pre-trained models are openly available for public use through various means. Python library like Keras provides a platform for downloading some existing popular models. The pre-trained model that we are implementing here is Inception v3 (Fig. 6).

The goals of the project include:

- Identifying and collecting a large chest X-ray image dataset containing both normal and pneumonia cases.
- Preprocessing of the dataset to detect the features from the images.
- Proposing a deep learning model for performing the characterization purpose.
- Training the proposed model on the available training dataset to learn from them.
- Testing proposed model on the available test dataset in order to achieve the task of classification of images into normal or pneumonia cases with high validation accuracy rate.

Inception v3 with transfer learning implementation

Human beings have an inherited ability of transferring knowledge among different tasks. Knowledge acquired during learning about one particular task can be used to solve similar or related tasks. The more similarity in the tasks makes the knowledge transfer process easier. Conventional machine learning algorithms were previously designed for working in isolation. These types of algorithms are trained to solve particular or specific tasks. Transfer learning is a prominent idea to overcome from isolated learning methodology and for employing the knowledge gained from one task to solve similar tasks. The main motivation in context of deep learning is the factor that almost all models that solve complex problems need huge amount of data and acquiring that huge amounts of labeled data for deep learning models can be difficult, considering time required for labeling the data points.

In transfer learning, you can use knowledge (weights, features, etc.) gained from earlier trained models for purpose of training upcoming/newer models, and it can also handle problems like having smaller dataset for newer task. For computer vision field, several low-level features like corners, shapes, edges, and intensity can be used across different tasks which makes enable knowledge transfer possible among different tasks. In place of training a deep neural network from scratch for your problem:

- Take and use a network that is earlier trained on a different datasets for a different tasks.
- Employ it to your dataset and your target problem/task (Fig. 7).

Working Methodology

The main idea here is to employ the weighted layers of pre-trained model to extract features but not to modify the weights of layers of model during training with different datasets for the new task (Fig. 8).

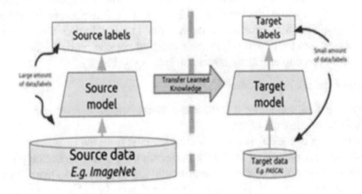

Fig. 7 Transfer learning ideas (taken from [25])

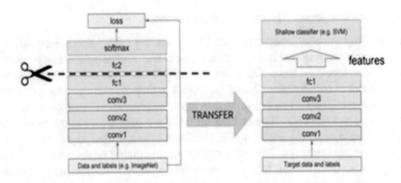

Fig. 8 Pre-trained model as feature extractor with transfer learning (taken from [25])

4 Materials and Methods

We present the point-by-point usage and assessment steps did for our proposed model to test its adequacy. Our examinations were done on a dataset containing X-beam picture of chest that is demonstrated as follows. We conveyed Keras which is a Python open-source library for profound learning system and utilized TensorFlow in backend for building and preparing motivation behind the convolutional neural system model. All tests identified with our venture were run on a PC.

- Dataset

The original dataset contains three key folders (training, testing, and validation) and two subfolders in each of them containing chest X-ray images of normal (N) and pneumonia (P) cases. A sum of 5840 X-ray pictures of lungs were taken from a repository available at Kaggle platform. This entire collection was made possible because of efforts of a number of medical personnel.

To distribute the available dataset [8, 26] into a balanced way into the training and testing folder, we modified the original data category. We rearranged a sum of 5216 images from the dataset into the training folder and the remaining 624 images in the testing folder for purpose of final testing of the model (Fig. 9).

- Preprocessing and Augmentation

We utilized different information increase techniques so as to falsely upgrade the quality and size of the accessible dataset. These stunts help in expelling overfitting-related issues and expands the model's speculation capacity during preparing process. The rescale activity is done for reason for picture amplification or decreases during the procedure of expansion.

The revolution extend tells about the range, in which the dataset pictures were haphazardly pivoted during the way toward preparing. Width move is utilized for even interpretation of the dataset pictures by 0.2%, and the tallness move means the vertical interpretation of the dataset pictures by 0.2%. Likewise, the shear go means a 0.2% of progress in picture points a standard counterclockwise way. The zoom go

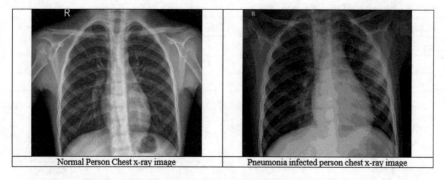

| Normal Person Chest x-ray image | Pneumonia infected person chest x-ray image |

Fig. 9 Chest X-ray of the normal and pneumonia-infected person

is utilized for zooming the dataset pictures to the proportion of 0.2%, and finally, the pictures were evenly flipped.

5 Results and Discussions

The complete design of the proposed convolutional neural system model comprises two fundamental parts that are the component extractor's part and a classifier part (sigmoid enactment work). Each layer in the element extraction process takes its past layer's yield as its information, and comparably, the yield created by this layer is passed as a contribution to the following upcoming layers: LeNet architecture, AlexNet architecture, and VGG architecture as depicted in Tables 1, 2, and 3, respectively.

The proposed engineering appeared in underneath comprises various layers, for example, convolution, max pooling, and characterization layers consolidated one after one. The element extractors comprise conv3 × 3, 32; conv3 × 3, 32; conv3 × 3, 64; max pooling layer of size 2 × 2; and a ReLU activator between these layers.

Here, outputs of convolution and max pooling layer operations are combined into form of feature maps. For the convolution operations and the pooling operations, sizes of feature maps are 34 × 34 × 128 and 17 × 17 × 64, respectively, whereas the input image is of size 150 × 150 × 3 as shown in table below. Here, point of notice is that each plane of a particular layer in network was produced by joining one or more than one planes of previous layers.

The classifier work is put toward the finish of the proposed CNN model. It is commonly alluded as a thick layer. This classifier takes a shot at singular highlights (vectors) to do computations like some other classifiers. The yield got by the element extractor part is then changed over into a one- dimensional (1D) highlight vector for the classifiers. This procedure is called as "straightening" where the convolution activity yield is smoothed to deliver single protracted component vector for usage by the thick layer in its last grouping task.

Table 1 Summary of LeNet architecture

Layer		Feature map	Size	Kernel size	Stride	Activation
Input	Image	1	32 × 32			
1	Convolution	6	28 × 28	5 × 5	1	tanh
2	Average pooling	6	14 × 14	2 × 2	2	tanh
3	Convolution	16	10 × 10	5 × 5	1	tanh
4	Average pooling	16	5 × 5	2 × 2	2	tanh
5	Convolution	120	1 × 1	5 × 5	1	tanh
6	FC		84			Tanh
Output	FC		10			Softmax

Table 2 AlexNet architecture summary

Layer		Feature map	Size	Kernel size	Stride	Activation
Input	Image	1	$227 \times 227 \times 3$			
1	Convolution	96	$55 \times 55 \times 96$	11×11	4	ReLU
	Max pooling	96	$27 \times 27 \times 96$	3×3	2	ReLU
2	Convolution	256	$27 \times 27 \times 256$	5×5	1	ReLU
	Max pooling	256	$13 \times 13 \times 256$	3×3	2	ReLU
3	Convolution	384	$13 \times 13 \times 384$	3×3	1	ReLU
4	Convolution	384	$13 \times 13 \times 384$	3×3	1	ReLU
5	Convolution	256	$13 \times 13 \times 256$	3×3	1	ReLU
	Max pooling	256	$6 \times 6 \times 256$	3×3	2	ReLU
6	FC		9216			ReLU
7	FC		4096			ReLU
8	FC		4096			ReLU
Output	FC		1000			Softmax

Table 3 Summary of VGG architecture

Layer		Feature map	Size	Kernel size	Stride	Activation
Input	Image	1	$224 \times 224 \times 3$			
1	$2 \times$Convolution	64	$224 \times 224 \times 64$	3×3	1	ReLU
	Max pooling	64	$112 \times 112 \times 64$	3×3	2	ReLU
3	$2 \times$Convolution	128	$112 \times 112 \times 128$	5×5	1	ReLU
	Max pooling	128	$56 \times 56 \times 128$	3×3	2	ReLU
5	$2 \times$Convolution	256	$56 \times 56 \times 256$	3×3	1	ReLU
	Max pooling	256	$28 \times 28 \times 256$	3×3	2	ReLU
7	$3 \times$Convolution	512	$28 \times 28 \times 512$	3×3	1	ReLU
	Max pooling	512	$14 \times 14 \times 512$	3×3	2	ReLU
10	$3 \times$Convolution	512	$14 \times 14 \times 512$	3×3	1	ReLU
	Max pooling	512	$7 \times 7 \times 512$	3×3	2	ReLU
13	FC		25,088			ReLU
14	FC		4096			ReLU
15	FC		4096			ReLU
Output	FC		1000			Softmax

The classification layer consists of a flattened layer with a dropout size of 0.5 and then two dense layers of size 64 and 2, respectively, and at the end, a sigmoid classifier function performs the task of classification of input images. To approve and look at the adequacy of the proposed model, we completed trials for different number of times for a considerable length of time. Parameter alongside hyperparameters were exceptionally tuned to build the general execution of the model. Various outcomes were accomplished during each test yet we spared the most exact outcomes that we got at definite run starting at now. So as to help the little dataset into profound learning convolutional neural system model design 2 techniques that is learning rate variety and information growth were utilized in the model.

The implementation of pre-trained inception v3 model with help of transfer learning on our domain dataset and target task gives approximately **85% of validation accuracy** for the purpose of classification of chest X-ray images into a particular category of normal or pneumonia-infected case. Plots representing the model accuracy and model loss for both training and testing processes are shown in Figs. 10 and 11.

The final results achieved are training loss of 0.5521, training accuracy of 0.7492, validation loss of 0.5247, and validation accuracy of 0.7500 (This state of model

Fig. 10 Representation of model accuracy

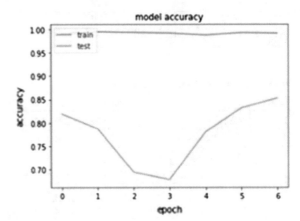

Fig. 11 Representation of model loss

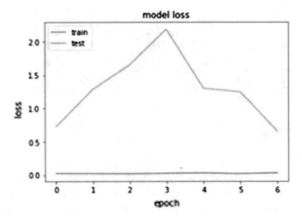

is saved finally.). Lesser validation accuracy is achieved due to large size of transformed images, whereas validation accuracy shows improvement in case of small-size images. However, the little changes in validation accuracy do not make large impact on proposed model's overall classification performance. The computation cost and training time required by larger images are also high. Finally, we proposed $150 \times 150 \times 3$ model because it achieved better validation accuracy of 75% with a training loss of 0.5521.

6 Conclusion and Future Work

As discussed, the strategy is to complete the order of pneumonia and typical cases through our proposed model from an assortment of chest X-beam pictures. Our proposed model is worked without any preparation that separates it from different models that depend to a great extent on move learning approach. Our findings during this work support the belief that deep learning methods can be employed in purpose of diseases' management and for simplify the process of diagnosis of a disease. Commonly, the pneumonia diagnosis cases are confirmed or detected by a single doctor, and this process allows error possibility at this state, but the use of deep learning models can be seen as a method of two-way confirmation system.

In this case, the proposed model which is based on chest X-ray images can provide a system that helps in making decision during diagnosis process, and later, the attending physician can confirm the case which ultimately decreases the error possibility by human and computer to a great extent. Our experiment results support the notion that use of deep learning models can enhance the treatment quality as it can better diagnose a diseases comparison to traditional methods. When compared with earlier traditional methods, our proposed model can effectively detect normal cases or pneumonia-infected cases in chest X-ray images. Later on, this work can be stretched out for motivation behind distinguishing and grouping of X-beam pictures comprising lung disease and pneumonia. As of late, the issue of qualification between X-beam pictures that contain pneumonia and lung disease has come out in this way; in the future, our work can likewise be stretched out to this tackle this issue.

References

1. Andre E, Brett K, Roberto A et al (2017) Dermatologist-level grouping of skin malignant growth with profound neural systems. Nature 542(7639):115–118
2. Naicker S, Plange-Rhule J, Tutt RC, Eastwood JB (2009) Deficiency of human services laborers in creating nations. Afr Ethn Dis 19:60
3. https://medium.com/analytics-vidhya/cnns-architectures-lenet-alexnet-vgg-googlenet-resnet-and-more-666091488df5

4. Grewal M, Srivastava MM, Kumar P, Varadarajan S (2017) Radiologist level precision utilizing profound learning for drain recognition in CT examines. http://arxiv.org/abs/1710.04934. View at: Google Scholar

5. Rudan I, Tomaskovic L, Boschi-Pinto C, Campbell H (2004) Worldwide gauge of the rate of clinical pneumonia among youngsters under five years old. Bull World Health Organ 82:85–903

6. Huang P, Park S, Yan R et al (2017) Included estimation of PC supported CT picture highlights for early lung malignant growth finding with little aspiratory knobs: a coordinated case-control study. Radiology 286(1):286–295

7. Demner FD, Kohli MD, Rosenman MB et al (2015) Setting up an assortment of radiology assessments for conveyance and recovery. J Am Med Inform Assoc 23(2):304–310

8. Shin HC, Lu L, Kim L, Seff A, Yao J, Summers RM (2015) Interleaved content/pictureprofound mining for a huge scope radiology database. In: Proceedings of the conference on computer vision and pattern recognition (CVPR), Boston, MA, USA, June 2015

9. Barret Z, Quoc VL (2016) Neural design search with fortification learning. http://arxiv.org/abs/1611.01578

10. Shin HC, Lu L, Kim L, Seff A, Yao J, Summers RM (2015) Interleaved content/pictureprofound mining for an enormous scope radiology database. In: Proceedings of the conference on computer vision and pattern recognition (CVPR), Boston, MA, USA, June 2015

11. Narasimhan, H. Earthy colored, Pablos-Mendez A et al. (2004) Reacting to the worldwide HR emergency. The Lancet **363**(9419):1469–1472

12. Melendez J, Van GB, Maduskar P et al (2015) An epic different example learning-based way to deal with PC supported identification of tuberculosis on chest x-beam. IEEE Trans Med Imaging 34(1):179–192

13. Avni U, Greenspan H, Konen E, Sharon M, Gold-berger J (2011) X-beam order and recovery on the organ and pathology level, utilizing patch-based visual words. IEEE Trans Med Imaging **30**(3)

14. Guo R, Passi K, Jain CK (2020) Tuberculosis diagnostics and localization in chest X-rays via deep learning models, 05 October 2020. https://doi.org/10.3389/frai.2020.583427

15. Rahman T, Khandakar A, Kadir MA et al Reliable tuberculosis detection using chest X-ray with deep learning, segmentation and visualization. IEEE Access. https://doi.org/10.1109/ACCESS.2020.

16. Masud M, Alshehri MD, Alroobaea R, Shorfuzzaman M Leveraging convolutional neural network for COVID-19 disease detection using CT scan images. Intell Autom Soft Comput. https://doi.org/10.32604/iasc.2021.016800

17. Hashmi MF, Katiyar S, Hashmi AW, Keskar AG (2021) Pneumonia detection in chest X-ray images using compound scaled deep learning model. J Control Meas Electron Comput Commun 62(3–4):397–406

18. Maeda-Gutiérrez V, Galván-Tejada CE, Zanella-Calzada LA et al (2020) Comparison of convolutional neural network architectures for classification of tomato plant diseases. Appl Sci 10:1245. https://doi.org/10.3390/app10041245

19. Lakhani P, Sundaram B (2017) Profound learning at chest radiography: computerized characterization of pneumonic tuberculosis by utilizing convolutional neural systems. Radiology 284(2):574–582

20. https://www.datasciencecentral.com/profiles/blogs/lenet-5-a-classic-cnn-architecture

21. https://towardsdatascience.com/convolutional-neural-network-champions-part-2-alexnet-ten sorflow-2-x-de7e0076f3ff

22. https://blog.paperspace.com/popular-deep-learning-architectures-alexnet-vgg-googlenet/

23. Choi H, Ryu S, Kim H (2018) Short-term load forecasting based on ResNet and LSTM. In: 2018 IEEE international conference on communications, control, and computing technologies for smart grids (SmartGridComm). https://doi.org/10.1109/SmartGridComm.2018.8587554

24. https://blog.paperspace.com/popular-deep-learning-architectures-resnet-inceptionv3-squeez enet/

25. Sarkar DJ (2018) A comprehensive hands-on guide to transfer learning with real-world applications in deep learning. Deep learning on steroids with the power of knowledge transfer,

15 November 2018. https://towardsdatascience.com/a-comprehensive-hands-on-guide-to-tra nsfer-learning-with-real-world-applications-in-deep-learning-212bf3b2f27a

26. Xiaosong W, Yifan P, Le L, Zhiyong L, Mohammadhadi B, Ronald MS (2017) Chest X-ray8: hospital-scale chest X-beam database and benchmarks on pitifully managed order and confinement of normal chest maladies. http://arxiv.org/abs/1705.02315. View at: Google Scholar

27. Varun G, Lily P, Marc C et al (2017) Improvement and approval of a profound learning calculation for discovery of diabetic retinopathy in retinal fundus photos. JAMA 316(22):2402–2410

28. Hermann S (2014) Assessment of sweep line streamlining for 3D clinical picture enlistment. In: Proceedings of conference on computer vision and pattern recognition (CVPR), Columbus, OH, USA, June 2014

29. Mohammad TI, Md AA, Ahmed TM, Khalid A (2017) Variation from the norm discovery and confinement in chest X-beams utilizing profound convolutional neural systems. http://arxiv.org/abs/1705.09850

30. Xue Z, You D, Candemir S et al. (2015) Chest X-beam picture see arrangement. In: Proceedings of the computer-based medical systems IEEE 28th international symposium, São Paulo, Brazil, June 2015

31. Boussaid H, Kokkinos I (2014) Quick and careful: ADMM-based discriminative shape division with loopy part models. In: Proceedings of conference on computer vision and pattern recognition (CVPR), Columbus, OH, USA, June 2014

Comparative Analysis to Classify Human Brain Anomalies for Brain Tumour

Nitu Singh and Jitendra Agrawal

1 Introduction

It is challenging to detect and treat brain tumour disease in the skull for different reasons. Early diagnosis makes treatment more accessible. The rapid growth of the brain tumour in a limited area such as the skull makes early diagnosis systems even more critical [3]. Brain tumour causes different symptoms in the patient depending on its location, type, and size. Most patients with brain tumours complain of headache [2]. However, for detecting this disease, it is necessary to have a brain MRI and can be examined by healthcare personnel and diagnosis.

In the field of computer vision and image processing, many scientific studies have been conducted on tumour detection in MR images. For example, in 2017, two different studies conducted by Harshavardhan et al. [6] and Raj et al. [15] examined the scientific studies conducted to detect brain tumours with MRI images until that year. Especially in the research conducted by the Bauer team, tumour types and types of reflections in two- and three-dimensional images were examined and determined how the clinical decision support system should detect these types. New developments have been included in feature selection, segmentation, and classification methods to be used in tumour detection. In addition, they stated that the texture properties in multimode images are advantageous compared to the others according to density or shape properties. Studies in this area have been classified in both studies, and the differences between them have been revealed. In addition, there is scientific

N. Singh (✉) · J. Agrawal
School of Information Technology, Rajiv Gandhi Proudyogiki Vishwavidyalaya, Bhopal, MP, India
e-mail: nitusingh.sept5@gmail.com

J. Agrawal
e-mail: jitendra@rgtu.net

© The Author(s), under exclusive license to Springer Nature Singapore Pte Ltd. 2023
R. Agrawal et al. (eds.), *International Conference on IoT, Intelligent Computing and Security*, Lecture Notes in Electrical Engineering 982,
https://doi.org/10.1007/978-981-19-8136-4_32

knowledge about the accuracy of the results obtained in tumour detection. The criteria used in the studies are listed.

Various studies have been conducted on the detection of a brain tumour in the literature. Tumour detection from MR images was performed using image segmentation methods and semi-supervised and supervised learning algorithms [7]. In a study conducted by Wulandari et al. [17], many segmentation methods were analysed, and their performance was evaluated [17].

In the study presented by Papezova [14], they tried to obtain precise and automatic information about the location, size, and margins of brain tumours from brain MRI images. Hybrid geodesic region-based curve evolutions for image segmentation was semi-automatically adapted to the system and demonstrated 70% success in the obtained MRI images.

In a 2018 study conducted by Abdalla et al. [1], using artificial neural networks, they made detection studies using a new architecture and achieved successful results. The proposed method stated that they developed an automatic process regardless of the shape, size, and visual features of the tumour. Some theoretical studies have been done on optimising the segmentation technique [3].

In a study conducted by Kumar et al. [10], segmentation was desired to be performed on high-resolution and two-dimensional MR images using firefly optimisation algorithms [10]. It is said that accurate segmentation is made in a shorter time than the calculation time. In a study conducted by Dandıl et al. in 2017, they surveyed to detect cancer cells in brain cells with the Otsu's algorithm [5]. Otsu's thresholding value and watershed segmentation method were applied on computed tomography (CT) images for image decomposition in this study. The primary purpose of this paper is to apply brain nodule segmentation and feature extraction using digital image processing to reduce the percentage distribution of cancer and classify disease stages.

The Kittler's method, also called the minimum error thresholding, basically works on the grey colour puck. However, current threshold algorithms are known not to work efficiently for noisy greyscale images. Three-dimensional minimum error thresholding (3D-MET) method was proposed by Jagan et al. in 2018 for dividing 3D photos, and it is claimed to be more successful in grey scales [8]. The proposed 3D-MET approach is applied with an optimal threshold separation based on the relative entropy theory and 3D histogram. A histogram consists of the grey distribution information of pixels and related information of neighbouring pixels in an image. Also, a fast recursive method is proposed to reduce the time complexity of 3D-MET; grey levels are involved here. Experimental results show that the proposed approach can provide superior segmentation performance compared to other methods for grey image fragmentation.

Bhandari et al. performed multilevel segmentation (CS) operation with Cuckoo search algorithm and wind-driven optimisation (WDO) method in 2014 using Kapur's entropy method [5]. In this study, the best solution as a fitness function has been obtained by using the CS and WDO algorithm and Kapur's optimum multilevel thresholding entropy. A new CS and WDO algorithm approach was used for the selection of the most suitable threshold value. In this technique, the correlation function is used to obtain the best solution or best fit value from the first random threshold values. They stated that they achieved successful results by applying their method to a series of satellite images.

In recent years, the MRF technique, a successful segmentation algorithm, has been used in different health fields. The imaged white lesions (dead brain cells) were segmented using the MRF technique with contrast enhancement [12]. It has been proven that the method used in the comparisons gives more successful results than other methods. In another study conducted in recent years, the segmentation of neuron interactions from MR images was again obtained using the MRF technique. In this study, graph-based methods that make automatic decomposition are also used. Likewise, in another study, ultrasound images' Tessellation of similar structures has also been successfully performed using the MRF technique [11].

This paper aimed to gives an review to detect brain tumour present in MR images with four different image separation algorithms such as MRF, Kapur, Kittler, and Otsu.

There are three more sections in the remainder of the article. In the second section that follows, the explanation of the methods used in the study, and the discussions and results are shown in the last section.

2 Segmentation Methods and Performance Criteria

In the study, the method for tumour detection consists of four main stages:

1. The first step is to prepare the image for use (image preprocessing). At this stage, unnecessary parts are trimmed. It is ensured to be in a specific contrast range, the density is normalised, and the areas that create noise are cleaned.
2. The second stage is a segmentation of brain tumours (segmentation of brain tumour images). The features used for segmentation depend primarily on the tumour type and class, as different tumour types and levels can differ significantly in appearance (e.g. contrast uptake, shape, regularity, location, etc.). In addition, feature selection will also depend on the lower compartment of tumour to be segmented. The most common property used for brain tumour disruption is image density. This is based on the assumption that different textures have different levels of grey. The same method was used in this study.
3. The third stage is the implementation of segmentation algorithms. There are two different methods in the segmentation process. These are region- and edge-based tumour analysis methods. Various algorithms can be preferred in the field of classification and clustering in segmentation operations. Region-based methods were selected in this study.
4. The last stage is to determine the accuracy of the tumour detection algorithms. Segmentation algorithms are also used here for tumour detection purposes. Normally, the segmentation process is used to analyse regions of different densities to decide whether the tumour is present and extract the tumour location.

Image segmentation is frequently used in pattern recognition, machine learning, and medical image processing. Image decomposition techniques generally use the greyscale histogram information of MR images. Colour MR images in RGB format are used by converting them to greyscale (Grey Scale) in the range of [0, 255].

Generally, the tumour region found in MR images has a minimal ratio compared to the background. This situation makes the detection of the tumour difficult. For this purpose, the probable tumour region was taken into a convex area with region of interest (ROI). In applications, the multigene region that is not more concave than rectangular is determined manually. In this way, higher success will be achieved since the region outside the ROI region will be ignored [4]. Descriptions of the four different image separation techniques are given below.

2.1 Otsu

Nobuyuki Otsu first proposed it in 1979 [13]. The Otsu's algorithm, a histogram-based thresholding method, is used to find the most suitable location on the histogram to be thresholded. In the Otsu's method, the variance between classes is taken as a basis. It is a cluster-based thresholding method. The herbaceous method assumes that the digital image consists of two objects, foreground and background. Calculates the variance value of each class. The optimum grey threshold value is found by minimising the weighted interclass variance value of the objects in the image. This method is very successful if the pixel numbers of the objects are close to each other [18]. The weighted interclass variance is found as follows:

$$\mu = \frac{\rho(r)[1 - \rho(r)]\big[\gamma_{\text{front}}(r) - \gamma_{\text{back}}(r)\big]^2}{\rho(r)\sigma_{\text{front}}^2(r) + \big[1 - \rho(r)\sigma_{\text{front}}^2(r)\big]}$$

Variance μ is the square root of the standard deviation. Here, the γ_{front} and the γ_{back} show the variance value between the classes of the front and back object in the picture, respectively. σ_{front} and σ_{back} show only the variance values of the front and back objects in the picture. "r" is a randomly chosen threshold value in the [0, 255] grey colour range. $\rho(r)$ indicates the probability of the corresponding grey level in the picture.

2.2 Kapur

The Kapur's method is based on the entropy of the greylevel histogram. The most appropriate threshold value is determined by considering the entropy values between classes. It is an entropy-based image thresholding method [16]. Kapur's method treats the digital image as two objects as foreground and background. It takes the entropy of the objects and calculates the sum for each grey level, where 0 is black and 255 is white. The grey level at which the total entropy of both objects reaches the maximum point is called the Kapur's threshold value. The image can be decomposed into two different objects from the Kapur's threshold value found. The total entropy $(E_{front} + E_{back})$ is found as:

$$E = -\sum_{c=0}^{r} \frac{\rho(c)/}{\rho(r)} \log \frac{\rho(c)}{\rho(r)} - \sum_{c=r+1}^{C} \rho(c)/\rho(r) \log \rho(c)/\rho(r)$$

Here, c is the grey level in the range of $[0, 255]$, and C is the maximum value of grey level in the image.

2.3 Kittler

It is a cluster-based thresholding method developed by Kittler and Illingworth [9]. Unlike most thresholding methods in the literature, Kittler acknowledged that two objects of a digital image can have very different standard deviation and variance values and stated that they should be considered Gaussian curves. Therefore, he transformed the thresholding problem into a Gaussian fitting problem.

$$\rho_k(r) = \sum_{c=p}^{q} j(c)$$

$$\mu_k(r) = 1/\rho_i(r) \sum_{c=a}^{b} c.j(c)$$

$$\sigma_k^2(r) = 1/\rho_i(r) \sum_{c=p}^{q} c - \mu_k(r)^2.j(c)$$

Classification rule has determined the threshold value with a Bayesian function developed by it. Here, p and q are:

$$p = \begin{cases} 0, k = 1 \\ r + 1, k = 2 \end{cases}$$

$$q = \begin{cases} r, k = 1 \\ n, k = 2 \end{cases}$$

Fig. 1 Eight pixels adjacent
to I

1	2	3
4	*i*	5
6	7	8

2.4 Markov Random Field (MRF)

The probability-based MRF algorithm labels the pixels in the image according to their dispersion functions. In an S matrix of $M \, \ddot{O} \, N$, the pixels are indexed, and the target region is determined. The neighbour system defined for S is defined such that

$$N = \{N_i | \forall i \in S\}$$

Clique $c \in C$ is the set of adjacent points. Horizontal, vertical, and diagonal cliques are used in the application. In the $M \times N$ matrix, all pixels are tagged to the object they belong to. Assuming $L = 1, 2, \ldots, k$ as the object, then $k = 2$ in this scenario (0 represents the image; 1 represents the background). If the random variable X is taken while labelling S, the formula $XS \in L$ gives the energy value of the X in the S pixel. According to the Hammersley–Clifford theorem, the energy density of X is given by the following Gibbs density (Fig. 1):

$$p(x) = Z^{-1}.e^{-1/TU(x)}$$

T is the temperature coefficient to be reduced in MRF and is taken as 1 unless otherwise specified. The energy $U(x)$ of the pixel in the middle of the eight neighbours is found as follows:

$$U(x) = \left(\sum v(x) \right)$$

$c \in C$ the Clique energy $V(x)$ of the central pixel is as in the formula below and is calculated by finding the sum of the Clique energies of eight neighbours. The sum of two different energies can find the total energy of a pixel. The first of these is the logarithm of the Gaussian distribution formula that contains the mean and standard deviation of the object to which the pixel belongs. The second is the conditional probability of eight neighbours of the central pixel.

In this way, two different energy states are calculated according to the probability that each pixel belongs to the tumour or the background. While the MRF method is running, the class of the relevant pixel in each cycle is changed with the larger of the two energy mentioned above. Iterations are continued until the classes of pixels do not change. The final classification result distinguishes the brain tumour from the background. In this study, both the tumour and the pixels of the brain.

3 Performance Measurement Metrics

As shown in Fig. 2, in the histogram graph obtained after the segmentation process, the image to be separated is divided into two groups: tumour and background [22].

False-positive (FP), false-negative (FN), true-positive (TP), and true-negative (TN) values show the success of the parsing process. For example, suppose the tumour region is labelled as "Tumour" in the segmentation process. In that case, the result is TP; if the tumour region is marked as "Background", then FN; if the background is labelled as "Tumour", then FP; and finally, if the background is labelled as "background", the result is TN. In cases where TP, TN, FP, and FN values indicate the number of pixels, the success rate of the parsing process, that is, the accuracy (Accuracy), is found with the formula number 11.

$$\text{Accuracy} = (t_p + t_n)/(t_p + t_n + f_p + f_n)$$

TP, TN, FP, and FN values are transferred to a table defined as the confusion matrix as shown in Table 1. This table contains the accuracy of the prediction results given by a particular classifier in a specific dataset in a two class datasets.

As shown in Table 1, MRI with tumour and MRI without tumour have considered two different types of classes. TP gives the number of MR images that are correctly classified, that is, the actual tumour is detected.

Fig. 2 Eight pixels adjacent to *I*

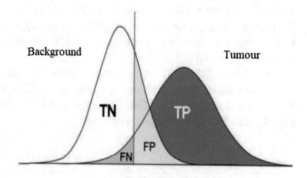

Table 1 Confusion matrix

	Tumour MRI	Tumour-free MRI
Tumour MRI	TP	FN
Tumour-free MRI	FP	TN

The FN number indicates that the MR image, which is a tumour, was detected as tumour-free. FP, on the other hand, shows that tumour-free MR images are predicted as tumourous. Finally, the TN number shows that tumour-free MR images can be detected accurately, that is, tumour-free.

Using the complexity table, the accuracy rate is obtained with formula number 11, true-positive rate (TPR) with formula number 12, and false-positive rate (FPR) with formula number 13.

TPR and FPR values varying in the range of [0, 1] are used to quantify the success of the algorithms. In addition, TPR and FPR values are also used to plot the ROC space. If necessary, success criteria such as precision, recall, and F1-score are also found using this table.

4 Comparative Result Analysis

Experimental results were obtained by using 23 MR images in four different algorithms. Eighteen were positive (cancer), and five were negative (not cancer). In the image separation process performed on 23 images, the relevant regions were marked with ROI and applied to four different algorithms.

A pixel-based performance evaluation in image parsing is a challenging task. For this reason, a performance evaluation method was preferred to determine whether tumour and non-tumour regions were detected correctly in 23 MR images obtained in our study. For this reason, the complexity matrix table is used. A separate complexity matrix was obtained for each algorithm used.

In Table 2, there are experimental results obtained from 23 MR images of Otsu's, Kapur's, Kittler's, and MRF algorithms, respectively. Complexity matrix, TPR, and FPR values were found for each technique. As is known, TPR shows the ratio of the number of correctly labelled tumour images within the tumour-labelled images. FPR, on the other hand, represents the ratio of MR images that are tumour but marked as tumour-free to those marked as tumour-free.

As shown in Table 2, if the algorithms were to be ranked from the most successful to the most unsuccessful with only the accuracy (ACC) criterion, MRF would be the first with a success rate of 87%. Others would be Kittler with 78.2%, Kapur with 60.8%, and Otsu with 65.2%, respectively.

Table 2 Comparative analysis

	True positive	False positive	True negative	False negative	TPR	FNR	Accuracy
Ostu	13	5	2	3	0.772	0.6	0.652
Kapur	11	7	3	2	0.611	0.4	0.608
Kittle	15	3	3	2	0.833	0.4	0.782
MRF	16	1	4	2	0.944	0.2	0.87

5 Conclusion

In this paper, a detail review was carried out on image processing technique to classify brain tumour, one of the most critical diseases. The study supports the decision of doctors in hospitals regarding tumour detection. As it is known, the final decision in disease detection belongs to the doctors. However, in the paper, it was aimed to minimise the margin of error of the doctors. For this reason, the study has been detailed to obtain the best detection system by considering different anomaly situations. As a result of the study, it was observed that the MRF algorithm was found to be better in the segmentation of medical data, and different results could be obtained by trying new algorithms. If an algorithm finds the output inconsistent, it has been seen that the work can be focused on the desired area with ROI. In the meantime, it has been stated that different ways and methods can be developed for the detection of brain tumour by imaging methods. As a result of the study, it was seen that the MRF algorithm is better than other methods in terms of segmentation of medical images.

References

1. Abdalla, HEM, Esmail MY (2018) Brain tumor detection by using artificial neural network. In: 2018 international conference on computer, control, electrical, and electronics engineering (ICCCEEE), pp 1–6. https://doi.org/10.1109/ICCCEEE.2018.8515763
2. Akbar S, Nasim S, Wasi S, Zafar SMU (2019) Image analysis for MRI based brain Tumor detection. In: 2019 4th international conference on emerging trends in engineering, sciences and technology (ICEEST), pp 1–5. https://doi.org/10.1109/ICEEST48626.2019.8981681
3. Arunmozhi S, Sivagurunathan G, Karpaga Meenakshi P, Karishma S, Rajinikanth V (2020) A study on brain tumor extraction using various segmentation techniques. In: 2020 international conference on system, computation, automation and networking (ICSCAN), pp 1–4. https://doi.org/10.1109/ICSCAN49426.2020.9262381
4. Babu AE, Subhash A, Rajan D, Jacob F, Kumar PA (2018) A survey on methods for brain tumor detection. In: 2018 conference on emerging devices and smart systems (ICEDSS), pp 213–216. https://doi.org/10.1109/ICEDSS.2018.8544353
5. Dandıl E (2017) Implementation and comparison of image segmentation methods for detection of brain tumors on MR images. In: 2017 international conference on computer science and engineering (UBMK), pp 1025–1029. https://doi.org/10.1109/UBMK.2017.8093425
6. Harshavardhan A, Babu S, Venugopal T (2017) An improved brain tumor segmentation method from MRI brain images. In: 2017 2nd international conference on emerging computation and information technologies (ICECIT), pp 1–7. https://doi.org/10.1109/ICECIT.2017.8453435
7. Hussain A, Khunteta A (2020) Semantic segmentation of brain tumor from MRI images and SVM classification using GLCM features. In: 2020 second international conference on inventive research in computing applications (ICIRCA), pp 38–43. https://doi.org/10.1109/ICIRCA 48905.2020.9183385
8. Jagan A (2018) A new approach for segmentation and detection of brain tumor in 3D brain MR imaging. In: 2018 second international conference on electronics, communication and aerospace technology (ICECA), pp 1230–1235. https://doi.org/10.1109/ICECA.2018.8474874
9. Kittler J, Illingworth J (1986) Minimum error thresholding. Pattern Recogn 19(1):41–47. https://doi.org/10.1016/0031-3203(86)90030-0
10. Kumar SB, Panda R, Agrawal S (2020) Brain magnetic resonance image tumor detection and segmentation using edgeless active contour. In: 2020 11th international conference on

computing, communication and networking technologies (ICCCNT), pp 1–7. https://doi.org/10.1109/ICCCNT49239.2020.9225296

11. Lin C, Wang Y, Wang T, Ni D (2019) Segmentation and recovery of pathological MR brain images using transformed low-rank and structured sparse decomposition. In: 2019 IEEE 16th international symposium on biomedical imaging (ISBI 2019), pp 1878–1881. https://doi.org/10.1109/ISBI.2019.8759441

12. Mathew AR, Babu Anto P (2017) Tumor detection and classification of MRI brain image using wavelet transform and SVM. In: 2017 international conference on signal processing and communication (ICSPC), pp 75–78. https://doi.org/10.1109/CSPC.2017.8305810

13. Otsu N (1979) A threshold selection method from gray-level histograms. IEEE Trans Syst Man Cybern 9(1):62–66. https://doi.org/10.1109/TSMC.1979.4310076

14. Papezova M, Faktorova D (2015) Automatic localization of epileptic seizures as a symptom of brain tumor. In: 2015 38th international conference on telecommunications and signal processing (TSP), pp 1–5. https://doi.org/10.1109/TSP.2015.7296427

15. Raj CPS, Shreeja R (2017) Automatic brain tumor tissue detection in t-1 weighted MRI. In: 2017 international conference on innovations in information, embedded and communication systems (ICIIECS), pp 1–4. https://doi.org/10.1109/ICIIECS.2017.8276094

16. Sen H, Agarwal A (2017) A comparative analysis of entropy based segmentation with Otsu method for gray and color images. In: 2017 international conference of electronics, communication and aerospace technology (ICECA), vol 1, pp 113–118. https://doi.org/10.1109/ICECA.2017.8203655

17. Wulandari A, Sigit R, Bachtiar MM (2018) Brain tumor segmentation to calculate percentage tumor using MRI. In: 2018 international electronics symposium on knowledge creation and intelligent computing (IESKCIC), pp 292–296. https://doi.org/10.1109/KCIC.2018.8628591

18. Öziç MÜ, Özbay Y, Baykan ÖK (2014) Detection of tumor with Otsu-PSO method on brain MR image. In: 2014 22nd signal processing and communications applications conference (SIU), pp 1999–2002. https://doi.org/10.1109/SIU.2014.6830650

Review on Customer Segmentation Methods Using Machine Learning

Rishi Gupta, Tarun Jain, Aditya Sinha, and Vishwas Tanwar

1   Introduction

There are several ways a business tries to attract new customers all the while trying to retain its current customers. A few decades back, these businesses practiced what was known as mass marketing—in which companies tried to sell the most popular product to all their customers or they practiced product differentiation—in this, companies offered a variety of products to a large market.

But, as technology rapidly evolved, companies moved to a newer approach—personalizing the products and targeting them to a specific market segment. Customer segmentation (or market segmentation as it is widely known) is the most approachable method for obtaining the above result. Understanding their customers and the market has allowed businesses to service each of their customers with care and personalization and proved the value of focusing on them. The oldest and most traditional form of segmentation is demographic segmentation, although with recent developments in big data, newer forms have emerged [1].

These approaches have also taken into consideration buyer attitudes, motivations, patterns of usage, and preferences [2]. The platform that utilizes segmentation the most is the e-commerce industry. Since the birth of e-commerce, sellers are constantly looking for ways to expand their reach while also maintaining a stable relationship with their frequent customers. And, in the age of booming social media, consumer data is as readily available as daily bread. Companies like Facebook, Google sell billions of dollars worth of consumer data to corporations which are then used as a basis of market segmentation. These e-commerce giants like Amazon, Flipkart target each segment with personalized promotional offers as well as personalized advertisements driving in clicks and eventually large amounts of profits. In this

R. Gupta (✉) · T. Jain · A. Sinha · V. Tanwar
Manipal University Jaipur, Rajasthan 303007, India
e-mail: genieousrishi@gmail.com

© The Author(s), under exclusive license to Springer Nature Singapore Pte Ltd. 2023   397
R. Agrawal et al. (eds.), *International Conference on IoT, Intelligent Computing and Security*, Lecture Notes in Electrical Engineering 982,
https://doi.org/10.1007/978-981-19-8136-4_33

Fig. 1 Market segmentation

article, we shall look at the currently available approaches to customer segmentation and their merits and demerits [14].

2 Background

The applications of customer segmentation are irreplaceable. It has now become the heart of product marketing and strategy in any industry. It is an indispensable tool for organizations to understand the market, whom to target with what product, and how to optimize the marketing strategy [13].

With the increasing popularity of social media, consumer data is readily available for businesses to use and profile the customers, to understand them better, and to act accordingly.

Let us say a website has a new user. The user would be asked to register an account on the website, for which they would be offered a discount coupon. Instead of offering every new user the same coupon, the website could—with the help of customer segmentation—give out targeted discount coupons. This will increase the likelihood of that user buying a product rather than window shop. It is also useful for planning customer communications as it allows for personalized recommendations for each group (Fig. 1).

3 Customer Segmentation

Cluster analysis or clustering is the task of grouping a set of objects in such a way that objects in the same group (called a cluster) are more similar (in some sense) to each other than to those in other groups (clusters) [7] (Fig. 2).

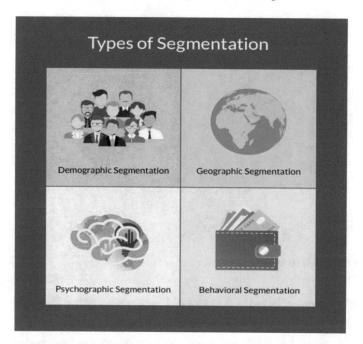

Fig. 2 Types of customer segmentation

Domains: Customer segmentation can be applied to many domains. These domains are defined by the type of data they are using for segmentation. The most prominent of these are listed below.

3.1 Demographic Segmentation

The oldest and the most traditional form of market segmentation, demographic customer segmentation, focuses on the structure of the population based on everyone's current living status. Age, gender, income, education, marital status, etc. are generally considered as demographic data. Segmentation performed on such data often yields basic groups which are the go-to for target marketing—if any other type of data is not available.

Most frequent segments formed after demographic segmentation include:

Students—this group involves unmarried, young adults/adolescents who have little or no income.

Parents—this group usually has married, middle-aged adults who have kids (dependents) and have a stable income.

Women restarting their careers—these middle-aged women tend to have a low academic background and low income.

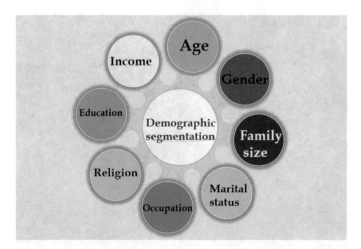

Fig. 3 Demographic customer segmentation

Senior citizens—this group includes old people who either live on a monthly pension or rely on their children, and they tend to live alone and so on.

These segments have different spending habits which can be directly related to their living scenario (Fig. 3).

3.2 Psychographic Segmentation

These are subjective attributes about a person that is not as easily available as other forms of data.

In this, data cannot be directly fed into an algorithm and is expected to yield results. The data and results must be interpreted by a psychological professional. Just as demographic segmentation emphasizes a person's current living situation, psychographic segmentation utilizes a person's real-life behavior which pertains to things like personality traits, values, attitudes, interests, lifestyles, subconscious, and conscious beliefs, motivations [2], etc. (Fig. 4).

3.3 Behavioral Segmentation

While demographic and psychographic segmentation focuses on who a customer is, behavioral segmentation focuses on how the customer acts [13]. The most widely used and most efficient form of segmentation—behavioral—relies on how the customer acts toward the company's services or services related to it. This involves

Fig. 4 Psychographic
customer segmentation

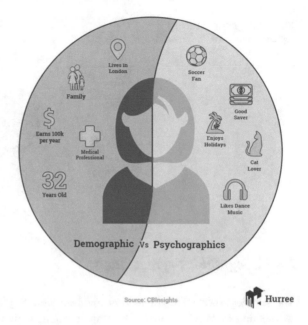

gathering data, performing analysis, and targeting the customers in real time. Therefore, this type requires higher computational power than others. Purchasing habits,
spending habits, user status, and brand interactions are the main attributes companies
look toward for behavioral analysis (Fig. 5).

Fig. 5 Behavioral customer
segmentation

Fig. 6 Geographic customer segmentation

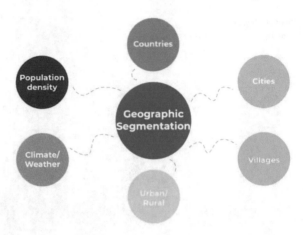

3.4 Geographic Segmentation

The simplest form of segmentation involves gathering the location of a user and segmenting them based on it. Geographic segmentation utilizes a customer's ZIP code, city, country, a radius around a certain location, climate, urban, or rural [14] attributes for segmentation.

Geographic segmentation is an effective methodology used by organizations with large national or international markets to better understand the location-based attributes that comprise a specific target market [11] (Fig. 6).

4 Pros and Cons

4.1 Demographic Segmentation

Pros:

- Demographic variables are quite easier to collect and measure when compared with other segmentation techniques.
- Targeting is usually more straightforward when using demographics as a metric—for example, you can target consumer groups, such as women or men between the ages of 40 and 50.
- Consumer profiles are more comprehensible across the board, making it easier to develop strategy among various departments (sales, customer service).

Cons:

- The model offers limited insight for marketers. Similar demographics among customers do not always mean that they all have the same needs.

- Because customers can have varying needs, a 'one-size-fits-all' approach to consumers based solely on broad demographics can make your marketing message ineffective.
- Skewed or problematic demographic data within a given region can produce unreliable assumptions, which can reduce the accuracy of your marketing methods.

4.2 Behavioral Segmentation

Pros:

- Marketers can build targeted consumer segments based on their responsiveness to certain product categories, promotion types, or path-to-purchase preferences.
- Monitoring and understanding the behavior of consumers online has become easier due to advances in data collection and tracking technologies.

Cons:

- While consumer behavior can be tracked, it is not always easy to pinpoint the motivations behind those behaviors with segmentation models, because they can vary greatly from person to person.
- Behavioral segmentation is often based on complex data constructs that are not always easy to understand without the help of a large team of data scientists and marketers.

4.3 Psychographic Segmentation

Pros:

- Marketers can get some insight into customer motivations.
- Psychographic segmentation can help brands in executing more emotive marketing to highly responsive segments.

Cons:

- Psychographic surveys which are self-reported can be inaccurate.
- Although marketers can use predictive modeling to create statistical projections, the accuracy of all these predictions firmly depends on the quality of the data used.
- Marketers need clear rules about how to interpret psychographic data to ensure a consistent approach among the individuals or departments that engage in customer segmentation analysis.

4.4 Geographic Segmentation

Pros:

- People in different communities have different needs. Something useful for someone in a more rural area, like gardening supplies, may not have any appeal to city dwellers. Separating consumers by where they live can ensure that marketers are only targeting those that may want or need your products or services.
- Marketers can adjust advertised pricing based on the cost of living of the customers that they are targeting.
- You can tailor products based on the local preferences of your customers. For example, certain clothes are more popular in some Canadian regions than others.
- You can use this segmentation to target customers in new areas where you want to grow your business.

Cons:

- This type of audience segmentation makes the assumption that people living in the vicinity of each other have got similar needs, which is not always true. This is true if you are targeting a wider geographic area.

5 Methodology

Although customer segmentation can be done through several methods, the structure is similar for all.

Prerequisites: Segmentation involves processing data; though any system can run the algorithms, larger data requires higher computational power.

5.1 Data Collection

There are many ways to collect consumer data, but since each domain requires a different type of data, data collection methods vary between them.

Demographic data can either be directly asked from the customer through surveys or be bought from data resources such as social media, census. Psychographic data is the least readily available data and contains sensitive personal data; therefore, it must be directly obtained from the users. Behavioral data is obtained through analysis of users' actions—be it clicks on advertisements or purchase habits, etc. Lastly, obtaining geographic data is a one-step process since it includes only the location.

5.2 Data Preprocessing

The raw data collected needs to be processed before analysis and segmentation. This involves removing outliers, null values, duplicates, and corrupted data. Principal component analysis also needs to be performed to compute the principal component and change the dimensionality of the data by selecting the first few principal components and ignoring the rest.

This ensures that machine learning algorithms are not overburdened and makes data visualization easier.

5.3 Data Analysis

In statistics, exploratory data analysis is an approach to analyzing datasets to summarize their main characteristics, often using statistical graphics and other data visualization methods [6]. EDA is a valuable tool that is used to analyze the data at hand and develop a model based on which the machine learning algorithm is performed.

Mainly performed in two ways: univariate analysis—for data consisting of a single variable—and multivariate analysis (aka relationship analysis)—for data consisting of more than one variable.

Data visualization is a major step involved in EDA, different forms of visualization of data include—box plots and histograms (for univariate analysis)—scatterplot, and bar charts (for multivariate analysis) (Fig. 7).

Fig. 7 Exploratory data analysis

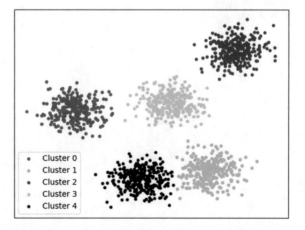

- Cluster 0
- Cluster 1
- Cluster 2
- Cluster 3
- Cluster 4

5.4 Segmentation

This is the main step in the whole process. This involves performing machine learning algorithms to obtain the segments and various fields attributed to it. There are several ways to perform segmentation (some do not involve machine learning), but to achieve the highest efficiency, some are more suitable to a type of segmentation than others.

But, before segmentation is performed, we need to find the optimal number of clusters or segments. This can be done using hyperparameter tuning and elbow method.

Parameters that define the model architecture are referred to as hyperparameters, and thus, this process of searching for the ideal model architecture is referred to as hyperparameter tuning [8]. When it comes to clustering algorithms, the hyperparameter in question is the number of clusters that need to be made. This is performed with the help of the elbow method.

These methods are discussed in the next section (Fig. 8).

6 Methods Available

There are many ways to perform customer segmentation; in this article, we will mainly focus on how machine learning algorithms compare in terms of usability and efficiency for segmentation.

Fig. 8 Customer segmentation process

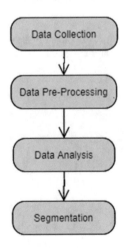

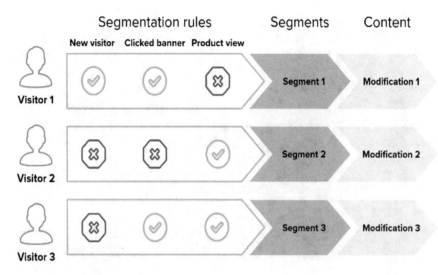

Fig. 9 Rule-based segmentation

6.1 Rule-Based

Rule-based segmentation is based on criteria such as Boolean logic or thresholds and is often two-dimensional; this is traditionally the easiest type of targeting for many professionals, as one can do it by filtering in Excel, e.g., marketers have traditionally segmented customers based on heuristics such as the industry, company size (in B2B) or age, income, etc. in B2C [3]. This follows an 'if A then B' type of rules and forms an algorithm (Fig. 9).

6.2 Supervised Clustering with Decision Tree

This method uses a specific target or dependent variable, and the target would predict differences in independent variables (input). Data utilized in this method is previous purchase patterns and customer demographic. The algorithm used is the decision tree with the target on their nodes. According to Baer, although this method connects the target with the other customer attributes, it shows only one aspect of customer behavior [4]. This is used for demographic and behavioral segmentation [13, 14].

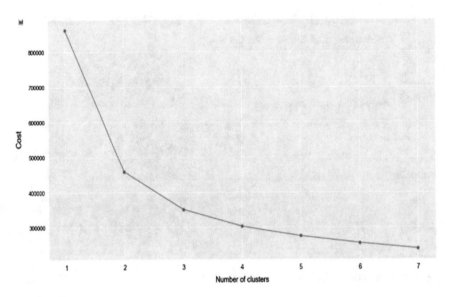

Fig. 10 Elbow method

6.3 k-means

First, for each centroid, the algorithm finds the nearest points (in terms of distance
that is usually computed as Euclidean distance) to that centroid and assigns them to
its category. Second, for each category (represented by one centroid), the algorithm
computes the average of all the points which has been attributed to that class. The
output of this computation will be the new centroid for that class [5]. To find the
optimal number of clusters, we use the elbow method (hyperparameter tuning). The
elbow method is a heuristic used in determining the number of clusters in a dataset.
The method consists of plotting the explained variation as a function of the number
of clusters and picking the elbow of the curve as the number of clusters to use [9].
This algorithm is the best algorithm for behavioral segmentation (Figs. 10 and 11).

6.4 k-prototype

K-prototype is a clustering method based on partitioning. Its algorithm is an improved
form of the k-means and k-mode clustering algorithm to handle clustering with mixed
data types [10]. It is used for datasets that have both numerical and categorical data
types. For numerical data types, it uses the k-means algorithm and calculates the
average after each iteration. For categorical data types, it uses the k-mode algorithm
which calculates Euclidean distance from each cluster center and calculates modes
after each iteration. To find the optimal number of clusters, we use the elbow method.

Fig. 11 k-means clustering

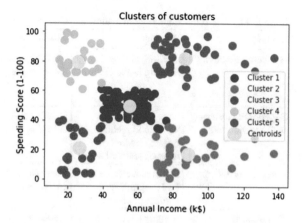

6.5 k-medoid (PAM)

A problem with the k-means and k-means++ clustering is that the final centroids are not interpretable. The idea of k-medoids clustering is to make the final centroids of the actual data points. This result makes the centroids interpretable [12].

The working is the same as the k-means, and the only difference comes in the updating centroids step in between iterations. Instead of computing the mean of points in a cluster (like in k-means), we swap the previous centroid with all other $(m-1)$ (if there are m-point in a cluster) points from the cluster and finalize the point as new centroid that has a minimum loss [12] (Table 1).

7 Conclusion

We discussed several ways a company profits from customer segmentation, its methodology, and various ways it can be performed. Even though there are many types of and any ways to perform customer segmentation, the purpose of each remains the same to personalize a business service and experience for its customers. For the last two decades, market segmentation has taken over every big and small company's marketing scheme. Every corporate has a data analyst in its marketing department. It is left to be seen what the future holds for artificial intelligence incorporated with business.

Table 1 Methods for customer segmentation

Method	Working	Advantage	Disadvantage
Business rule-based	Traditional targeting by filtering in Excel	Easy to apply, use database query	Tends to rely on heuristics developed over time and is slow to adapt to changes [3]
Supervised clustering with decision tree	Uses a dependent variable to predict differences in independent variables	Classify customers according to target	Uses one variable to cluster
k-means clustering	Uses unlabeled data to find a significant number of clusters	Uses' any number of customers' attributes	Speed of computation depends on k-values
Hierarchical clustering	Initially treats each observation as a cluster, then repeatedly merges two similar clusters	Relatively straightforward to program, no need to specify the number of clusters required	Very high time complexity compared to k-means
PAM clustering (k-medoid)	Finds a sequence of objects called medoids that are centrally located in clusters, and clusters are constructed by assigning each observation to the nearest medoid	Effectively deals with the noise and outliers present in data	Since the first k-medoids are chosen randomly, different results may be obtained on the same dataset
k-prototype	Combines working of k-means (for numerical values) and k-modes (for categorical values) algorithms	Can be applied to datasets with mixed data types, whereas k-means can be applied to only numerical data types	Unclear what weights have to be given to categorical variables

References

1. Marcus C (1998) A practical yet meaningful approach to customer segmentation, 1st ed. J Consumer Mark 15
2. Yesbeck J (2021) Types of market segmentation, 18th May 2021. https://blog.alexa.com/types-of-market-segmentation/
3. Elizabeth (2019) Customer segmentation: rules-based vs. K-means clustering, 1st ed. d3mlabs, Denmark
4. Sari J, Nugroho L, Ferdiana R, Santosa P (2011) Review on customer segmentation technique on E-commerce, 1st ed. American Scientific Publishers, Indonesia
5. Alto V (2021) Unsupervised learning: K-means vs hierarchical clustering, 18th May 2021. https://towardsdatascience.com/unsupervised-learning-k-means-vs-hierarchical-clustering-5fe2da7c9554
6. "Exploratory data analysis" Wikipedia. Wikimedia Foundation, 7 May 2021. en.wikipedia.org/wiki/Exploratory_data_analysis

7. "Cluster analysis" Wikipedia. Wikimedia Foundation, 2 June 2021. en.wikipedia.org/wiki/Cluster_analysis
8. Jordan J (2021) Hyperparameter tuning for machine learning models, 4th June 2021. https://www.jeremyjordan.me/hyperparameter-tuning/
9. "Elbow method (clustering)" Wikipedia. Wikimedia Foundation, 11 December 2020. en.wikipedia.org/wiki/Elbow_method_(clustering)
10. Aprilliant A The k-prototype as Clustering Algorithm for Mixed Data Type (Categorical and Numerical), 4th June 2021. https://towardsdatascience.com/the-k-prototype-as-clustering-algorithm-for-mixed-data-type-categorical-and-numerical-fe7c50538ebb
11. The Benefits of Geographic Segmentation. Alchemer, 7th June 2021. https://www.alchemer.com/resources/blog/benefits-of-geographic-segmentation/
12. Kumar S (2021) Understanding K-means, K-means++ and, K-medoids clustering algorithms, 7th June 2021. https://towardsdatascience.com/understanding-k-means-k-means-and-k-medoids-clustering-algorithms-ad9c9fbf47ca
13. Kaminskyi A (2021) Information technology model for customer relationship management of nonbank lenders: coupling profitability and risk, pp 234–237
14. Monil P (2020) Customer segmentation using machine learning. Int J Res Appl Sci Eng Technol 8(6):2104–2108. https://doi.org/10.22214/ijraset.2020.6344

Fish Species Classification Using Convolutional Neural Networks

Nishat Fatima and Vrinda Yadav

1 Introduction

Seafood is popular in a variety of meals, particularly in seaside countries, due to its flavor and nutritional value [1]. Numerous sectors depend on the ability to classify fishes based on their traits. Environmental pressures such as global warming, marine responses to climate change, and pollution, as well as cultural pressures such as unregulated and overfishing, and sustainable use of marine natural resources have an increasing impact on fish populations [2]. The ramifications of these events account for the urgency of developing a standardized, cost-effective, and reliable system for monitoring fishes across habitats.

Due to its importance in oceanography and marine research, fish recognition is one of the essential underwater object detection problems. Academic scholars, ocean scientists, and biologists can benefit from fish species recognition by conducting various experiments, analyzing the result, and concluding their findings.

Manual methods for classifying fishes can be time-consuming, necessitating substantial sampling efforts which are harmful to the marine environment. They may be expensive and produce limited data, and a scarcity of fish experts could lead to erroneous and subjective identification.

On the basis of scope, fish species classification can be classified into three application categories [3]:

- Identifying the species of dead fish. For example, in the industry, conveyor belt classification.

N. Fatima (✉) · V. Yadav
Department of Computer Science and Engineering, Centre for Advanced Studies,
Lucknow, Uttar Pradesh 226031, India

V. Yadav
e-mail: vrinda@cas.res.in

- Recognizing fish species in an artificial habitat. For example, aquariums, water tanks, etc.
- Recognize fish species in their natural habitat. For example, sea, ocean, etc.

Automated systems can assist in classifying fish species. For fish identification and to improve present approaches, there has been an increasing interest in utilizing electronic monitoring, electronic reporting, and artificial intelligence in this area of research. These techniques are portable, non-invasive, efficient, and non-destructive and they can produce high-resolution and high-quality images at a low cost. The recognition of fish species is a multiclass classification problem [4].

Machine learning algorithms may be applied to target effective fish species identification and segmentation. Fish classification is critical for understanding how species interact with one another and with the surrounding marine environment. While significant progress has been made in categorizing fish species in a controlled environment, there is still more work to be done.

When scientists use their previous knowledge to check spoiled fishes through their sense of sight, they frequently make subjective choices, which may result in spoiled seafood being sold. Hence, developing an automated system to solve the problem of spoiled fish detection while also reducing the financial strain on vendors is essential. Before differentiating fresh from ruined fish, an automated classification of fish types is required.

In the classification of fish species, our proposed model achieved a 99.89% accuracy against the mean accuracy of 98.74% for the large-scale fish dataset [1], in which they fed the gray-level co-occurrence matrice's (GLCMs) "contrast" and "energy" features to a support vector machine (SVM). GLCM presents how frequent distinct grayscale intensities of pixels appear in an image [5], whereas in our proposed model, a modified MobileNetV2 model which is a convolutional neural network is employed.

The paper is structured as follows: Sect. 2 summarizes previous research in this domain; Sect. 3 describes the methodology used in our work; and Sect. 4 summarizes the experimental findings by visualizing the results. The conclusion and future work are discussed in Sect. 5.

2 Related Work

This section describes the previous research work addressing the problem of fish species classification.

Rathi et al. [6] used a novel technique based on convolutional neural network (CNN), deep learning, and image processing to achieve an accuracy of 96.29% via Otsu's thresholding, morphological operation, pyramid mean shifting, and CNN.

Abdullah et al. [2] showed an improvement on a previously built model [6]. The performance of their model was demonstrated with real-world data from a research organization called The Nature Conservancy. The model achieved an accuracy of 3% higher than that of the Rathi et al. [6].

Vaneeda et al. [7] designed a novel training method based on realistic deep vision simulation. Using a deep vision trawl camera system, they automated the classification of species present in images through a deep learning neural network. On a dataset of images acquired from a conventional fisheries survey, using a commercially accessible camera system, the model was able to recognize fish species with up to 94% accuracy.

Aminul et al. [8] presented a hybrid local binary pattern (HLBP), which is a hybrid feature descriptor based on adaptive thresholds that extracts sign and magnitude from an image using a single channel. The hybrid local binary pattern (HLBP) achieved accuracy of 94.97%. Also, they used different kernels of support vector machines (SVMs) for classification and compared results with HLBP with well-known feature descriptors such as local binary pattern (LBP), local gradient pattern (LGP), noise adaptive binary pattern (NABP), CENTRIST, discriminative ternary census transform histogram (DTCTH), and local adaptive image descriptor (LAID).

Ricardus et al. [9] proposed fish species identification by an image enhancement model applied to the backpropagation neural network by choosing an appropriate interpolation method and an appropriate configuration of backpropagation neural network (BPNN), and accuracy obtained was 90.24%.

Deep et al. [10] proposed a hybrid convolutional neural network (CNN) framework that uses CNN for feature extraction, support vector machine (SVM), and k-nearest neighbor (k-NN) for classification. They used hybrid convolutional neural network (deep CNN-SVM) deep CNN-KNN. The framework using deep CNN-KNN achieved an accuracy of 98.79%.

Shubhi et al. [11] proposed a deep convolutional networks-driven classification algorithm (DCNDCA) which incorporated the efficiency of deep convolutional neural networks with the scalability and flexibility of other algorithms, making it a reliable and adaptable system in various scenarios. Preprocessing inputs data by locally linear embedding using Hessian eigenmaps. The accuracy achieved was 95.8%, with partially trained CNN models being trained on a multinomial Naive Bayes classifier in cascade.

An improved image-set matching approach is implemented by Snigdhaa et al. [4], in which graph-embedding discriminant analysis was used to achieve 91.66% accuracy. Implementation was done by computing the Gram matrix, the within class and between class graph similarity matrices. They solved the maximization problem by eigendecomposition, and the projection matrix obtained had eigenvectors sorted in a descending manner.

Shoaib et al. [12] attempted to compensate for the lack of labeled data by using pre-trained deep neural network models. The purpose of their research was to use deep learning techniques to solve fine-grained fish species classification, and they achieved a 94.3% of overall accuracy. A cross-layer pooling technique was devised that uses a pre-trained CNN for classification and a support vector machine for classification (SVM). They introduced an automatic method for fish species categorization in videos captured in uncontrolled underwater environments that uses current pre-trained deep neural network models.

The goal of Zhou et al. [13] was to investigate the object detection scheme under real sea water through an autonomous underwater vehicle (AUV) built-in digital camera. Their methods include data augmentation CNN, dropout algorithm for over fitting problems, and refine loss function by YOLO. An accuracy of 99% was achieved via their method.

The dataset used in the papers [2, 4, 6–13] is the Fish4Knowledge dataset [14], which includes underwater images of the fishes as well as binary masks to separate the fishes from the background. Our model and [1] were constructed on the images taken from the dataset "A Large-Scale Fish Dataset" [1].

Table 1 exhibits the comparison between the list of papers [1, 2, 4, 6–13] explained in this section. Table 1 is constructed against four columns consisting of serial number, list of papers, technique used, and the accuracy achieved in each paper and dataset used.

Table 1 Comparison with related work

S. No	List of papers	Technique used	Dataset used	Accuracy achieved (%)
1	Rathi et al. [6]	Otsu's thresholding, morphological operation, pyramid mean shifting with CNN	Fish4Knowledge dataset [14]	96.29
2	Abdullah et al. [2]	Convolutional neural network, deep learning, and image processing	Fish4Knowledge dataset [14]	98.6
3	Vaneeda et al. [7]	Deep learning neural network	Fish4Knowledge dataset [14]	94
4	Aminul et al. [8]	Hybrid local binary pattern (HLBP)	Fish4Knowledge dataset [14]	94.97
5	Ricardus et al. [9]	Backpropagation neural network (BPNN)	Fish4Knowledge dataset [14]	90.24
6	Deep et al. [10]	Hybrid convolutional neural network (CNN)	Fish4Knowledge dataset [14]	98.79
7	Shubhi et al. [11]	Deep convolutional networks-driven classification algorithm (DCNDCA)	Fish4Knowledge dataset [14]	95.8
8	Snigdhaa et al. [4]	Graph-embedding discriminant analysis	Fish4Knowledge dataset [14]	91.66

(continued)

Table 1 (continued)

S. No	List of papers	Technique used	Dataset used	Accuracy achieved (%)
9	Shoaib et al. [12]	Pre-trained deep neural network models	Fish4Knowledge dataset [14]	94.3
10	Zhou et al. [13]	Data augmentation CNN and refine loss function by YOLO	Fish4Knowledge dataset [14]	99
11	Ulucan et al. [1]	Gray-level co-occurrence matrices (GLCMs) "contrast" and "energy" features to a support vector machine (SVM)	A large-scale fish dataset [1]	98.74
12	Our work	Pre-trained MobileNetV2 convolutional neural networks' model	A large-scale fish dataset [1]	99.89

3 Methodology

CNNs are image processing systems based on the human visual system and the brain's biological structure. The main layers of a CNN include an input layer, a convolutional layer, a non-linearity layer, a pooling layer, and a fully connected layer. The convolutional and fully connected layers have parameters, but pooling and non-linearity layers do not have parameters [15]; the number of layers and their sequence are determined by the difficulty of the challenge at hand. MobileNetV2 is used as a pre-trained network because training an end-to-end network takes longer duration. The epoch number is set to 25, and the size of the batch is set at 32. The flowchart in Fig. 1 shows the steps that are followed in the research.

3.1 Dataset

The scarcity of publicly available datasets including regularly used fish image samples is a significant barrier in fish classification investigations. Existing databases include photos of fish captured underwater, as well as fish species that are not commonly consumed.

A large-scale fish dataset [1] is a collection containing the images of eight different fish species popular in Turkey's Aegean Region and shrimp purchased from a super-market's seafood department. There are nine different types of seafood (eight types

Fig. 1 Approach used

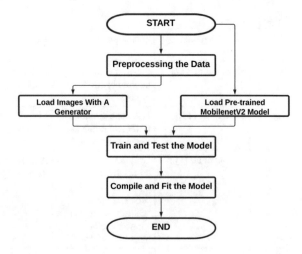

Fig. 2 MobileNetV2
architecture [16]

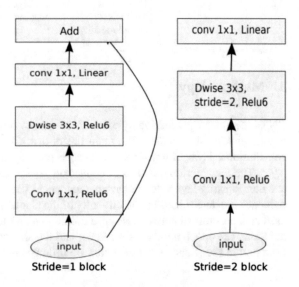

of fishes and shrimp) in the dataset. There are 1000 augmented images and their
pairwise augmented ground truths for each class.

The ratio of training to test data is 7:1. Figure 3 shows randomly selected images
of fishes from the dataset which includes trout, red mullet, Black Sea sprat, hourse
mackerel, gilt-head bream, sea bass, red sea bass.

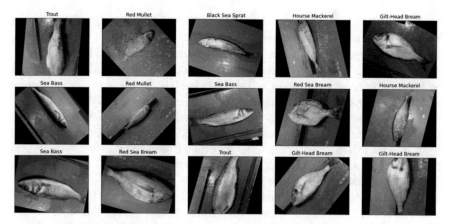

Fig. 3 Large-scale fish dataset [1]

3.2 Approach Used

The steps in Fig. 4 are explained next.

Step 1: Preprocessing the Data

The images within the dataset were pre-processed before being fed to the CNNs. The steps were:

Load the dataset in directory and get its file path and labels in a list form. Convert the list into series for both file path and labels, combine them into one series, and store it under a directory. Drop the ground truth (GT) images from the dataset. Shuffle the data frame, reset the index so as to prevent index from creating columns containing old index entries. Show the result of the data frame through head function. Display 15 images from the dataset with their respective labels.

Step 2: Load the Images with a Generator

Loading the images with a generator, and separating the entire dataset into batches for training, testing, and validating the data. Training set has 5085 images belonging to nine classes, the testing set has 2725 images, and the validation set has 1271 images belonging to nine classes. The output results in a tuple of (x, y), where x is a numpy array containing a batch of images and y is a numpy array of corresponding labels.

Step 3: Load the Pre-trained MobileNetV2 Model

Convolutional neural networks' features (CNNsFs) are used to extract numerous characteristics such as edges, blobs, and microscopic details in the paper. It is centered on an inverted residual structure, with bottleneck levels connected by residual connections. The intermediate expansion layer filters features with lightweight depthwise convolutions as a source of non-linearity. The architecture of MobileNetV2 consists of a fully convolutional layer with 32 filters, followed by 19 residual bottleneck layers.

The MobileNetV2 model as shown in Fig. 2 comprises 53 convolution layers and a single AvgPool with a giga floating point operations per second (GFLOP) of 350. It is made up of two primary parts: residual block inverted and residual block bottleneck. In the MobileNetV2 used in the paper, there are two types of convolutional layers: 3 × 3 depthwise convolution and 1 × 1 convolution [16].

Each block is made up of three layers: 1 × 1 convolution (CONV2D) and depthwise convolution (DWCONV2D) with ReLU6 and 1 × 1 convolution without any linearity. The model has weights trained on ImageNet with global average pooling applied to the output layers of the last convolutional block. Freeze the convolutional base and use it as a feature extractor, so as to prevent the weights in a given layer from being updated during training the data.

Figure 4 explains the layers and parameters of each layer in the modified version of MobileNetV2 pre-trained model with layers being: convolution (CONV2D), batch normalization (BN), ReLU, depthwise convolution (DW-CONV2D), depthwise batch normalization (DW-BN), depthwise ReLU (DW-ReLU). The main difference between our model and the pre-trained model is the addition of two dense layers (128 as dimensionality of the output space) with activation function "ReLU" as shown in Fig. 4.

Step 4: Test and Train the Model

Seventy percent of the images are randomly used for training the algorithm, while the remaining 30% are used to test the algorithm. We train the pre-trained model with two dense layers of 128 filters and ReLU activation function and then the final layer with a Softmax activation function for nine classes (Black Sea sprat, trout, striped red mullet, shrimp, gilt-head bream, sea bass, red mullet, hourse mackerel, and red sea bream) and specify the input and output. The trained model is tested against the 30% of the randomly selected images from the dataset.

Step 5: Compile and Fit the Model

Compile the entire model with ADAM optimizer and categorical cross-entropy loss. Fit the model with epoch as 25 with early stopping so as to avoid the overfitting situation.

4   Results and Discussions

The test accuracy achieved by our model is 99.89% and the test loss is 0.0038%.

The loss graph and accuracy graph plotted for our model are shown in Figs. 5 and 6, respectively.

A confusion matrix is a (N * N) table, where N is number of target classes, that is used for evaluating a classification model (or "classifier's") performance on a set of test data, for which the true values are known. Figure 7 shows the confusion (9 * 9) matrix for our model for fish species classification.

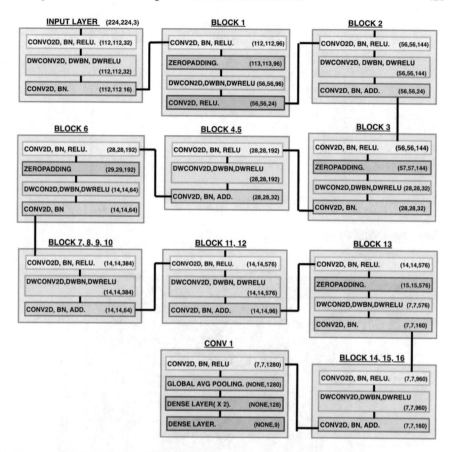

Fig. 4 Modified MobileNetV2 architecture

Fig. 5 Loss graph

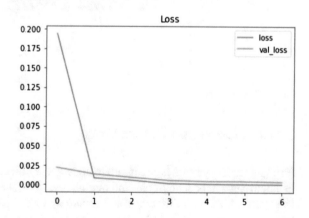

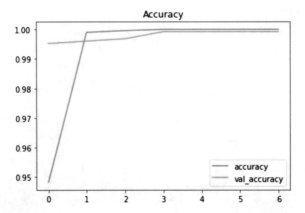

Fig. 6 Accuracy graph

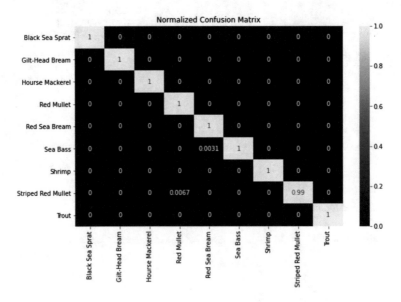

Fig. 7 Confusion matrix

5 Conclusion and Future Work

Fish species recognition is an important underwater object detection challenge due to its importance in oceanography and marine research. The existing model to classify fish species into nine different classes, namely trout, red mullet, Black Sea Sprat, hourse mackerel, gilt-head bream, sea bass, red sea bass for the large-scale fish dataset achieved an accuracy of 98.74% using a support vector machine (SVM). In this paper, we modified MobileNetV2 architecture by adding two dense layers (128 as dimensionality of the output space) with activation function "ReLU" and achieved an accuracy of 99.89% for the large-scale fish dataset.

We are optimistic about the transferability of this model to other real-world datasets and the potential of using it for high-level classification of fishes. In the future, more images and classes can be added in the dataset for better accuracy.

References

1. Ulucan O, Karakaya D, Turkan M (2020) A large-scale dataset for fish segmentation and classification. In: 2020 innovations in intelligent systems and applications conference (ASYU). IEEE, pp 1–5
2. Albattal A, Narayanan A Classifying fish by species using convolutional neural networks
3. Shafait F, Mian A, Shortis M, Ghanem B, Culverhouse PF, Edgington D, Cline D, Ravan-bakhsh M, Seager J, Harvey ES (2016) Fish identification from videos captured in uncontrolled underwater environments. ICES J Mar Sci 73(10):2737–2746
4. Hasija S, Buragohain MJ, Indu S (2017) Fish species classification using graph embedding discriminant analysis. In: 2017 international conference on machine vision and information technology (CMVIT). IEEE, pp 81–86
5. Haralick RM, Shanmugam K, Dinstein IH (1973) Textural features for image classification. IEEE Trans Syst Man Cybern 6:610–621
6. Rathi D, Jain S, Indu S (2017) Underwater fish species classification using convolutional neural network and deep learning. In: 2017 ninth international conference on advances in pattern recognition (ICAPR). IEEE, pp 1–6
7. Allken V, Handegard NO, Rosen S, Schreyeck T, Mahiout T, Malde K (2019) Fish species identification using a convolutional neural network trained on synthetic data. ICES J Mar Sci 76(1):342–349
8. Islam MA, Howlader MR, Habiba U, Faisal RH, Rahman MM (2019) Indigenous fish classification of Bangladesh using hybrid features with SVM classifier. In: 2019 international conference on computer, communication, chemical, materials and electronic engineering (IC4ME2). IEEE, pp 1–4
9. Pramunendar RA, Wibirama S, Santosa PI (2019) Fish classification based on underwater image interpolation and back-propagation neural network. In: 2019 5th international conference on science and technology (ICST), vol 1. IEEE, pp 1–6
10. Deep BV, Dash R (2019) Underwater fish species recognition using deep learning techniques. In: 2019 6th international conference on signal processing and integrated networks (SPIN). IEEE, pp 665–669
11. Sareen S, Thukral A, Indu S Classification of fish species using deep learning based hybrid model
12. Ahmed SS, Salman A, Malik MI, Shafait F, Mian A, Shortis MR, Harvey ES (2018) Automatic fish species classification in underwater videos: exploiting pre-trained deep neural network models to compensate for limited labelled data. ICES J Marine Sci 75(1):374–389
13. Cui S, Zhou Y, Wang Y, Zhai L (2020) Fish detection using deep learning. J Appl Comput Intell Soft Comput 2020
14. Huang P, Boom B, Fisher R (2013) Fish recognition ground-truth data. línea. Disponible en: http://groups.inf.ed.ac.uk/f4k/GROUNDTRUTH/RECOG
15. Albawi S, Mohammed TA, Al-Zawi S (2017) Understanding of a convolutional neural network. In: 2017 international conference on engineering and technology (ICET). IEEE, pp 1–6
16. Sandler M, Howard A, Zhu M, Zhmoginov A, Chen L-C (2018) Mobilenetv2: inverted residuals and linear bottlenecks. In: Proceedings of the IEEE conference on computer vision and pattern recognition, pp 4510–4520

Disease Detection in Tomato Leaves Using Raspberry Pi-Based Machine Learning Model

Jagdeep Rahul, Lakhan Dev Sharma, Rishav Bhardwaj, and Ram Sewak Singh

1 Introduction

India has the second-highest population in the world, and of this, nearly 70% of people rely on agriculture for their living, either directly or indirectly. Because of our wide climatic variations, they grow various types of cash crops and food crops, such as wheat, rice, and mangos [1]. Farmers have to face many difficulties, but one major difficulty is identifying plant disease. Plant disease detection by seeing is a more time-consuming and inaccurate process that can only be performed in specific domains. Using an automated detection approach, on the other hand, requires less efforts, less time, and improves accuracy. Some colour spots, early and late scorch, and various fungal, viral, and bacterial diseases are common in plants. Artificial intelligence-based approaches may be used to identify and classify plant diseases automatically as method used in signal processing with AI [2–6]. Today, most plant disease detection is done mostly by just seeing the plants and guessing the disease. Then, they use different pesticides for that disease. Sometimes it works, but many times it can't guess. It leads to the wrong and unnecessary use of chemicals, and the original disease is still present there. It results in more losses for farmers. Their crop is damaged. They have financial losses from buying pesticides, and their crops are also poisoned by these chemicals. We have some methods to detect diseases, but they

J. Rahul (✉)
Department of Electronics and Communication Engineering, Rajiv Gandhi University, Doimukh, India
e-mail: jagdeeprahul11@gmail.com

L. D. Sharma · R. Bhardwaj
School of Electronics Engineering, VIT-AP Campus, Amrawati, India

R. S. Singh
Department of Electronics and Communication Engineering, Adama Science and Technology University, Adama, Ethiopia

© The Author(s), under exclusive license to Springer Nature Singapore Pte Ltd. 2023
R. Agrawal et al. (eds.), *International Conference on IoT, Intelligent Computing and Security*, Lecture Notes in Electrical Engineering 982,
https://doi.org/10.1007/978-981-19-8136-4_35

are costly, take time, and one needs to go to laboratories for testing. So, we have used image processing and machine learning method to solve this problem. We utilized the k-nearest neighbour (KNN) algorithm with an implementation on the Raspberry Pi using a camera.

2 Literature Review

The method proposed in [7] was developed for plat disease detection which converts the RGB images to H, I3b, and I3a images. Method in [8] has used the k-means clustering algorithm and the Otsu method for analysis of infection in leaves. The extracted features such as texture and shapes from images can be used for classification of plant diseases [9]. The Gabor filter was utilized for extracting the features and performed classification using ANN (artificial neural network) in [10]. The method in [11] proposed a system which works with the help of an Android phone for detection of the diseases. In this, a person takes and collects pictures of wheat plants using their smart phone and sends those photos to the system for disease diagnosis. Sannakki et al. [12] has detected the disease in grape leaves which used a technique based on feeding forward and backward engendering neural networks. Good quality grape leaf photos were used by them for diagnosing the disease. Method in [13] proposed a system for the detection of anthracnose and downy mildew disease in watermelon leaves. That has used a framework based on the neural system. To establish the proposed system's accuracy, the real positive and negative rates were calculated. In [14], researchers developed a technique for the automated classification of leaf diseases using high resolution multispectral and stereo pictures. The recognition and classification of disease on cereal plants based on visual symptoms such as colour and texture features which are affected by fungus-based diseases were proposed in [15]. For this, they used support vector machines (SVM) and artificial neural networks (ANN). In paper cotton leaf, disease detection was performed using pattern recognition algorithms in [16] has employed a back propagation neural network to identify the condition. The method in [17] has investigated several divisions and emphasized extraction methods that may be used to diagnose plant diseases using a photograph of their leaves. The disease identification in Malus domestica was implemented using an efficient technique such as k-mean clustering, texture, and colour analysis in [18]. It uses the texture and colour characteristics that usually appear in affected regions to distinguish and identify distinct agriculture. Method in [19] demonstrated that in cucumber plant leaves, two forms of fungus are found: *P. cubensis* and *S. fuliginea*. To describe their infection, they have used a 3-layered ANN model. The work proposed in [20] utilized deep learning to construct a classifier for disease detection. It also proposed the idea of localizing the regions affected by the disease and helping to recognize the disease. The author utilizes data sets from Good Fellow, Bengio, and other publications. For compact devices, further research is needed to minimize the processing and size of deep models. An image processing technique was developed to identify betel vine rot disease in [21]. To diagnose and observe peripheral disease characteristics, they used vision-based approaches. Identification is dependent on colour characteristics of the rotted leaf area disease. In this technique, authors selected the "Bangla desi" variants of betel vine. Mokhtar et al. [22] proposes feature extraction technique using Gabor wavelet transform techniques from tomato leaf. To detect leaf illnesses, they used SVM. For research purposes, images of actual tomato leaf samples were used, as well as the author's observations of two forms of tomato leaf disease, namely early blight and powdery mildew.

3 Proposed Methodology

Block diagram of the proposed methodology is given in Fig. 1. The image of the leaf is taken through a webcam, which is attached to a Raspberry Pi. The image is further processed by a Raspberry Pi-based machine learning model, and a decision is taken. The data set used in the proposed model for comparing images is taken from [23].

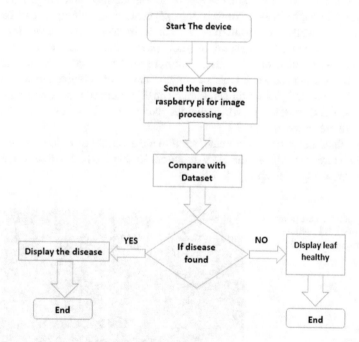

Fig. 1 Process flow diagram of the proposed model

3.1 Image Acquisition

Image acquisition is the operation of acquiring an image from a source, often hardware devices such as cameras, sensors, and so on, in image processing and machine vision. It is the first and most critical stage in the workflow sequence since the system cannot process anything without a picture.

In this step, we have captured an image using a camera and large number of different infected and healthy leaves of that particular plant to create a big data set which has been used for training purposes. The image should be good, and there should be ample lighting and good resolution.

3.2 Image Segmentation

In image segmentation, we removed all the background from the image and keep only the leaf part. The numerous segmentation techniques that range from basic thresholding to sophisticated colour image segmentation techniques can be used. These pieces typically refer to anything that can be isolated and treated as individual objects by humans easily. In general, the more precise the segmentation is, the more likely it is to succeed with recognition. We have used the segmentation method of Otsu [24] in the proposed work. Otsu's technique, named after Nobuyuki Otsu, is used to conduct automated picture thresholding. The algorithm returns a single intensity threshold in the simplest form, which divides pixels into two classes: foreground and background are shown in Fig. 2.

To eliminate noise from the image, we utilized a variety of functions. To obtain an HSV picture, we utilized the HSV function. In this method, we used Gaussian blur and canny edge detection [25–27].

Fig. 2 **a** Infected leaf (after segmentation). **b** Healthy leaf (after segmentation)

(a)   (b)

3.3   *Feature Extraction*

This is one of the most crucial techniques. First, we transform the RGB picture to a greyscale image. For this technique, we employed a grey-level co-occurrence matrix (GLCM) [23]. This matrix provides us with colour and texture characteristics for recognition and categorization. These colour and texture features were then supplied into the classifier for training. The GLCM consists of three key phases. The first RGB colour image is converted to an HSI image. Secondly, GLCM is generated from the HIS image. And lastly, texture features from GLCM are generated [28]. Following features were used in this study.

Shape Based Features

The area of a segmented picture is the total number of pixels. The number of border pixels on the leaf margin is referred to as the leaf perimeter. Aspect ratio refers to the relationship between a leaf's physiological length and physiological breadth. Rectangularity is the resemblance of a rectangle to a leaf, where L is the length, W is the breadth, and A is the leaf area; the formula is $(L*W)/A$.

Features that are dependent on texture

The angular second moment (ASM) is a metric that quantifies orderliness, or how regular or orderly the pixel values in a window are. The sum of squares of all elements in the GLCM is obtained by using energy. The image entropy is a metric for determining the image's degree of uncertainty (variations) which determines the grey-level co-occurrence matrix's local variations. The features equations are represented from Eqs. (1) to (5).

Features Equations

$$\text{Energy} = \sum_{i,j=0}^{N-1} \left(P_{ij}\right)1*^2 \tag{1}$$

$$\text{Entropy} = \sum_{i,j=0}^{N-1} \ln\left(P_{ij}\right)P_{ij} \tag{2}$$

$$\text{Contrast} = \sum_{i,j=0}^{N-1} P_{ij}(i-j)^2 \tag{3}$$

$$\text{Homogeniety} = \sum_{i,j=0}^{N-1} \frac{P_{ij}}{1-(i-j)^2} \tag{4}$$

$$\text{Correlation} = \sum_{i,j=0}^{N-1} \frac{P_{ij}}{1-(i-j)^2} \tag{5}$$

P_{ij} = Element i, j of the normalized GLCM
N = Number of grey levels in the image as specified by number of levels in under quantization
U = GLCM mean
σ = Varience of all intensities.

3.4 Data Set Used

The tomato is the plant that has been considered (Solanum lycopersicum). We collected 341 bacteria-infested leaves, 355 healthy leaves, and 269 infected leaves. In total, we examined 965 leaves.

3.5 Hardware Implementation

Raspberry Pi: In order to connect the Raspberry Pi to a computer display or television, a conventional keyboard and mouse are used. It's a small, powerful computer that lets anybody learn Scratch and Python programming. It can do everything a laptop can do, including browsing the Internet, streaming HD video, spreadsheets, word processing, and football. It runs on Broadcom BCM. It has many GPIO to interact with the outer world through sensors. It has an inbuilt Wi-Fi adapter, a USB jack, and other peripheral devices. The function of the Raspberry Pi is to do image processing and classification. An image of the working model is shown in Fig. 3.

Fig. 3 Demonstration of working model

3.6 Classification

K-nearest neighbour (KNN): It is one of the easiest learning algorithms for computers based on supervised learning methods. This is a nonparametric algorithm, which means that the underlying data [29] does not require any predictions. This is sometimes referred to as a lazy learner method, because rather than storing the information, it learns directly from it and performs an action on it during classification [30–33]. It makes the assumption that the new case/data and the existing cases are equivalent and assign the new case to the category that is closest to the existing categories. This algorithm stores all the data available and, depending on the similarities, classifies a new data point. This implies that when new data becomes available, it may be promptly sorted into a well-suited group using the KNN algorithm [34]. At the training stage, the algorithm simply keeps the data set and then classifies the data as it gathers new data into a group that is fairly similar to the present data.

4 Results

The algorithm and methodology used in this work were implemented using Python on the Raspberry Pi. We took 80% of the data as training and 20% as testing data. We tested for various K values, such as 5, 7, and 3. We got the best precision and accuracy at $K = 5$. This gave precision and accuracy of the algorithm. The precision for bacterial detection is 98%, 97% for healthy detection, and 93% for virus detection.

Table 1 shows that we collected 341, 355 samples of bacterial, healthy, and viral infection, respectively. We achieved precision 93% on virus, 97% in healthy, and 95% in bacterial. Recall was obtained 99% in bacterial, 97% in healthy, and 98% in virus. F1 score was 97% in bacterial and healthy and 95% in virus. The result shows overall Precision was 96%. The proposed method is compared with the existing models in Table 2.

Table 1 Result using KNN-based model

	Precision (%)	Recall (%)	F1 score (%)	Support
Bacterial	95	99	97	341
Healthy	97	97	97	355
Virus	93	98	95	269
Accuracy	96			

Table 2 Comparison with existing techniques

Name of project	Method	Accuracy (%)
Neural Network Analysis was used to classify watermelon leaf diseases [13]	SPSS and MATLAB's neural network pattern recognition toolbox	75
Quick and accurate plant disease detection and classification [14]	ANN	83
Using colour texture features, classification of fungal disease symptoms on cereals [15]	Recognition using support vector machines (SVM) and artificial neural networks (ANN)	77.5
Cotton leaf disease pattern recognition [16]	Back propagation neural networks	85
Our proposed system	Use of KNN for disease detection	96

5 Conclusion

In this work, we proposed an image processing and k-nearest neighbours (KNN)-based algorithms to detect plant disease with an implementation on the Raspberry Pi. From this research, we can conclude that implementing this approach will have a positive impact on the lives of our farmers and their families. We will save labour and time that would have been squandered due to plant disease if we used this method. The work done here incorporates both image processing and pattern recognition approaches to classify crop diseases. We can also add features such as suggesting preventive measures to stop that disease or showing pesticides for that disease.

References

1. Food Wastage in India. https://www.chintan-india.org. Last accessed 2021/10/12
2. Rahul J, Sharma LD, Bohat VK (2021) Short duration Vectorcardiogram based inferior myocardial infarction detection: class and subject-oriented approach. Biomed Eng/Biomedizinische Technik
3. Rahul J, Sora M, Sharma LD (2021) Dynamic thresholding based efficient QRS complex detection with low computational overhead. Biomed Sig Process Control 67:102519
4. Rahul J, Sora M, Sharma LD (2021) A novel and lightweight P, QRS, and T peaks detector using adaptive thresholding and template waveform. Comput Biol Med 132:104307
5. Rahul J, Sora M (2020) A novel adaptive window based technique for T wave detection and delineation in the ECG. Bio-Algorithms Med-Syst 16(1)
6. Rahul J, Sharma LD (2021) An enhanced T-wave delineation method using phasor transform in the electrocardiogram. Biomed Phys Eng Exp 7
7. Camargo A, Smith JS (2009) An image-processing based algorithm to automatically identify plant disease visual symptoms. Biosys Eng 102(1):9–21
8. Badnakhe MR, Deshmukh PR (2012) Infected leaf analysis and comparison by Otsu threshold and k-means clustering. Int J Adv Res Comput Sci and Softw Eng 2(3):449–452
9. Naikwadi S, Amoda N (2013) Advances in image processing for detection of plant diseases. Int J Appl Innov Eng Manam 2(11)

10. Kulkarni AH, Patil A (2012) Applying image processing technique to detect plant diseases. Int J Modern Eng Res 2(5):3661–3664

11. Xia Y, Li Y, Li C (2015) Intelligent diagnose system of wheat diseases based on android phone. J Inform Comput Sci 12(18):6845–6852

12. Sannakki SS et al (2013) Diagnosis and classification of grape leaf diseases using neural networks. In: 2013 fourth international conference on computing, communications and networking technologies (ICCCNT). IEEE

13. Kutty SB, Abdullah NE, Hashim H, Sulinda A (2013) Classification of watermelon leaf diseases using neural network analysis. In: 2013 IEEE business engineering and industrial applications colloquium (BELAC), pp 459–464. IEEE

14. Reyalat M, Al Hiary B, Ahmad B, Rahamneh Z (2017) Fast and accurate plant disease detection and classification. Int J Comput Appl 17(1). ISSN 0975-8887

15. Pujari JD, Yakkundimath R, Byadgi AS (2013) Grading and classification of anthracnose fungal disease in fruits. Int J Adv Sci Technol 52

16. Rothe PR, Kshirsagar RV (2015) Cotton leaf disease identification using pattern recognition techniques. In: 2015 international conference on pervasive computing (ICPC). IEEE, pp 1–6

17. Khirade SD, Patil AB (2015) Plant disease detection using image processing. In: 2015 international conference on computing communication control and automation (ICCUBEA). IEEE, pp 768–771

18. Sabah B, Navdeep S (2012) Remote area plant disease detection using image processing. IOSR J Electron Commun Eng 2(6)

19. Asefpour Vakilian K, Massah J (2013) An artificial neural network approach to identify fungal diseases of cucumber (*Cucumis sativus L.*) plants using digital image processing. Arch Phytopathol Plant Prot 46(13):1580–1588

20. Brahimi M, Boukhalfa K, Moussaoui A (2017) Deep learning for tomato diseases: classification and symptoms visualization. 31(4):299–315

21. Dey AK, Sharma M, Meshram MR (2016) Image processing based leaf rot disease, detection of betel vine (*Piper BetleL.*). Procedia Comput Sci 748–754

22. Mokhtar U, Ali MAS, Hassenian AE, Hefny H (2015) Tomato leaves diseases detection approach based on support vector machines. IEEE

23. Mark Nixon AS (2012) Feature extraction and image processing for computer vision

24. Fan JL, Zhao F, Zhang XF (2007) Recursive algorithm for three-dimensional Otsu's thresholding segmentation method. Acta Electron Sinica 35(7):1398–1402

25. Sekehravani EA, Babulak E, Masoodi M (2020) Implementing canny edge detection algorithm for noisy image. Bull Electr Eng Inform 9(4):1404–1410

26. Singh, Misra AK (2017) Detection of plant leaf diseases using image segmentation and soft computing techniques. Inform Process Agric 4(1):41–49

27. Sethy PK, Negi B, Behera SK, Barpanda NK, Rath AK (2017) An image processing approach for detection, quantification, and identification of plant leaf diseases—a review. Int J Eng Technol 9(2):635–648

28. Asery R, Sunkaria RK, Sharma LD, Kumar A (2016) Fog detection using GLCM based features and SVM. In: Conference on advances in signal processing (CASP), 9 June 2016. IEEE, pp 72–76

29. Sharma LD, Sunkaria RK (2018) Inferior myocardial infarction detection using stationary wavelet transform and machine learning approach. Sig Image Video Process 12(2):199–206

30. Sharma LD, Sunkaria RK (2019) Detection and delineation of the enigmatic U-wave in an electrocardiogram. Int J Inform Technol 1–8

31. Rahul J et al (2021) An improved cardiac arrhythmia classification using an RR interval-based approach. Biocybern Biomed Eng 41(2):656–666

32. Rahul J, Sora M (2020) Premature ventricular contractions classification using machine learning approach. In: 2020 international conference on smart electronics and communication (ICOSEC). IEEE

33. Rahul J, Sharma LD (2022) Artificial intelligence-based approach for atrial fibrillation detection using normalised and short-duration time-frequency ECG. Biomed Sig Process Control 71:103270
34. Sharma LD, Sunkaria RK, Kumar A (2017) Bundle branch block detection using statistical features of QRS-complex and k-nearest neighbors. In: 2017 conference on information and communication technology (CICT), IEEE, pp 1–4

A Review on Crop Disease Detection Techniques

V. Praba and K. Krishnaveni

1 Introduction

Agriculture is the most important source of country's livelihood and development. According to the United Nations Food and Agriculture Organization (UNFAO) report, the world's population will quadruple by 2050 [1]. The developing country like India has to increase its agricultural production to manage its rapid rise of birth rate. To raise the productivity and efficiency and to minimize the challenges that farmers face in agriculture, tremendous mechanical and chemical progress has been made. One of the main causes of productivity reduction is crop diseases. The growth of crop diseases increases in recent times due to drastic climate variations and lack of immunity in crops. This may ultimately decrease the cultivation, cause large-scale demolition of crops and eventually lead to financial loss of farmers. The identification and treatment of crop diseases have become a major challenge due to hasty growth in variety of diseases and inadequate knowledge of farmers. In traditional farming, a huge team of experts with expensive time-consuming crop disease diagnosis process is required owing to texture and visual similarities of plant leaves.

To reach excessive yields and healthy plants, farmers throughout the world struggle to prevent and eliminate numerous diseases from their crops. Each crop is susceptible to specific sicknesses which have an impact at the excellent and final yield capacity [2]. Recent technological advancements have had a great impact in enhancing the quality of agriculture. The Internet of Things (IoT)-based solutions

V. Praba (✉) · K. Krishnaveni
Department of Computer Science, Sri. S. Ramasamy Naidu Memorial College (Affiliated to Madurai Kamaraj University, Madurai), Sattur, Virudhunagar District, Tamil Nadu 626203, India
e-mail: praba@srnmcollege.ac.in

K. Krishnaveni
e-mail: kkrishnaveni@srnmcollege.ac.in

© The Author(s), under exclusive license to Springer Nature Singapore Pte Ltd. 2023   435
R. Agrawal et al. (eds.), *International Conference on IoT, Intelligent Computing and Security*, Lecture Notes in Electrical Engineering 982,
https://doi.org/10.1007/978-981-19-8136-4_36

can automatically maintain and monitor the agricultural farms with minimal human intervention.

The synergistic evolution and application of techniques like artificial intelligence (AI), deep learning (DL) and IoT have gathered the attention of researchers to make predictions for agricultural applications. Healthy crops can increase the productivity to an extent. Hence, computer vision-based models are in need to monitor the crops, detect and predict the crop disease and to provide the way to solve this problem.

In this research work, a variety of crop diseases and various existing smart farm techniques for identifying them are reviewed. This paper is organized into the following sections. Section 2 discusses various crop diseases. In Sect. 3, the existing crop disease detection techniques are reviewed and analyzed. Finally, the conclusion and the future direction of the research work are made in Sect. 4.

2 Crop Diseases

Crop diseases are a vast yield and quality constraint for growers of crops. Crop disorder includes any dangerous deviation or alteration from the everyday functioning of the physiological approaches. Therefore, diseased crop suffers disturbances from normal process and their essential capabilities. The crop diseases are grouped by the type of causative agent as follows [3, 4].

2.1 Fungal Diseases

Fungal represents a wide variety of plant pathogens and responsible for serious plant diseases. Most vegetable illnesses are induced with the aid of fungal. They damage plant life via killing cells or inflicting plant stress. The sources of fungal infections are infected seeds, soil, crop debris, nearby vegetation and weeds. Fungi are unfolding by wind and water splash and through the motion of contaminated soil, animals, workers, machinery, tools, seedlings and other plant materials. Various fungal diseases, crops to be affected and the factors conducive to spread are listed in Table 1.

2.2 Bacterial Diseases

Pathogenic bacteria which are capable of multiplying quickly can spread many serious diseases on crops. They penetrate through the plant via insects and different pathogens. Infection includes excessive humidity, crowding, and terrible air circulation, plant stress prompted by the way of over-watering, under-watering, or irregular

Table 1 Crop fungal diseases

Fungal disease	Image	Crops affected	Factors conducive to spread
Black spot		Rose leaves	Temperature range (20–27 °C)
Rust		Snapdragons, beans, tomatoes and lawns	Epidemiological factors
Botrytis blight		Celery, lettuce, beans, brassicas, cucumber, capsicum, tomato	1. Temperature range (50°–60 °F) 2. High relative humidity 3. Dead or dying plant tissue
Powdery mildew		Cucumber, pumpkin, beetroot, potato, herbs, peas, bitter melon, tomato, capsicum, cabbage	Low relative humidity
Anthracnose		Lettuce, celery, beans, cucurbits, tomato, capsicum, potato	Hot and humid environmental conditions

watering, negative soil health, and poor or extra nutrients. The bacterial diseases, crops to be affected and the factors conducive to spread are listed in Table 2.

2.3 Viral Diseases

Viruses are usually transmitted from one plant to another by means of living organism referred as vector or carrier. The most sizeable vectors of plant viruses consist of aphids, whiteflies, thrips and leafhoppers, which have piercing sucking mouth parts that permit the bugs to access and feed on the contents of the plant cells. Virus can also be transmitted by other insects, mites, nematodes, fungi, infected pollen

Table 2 Crop bacterial diseases

Bacterial disease	Image	Crops affected	Factors conducive to spread
Black rot		Brassicas	Warm and wet conditions
Bacterial canker		Tomato, capsicum, chilli	1. Moderate temperatures 2. High humidity
Bacterial soft rot		Tomato, capsicum, potato, sweet potato, carrots, herbs	Warm and wet conditions
Bacterial leaf spot/bacterial spot		Lettuce, cucurbits, tomato, capsicum	1. Overhead irrigation 2. Windy conditions
Bacterial speck		Tomato	1. Humidity 2. Overhead irrigation
Bacterial brown spot		Beans	Cool, wet and windy conditions

or vegetative propagating material, contact between plants and contaminated seeds. The viral diseases, crops to be affected and the factors conducive to spread are listed in Table 3.

3 Crop Disease Detection and Classification Techniques—Review

A literature review carried out on existing crop disease detection and classification is presented in this section. This review is grouped as:

- Crop disease detection methods with IoT
- Crop disease detection methods without IoT.

3.1 Crop Disease Detection Methods Without IoT

Various existing crop diseases' detections without IoT techniques reviewed are described in this section.

Bhimte and Thool [5] proposed an automatic cotton leaf disease diagnosis system using SVM classifier. Cotton leaf images are received by digital camera. Then, preprocessing techniques such as image cropping, resizing, color transformation, contrast enhancement and filtering are applied to enhance the quality of images. After preprocessing, the images are segmented by k-means clustering using color. Texture statistical features are extracted from segmented image using gray-level co-occurrence matrix (GLCM). Finally, the two diseases, namely bacterial blight and magnesium deficiency, are identified using multiclass SVM classifier based on the extracted features with 98.46% accuracy.

Das et al. [6] proposed a model to identify different types of diseases using support vector machine (SVM) classifier. The tomato leaf images are taken from Kaggle dataset and web. Initially, the dimension of each image is reduced and converted to grayscale image. Then, the leaf images are segmented by masking and thresholding techniques, and the texture features are extracted using Haralick algorithm. Finally, three different machine learning algorithms such as logistic regression, random forest and SVM have been used to classify the leaf images. Out of these three models, SVM produces better result in identifying healthy leaf from an unhealthy one with 87.6% accuracy.

Padol and Yadav [7] acquired the images using digital camera from different regions like Pune, Nasik and some of the images are taken from internet. The noise particles of the leaf images are removed by using Gaussian filter and resized, and the quality of the image is increased using thresholding technique. The preprocessed images are segmented by k-means cluster algorithm. Then, both texture and color features are extracted, and finally, SVM classification technique is used to detect the

Table 3 Crop viral diseases

Viral disease	Image	Crops affected	Factors conducive to spread
Tobacco mosaic virus		Tomato, bitter melon, long melon, snake bean	Direct contact of plants
Potato leafroll virus		Potato	Circulative, persistent, non-propagative manner by several aphid species
Sugarcane mosaic virus		Sugarcane	Mechanical transmission and other equipment
Pepper mild mottle virus		Capsicum, including chillies	Mechanical transmission and by contaminated seeds
Leaf curl virus		Cotton, papaya, bhendi, chilli, capsicum, tomato, tobacco	Both biotic and abiotic factors
Capsicum chlorosis virus		Capsicum, tomato, chillies	Thrips

disease of the grape leaves such as downy mildew and powdery mildew with 88.89% average accuracy.

Agarwal et al. [8] proposed a convolution neural network model to identify the disease in corn crop. Corn crop images are acquired from PlantVillage dataset. CNN model with three convolution and max pooling layers followed by fully connected layers detects corn crop disease with 94% accuracy after running 1000 epochs.

Patel and Vaghela [9] proposed a system to detect crop diseases as well as pests using deep learning techniques. This system consists of the following phases: (i) Image acquisition: acquires the parts of infected crop images from PlantVillage dataset and internet. (ii) Image preprocessing: images are resized. Then, the images are augmented by random rotation, zooming, shearing and flipping approaches. (iii) Feature extraction: Features are extracted and classified using transfer learning.

Sharath et al. [10] proposed an android application to detect the disease in plants like orange, pomegranate, papaya, grapes, etc., using convolutional neural network. In this system, images are captured using mobile. Next, the images are resized to increase the quality. Then, applied GrabCut technique for image segmentation and morphological processing are used to remove islands and small objects present in the segmented image. Finally, the diseases are classified using CNN algorithm with 91% accuracy.

Sardogan et al. [11] proposed convolutional neural network (CNN) model and learning vector quantization (LVQ) algorithm to detect and classify the tomato leaf disease. The images are acquired from PlantVilliage dataset containing 500 images of tomato leaves with four symptoms of diseases. CNN algorithm is used to extract the features automatically. The LVQ has been fed with the output feature vector of convolution part for training the network. The proposed method effectively recognizes the healthy leaf and four different types of tomato leaf diseases such as late blight, bacterial spot, yellow leaf curl and septoria leaf spot.

The comparative analysis of various crop diseases' detection techniques without IoT has been carried out and is shown in Table 4.

The limitations of the existing crop disease detection methods emphasized are listed below:

- Monitoring large fields leads to additional labor cost.
- Images under real conditions are not used.
- Detection of unhealthy crops consumes more time that leads to productivity loss.

3.2 Crop Disease Detection Methods with IoT

Various existing IoT-based crop diseases' detection techniques reviewed are described in this section.

Sarangdhar and Pawar [12] proposed two Android applications: one for displaying soil parameters and the other for presenting disease information, as well as for turning external devices like sprinklers or motors ON/OFF and for moving the entire system from one location to another to check soil parameters. In soil quality monitoring

Table 4 Comparative study on different works without IoT

References	Dataset	Algorithms	Crop	Pros	Cons
Bhimte and Thool [5]	Digital camera	Segmentation: k-means clustering Feature extraction: GLCM Classification: SVM	Cotton leaf	Bacterial blight and magnesium deficiency are identified with good accuracy	The input dataset contains only 130 images Not applicable for all kind of diseases
Das et al. [6]	Kaggle dataset and web	Segmentation: masking and thresholding Feature extraction: Haralick algorithm Classification: SVM	Tomato leaf	Solve both linear and nonlinear problems using SVM	Did not classify the diseases
Padol and Yadav [7]	Digital camera and internet	Segmentation: masking and thresholding Feature extraction: Haralick algorithm Classification: SVM	Grape leaves	Features are extracted by both texture and color features	Only detects downy mildew and powdery mildew diseases, and dataset contains only 137 images
Agarwal et al. [8]	PlantVillage dataset	CNN model	Corn crop	Better than VGG 16 in terms of inference time and/or storage space	Accuracy of discussed model is not better than VGG 16
Patel and Vaghela [9]	PlantVillage aataset and internet	Feature extraction and classification: transfer learning techniques	Tomato crop	It processes all parts of the crop like upper and lower sides of the leaf, stem, root, fruit images	Only suitable for tomato crop
Sharath et al. [10]	Mobile	Segmentation: GrabCut technique and morphological processing Classification: CNN	Orange, pomegranate, papaya, grapes	Solutions are provided through Android application	Solutions are common to all the farmers irrespective of the region

(continued)

Table 4 (continued)

References	Dataset	Algorithms	Crop	Pros	Cons
Sardogan et al. [11]	PlantVillage	Feature extraction: CNN Classification: LVQ	Tomato leaf	Recognized the four types of tomato leaf diseases	Not applicable for all kind of diseases

application, temperature, moisture, humidity and water sensors are employed and interfaced with a Raspberry Pi. It shows the values of soil characteristics, as well as the level of water in a tank. Farmers may operate the motor and sprinkler assembly using this application by turning on and off the relay. Another application to detect and control the five different cotton leaf diseases Bacterial Blight, Alternaria, Gray Mildew, Cereospra, and Fusarium wilt on the Raspberry Pi using an SVM-based regression technique was developed in Python code. Initially, the input image is selected from database. Then, the images are resized, and median filter is applied for preprocessing to remove noise and to increase quality of an image. Image segmentation has the following processes: First, the RGB color format image is transformed to YCbCr color format. After color conversion, bi-level thresholding is applied to get logical black and white image. Next, the image is converted to RGB-masked image using bitwise operation. Then, RGB-masked image is converted to gray image. In feature extraction step, extract the two texture features using Gabor filter with parameters such as frequency, angle in 16 orientations with ten different frequencies and color features using mean and standard deviation to identify the disease. Finally, SVM-based regression technique with nonlinear Gaussian kernel is used to classify the diseases. The overall accuracy of cotton leaves' disease detection is 83.26% accuracy. After successful detection and classification, the disease name and remedies are displayed on the application. This method is effective in big farms for disease identification and control.

Gupta et al. [13] developed a system based on IoT and deep learning. Images are captured at regular intervals using drones equipped with cameras. These images are tested with trained machine (uses CNN) and generate the appropriate prediction. This system would be tested on datasets of bottle gourd, maize and papaya leaves that were captured in a controlled environment and included both healthy and damaged leaves. The system would be able to tell whether the leaves were infected or not. And, the sensors are present in the soil capture moisture content, salinity and other factors. If any environmental unbalance is observed, then the crop owner would be notified through the web AI.

Thora et al. [14] proposed a smart solution to detect leaf disease through various sensors put at agricultural areas. These all sensor operations have been employed on single controller known as the Raspberry Pi (RPI). The leaf images are captured by camera. And, the captured images are processed using OpenCV. Then, leaf disease state and environmental elements impacting the crop are sent to the farmers immediately via WIFI server and RPI.

Khan et al. [15] proposed Internet of Things (IoT) to apply automation in a greenhouse environment. The images are acquired from greenhouse environment to prepare dataset. The image preprocessing process is applied on acquired images for color transformation. The color transformed images are segmented by k-means clustering algorithm. Then, features are extracted from segmented image using color co-occurrence method (CCM). Finally, detects the disease on plant leaf by applying the convolutional neural network with AlexNet algorithm.

Fuke and Raut's [16] proposed an IoT-based solution to predict some leaf diseases such as brown and yellow spots, early scorch, late scorch and others are fungal, viral and bacterial diseases. Deploys various sensors like color sensor, temperature–humidity sensor and camera for predicting the diseases on the plant leafs. The proposed IoT model consists of hardware module for measuring the temperature and humidity color of leaf image and software modules which are used to identify the affected area in leaf by the image processing. Initially, data's are collected from sensors and send to Arduino Uno through wired or wireless devices. Ardunio Uno monitors and records the values of temperature, humidity, soil moisture and sunlight of the natural environment and continuously updated. Then, RGB images are acquired by digital camera. Apply the preprocessing method to remove distortion and color transformation; clipping and smoothing techniques are used to enhance the image quality. After preprocessing, images are segmented using k-means algorithm. Finally, classifies the disease using mask and unmask pixels' technique.

Yakkundimath et al. [17] proposed a system to determine whether the plant is normal or diseased. This system detects the crop diseases based on sensor values captured. The values based on temperature, humidity and color parameters are used to identify the presence of plant disease.

Ramesh and Rajaram [1] suggested image processing techniques to cope with IoT-based systems for early disease identification in rice crops based on visual symptoms. Initially, the images are captured with a web camera and sent to Raspberry Pi. Then, images are resized and enhanced, and the RGB images are converted to grayscale. Then, compare the images with existing database by optimization techniques using OpenCV, and image will be sent to the cloud. Finally, cloud prediction framework identifies the rice crop disease.

Thangadurai et al. [18] proposed an automated robot to detect illness in sugarcane leaves. This system focuses on the disease-affected parts, amount of pesticide and assessing disease seriousness using image handling procedure. Initially, the images are taken from controlled condition. Then, images are segmented by straightforward edge and triangle thresholding algorithms. Finally, computing the remaining injury region and leaf territory, flaws are classified using CNN.

Yoganand et al. [19] developed a sensor-based system for crop disease monitoring. The plant's humidity and air temperature are measured using humidity and temperature sensors. A soil moisture sensor is also used to determine the state of the soil. The data from the groundnut farm are received using sensors, web cameras, GSM and controllers. The data are analyzed using machine learning models (XG boost), and crop disease predictions are made. This method of preventing crop disease (groundnut crop) is offered, and farmers are notified the prediction by means of SMS/email.

Thakur and Mittal [20] proposed real-time IoT application for classification of crop diseases using machine learning in cloud environment. The images of the crops are captured by sensor cameras. And, captured images are sent to the cloud server via Raspberry Pi 3 model. The RGB color image was converted to graycolor using Java histogram equation for preprocessing. Then, infected part of the image was segmented by k-means clustering method. Features are extracted from leaf image such as color, texture and morphology. Finally, detects and classifies the diseases by artificial neural network (ANN).

Comparative analysis of various crop diseases' detection techniques with IoT has been carried out and is shown in Table 5.

The advantages of the IoT-based crop disease detection methods are listed below:

- Monitor large field crops with low cost and time.
- Automatically detect unhealthy crops at very early stage itself.
- Generate crop disease datasets using images acquired in real conditions with the help of IoT sensors.

4 Conclusion and Suggestions

In this paper, various crop diseases and factors affecting the production of the agriculture crops are discussed initially. Then, the existing crop disease detection methods utilizing IoT sensors for capturing field data and existing datasets without using IoT are reviewed, compared and discussed. From this review, it has four main phases

Table 5 Comparative study on different works with IoT

References	IoT techniques	Algorithms	Crop	Pros	Cons
Sarangdhar and Pawar [12]	Sensors and Raspberry Pi	Classification: SVM	Cotton	Reduce the manual work	Cannot detect all types of cotton leaf diseases
Gupta et al. [13]	Sensors and web API's	Classification: CNN	Bottle gourd, maize, papaya	Observes each and every aspect of the plant	Cannot process any plant images
Thora et al. [14]	Raspberry PI, camera, WiFi server and RPI	OpenCV	Common	Can be successfully interfaced with the devices using wireless communication	Sunlight is the main factor which affects the result

(continued)

Table 5 (continued)

References	IoT techniques	Algorithms	Crop	Pros	Cons
Khan et al. [15]	Sensors and fixed camera	Segmentation: k-means clustering Feature extraction: GLCM Classification: CNN with AlexNet	Not mentioned	It is an effective method for greenhouse environment	Difficult to implement on normal field. It may be a costly
Fuke and Raut [16]	Arduino Uno and sensors	Segmentation: k-means clustering	Not mentioned	It also finds out and informs the weather changes	Did not classify the diseases
Yakkundimath et al. [17]	Temperature, color and humidity sensors, Arduino Uno, ThingSpeak	Not available	Not mentioned	To determine the quality of the leaves	Result based on sensor values only
Ramesh and Rajaram [1]	Sensors, Raspberry Pi	Optimization techniques with OpenCV	Rice crops	It identifies rice crop diseases	Not applicable to all kind of crops
Thangadurai et al. [18]	Sensors and Arduino Uno	Segmentation: straightforward edge and triangle thresholding algorithms Classification: CNN	Sugarcane leaves	Identify the affected portion of the leaf zone and sore district region individually	Fungal disease only be detected
Yoganand et al. [19]	Web camera, sensors, Arduino, cloud, GSM controller	XG boost for data analysis	Groundnut	It also has irrigation system	Only consider the groundnut plant diseases
Thakur and Mittal [20]	Sensors, web cameras, GSM and controllers			Automatically notify via the SMS/email	Not applicable to all kind of crops

which are image dataset acquisition, segmentation, features extraction and classification. Most of the existing works applied the k-means clustering algorithm for segmentation. Then, the GLCM and masking and thresholding algorithms are implemented for features' extraction. The SVM and CNN algorithms are yielding better accuracy for crop disease classification. This review concludes that there is a need for efficient image capturing technique to reduced human intervention compared to existing traditional methods. Since deep learning techniques play a vital role in classification, implementation of deep neural networks utilizing real-time IoT-sensed data to identify all possible crop diseases with more accuracy and low computation time is considered as future research work.

References

1. Ramesh S, Rajaram B (2018) IoT based crop disease identification system using optimization techniques. ARPN J Eng Appl Sci 13(4)
2. IoT in Agriculture|Global Trends of IoT Applications in Agriculture (educba.com).
3. https://www.agric.wa.gov.au/pests-weeds-diseases/diseases/crop-diseases
4. https://www.britannica.com/topic/list-of-plant-diseases-2033263
5. Bhimte NR, Thool VR (2018) Diseases detection of cotton leaf spot using image processing and SVM classifier. In: Second international conference on intelligent computing and control systems (ICICCS 2018). IEEE Xplore
6. Das D, Singh M, Mohanty SS, Chakravarty S (2020) Leaf disease detection using support vector machine. In: International conference on communication and signal processing, India, 28–30 July 2020. IEEE
7. Padol PB, Yadav AA (2016) SVM classifier based grape leaf disease detection. In: Conference on advances in signal processing (CASP), Cummins College of Engineering for Women, Pune, 9–11 June 2016. IEEE
8. Agarwal M, Sinha A, Bohat VK, Gupta SKr, Ansari MD, Garg D A Convolution Neural Network based approach to detect the disease in Corn Crop. In: 9th international conference on advanced computing (IACC). IEEE Xplore
9. Patel PP, Vaghela B (2019) Crop diseases and pests detection using convolutional neural network. IEEE
10. Sharath DM, Rohan MG, Suresh KV, Arun Kumar S, Akhilesh, Prathap C (2020) Disease detection in plants using convolutional neural network. In: Third international conference on smart systems and inventive technology (ICSSIT 2020). IEEE Xplore
11. Sardogan M, Tuncer A, Ozen Y Plant leaf disease detection and classification #ased on CNN with LVQ algorithm. In: 3rd international conference on computer science and engineering. IEEE
12. Sarangdhar AA, Pawar VR (2017) Machine learning regression technique for cotton leaf disease detection and controlling using IoT. In: International conference on electronics communication and aerospace technology, ICEC
13. Gupta AK, Gupta K, Jadhav J, Deolekar RV, Nerurkar A, Deshpande S (2019) Plant disease prediction using deep learning and IoT. In: 6th international conference on computing for sustainable global development
14. Thora A, Kumari S, Valakunde ND (2017) An IoT based smart solution for leaf disease detection. In: International conference on big data, IoT and data science (BID), Vishwakarma Institute of Technology, Pune, 20–22 December 2017
15. Khan FA, Ibrahim AA, Zeki AM (2020) Environmental monitoring and disease detection of plants in smart greenhouse using internet of things. J Phys Commun

16. Fuke RP, Raut NV IOT based solution for leaf disease prediction. Int J Innov Res Comput. ISBN 978-93-5416-539-9
17. Yakkundimath R, Saunshi G, Kamatar V (2018) Plant disease detection using IoT. Int J Eng Sci Comput
18. Thangadurai N, Vinay Kumar SB, Gayathri KM, Dhanashekaran R Detection of disease in sugarcane leaf using IoT. Int J Sci Technol 9(2)
19. Yoganand S, Narasingaperumal, Pratap Reddy P, Rahul S (2020) Prevention of crop disease in plants (Groundnut) using IoT and machine learning models. Int Res J Eng Technol (IRJET) 7(3)
20. Thakur TB, Mittal AK (2020) Real time IoT application for classification of crop diseases using machine learning in cloud environment. Int J Innov Sci Modern Eng (IJISME) 6(4). ISSN 2319-6386

Energy-Efficient Model (ARIMA) for Forecasting of Modal Price of Cod Pea Using Cloud Platform

Sachin Kumar, Saurabh Pal, Satya Singh, and Priya Jaiswal

1 Introduction

Price information plays an important role in the market arrival of agricultural commodities. Indian farmers often reach the market completely obvious of the prevailing prices and consequently fail to realize the best of price for their produce. The situation is more serious in the case of perishables such as fruits and vegetables, where storage and processing facilities are limited and option to sell in distant markets is rare.

Pea (*Pisum sativum L.*) is a good source of alimental protein to balance a cereal-predicated diet, mainly for vegetarian masses in the state. It is a highly productive crop and is grown for food, feed, and vegetables during the Rabi seasons. During 2010–11 in Sant Ravidas Nagar (a district of eastern Uttar Pradesh), field pea was grown in 1127 ha with a total production of only 1375 q and very poor productivity of 1220 kg/ha (Anonymous 2013). The leading cause for such meager yield in Uttar Pradesh and mainly in Sant Ravidas Nagar is cod pea farming in marginal areas and the adoption of enhanced agricultural techniques, minimal adoption of high-yielding improved varieties, and an overall lack of awareness among farmers about improved packages of practices. There is wide scope for extension machinery to educate the farmers of eastern Uttar Pradesh for higher adoption of improved and specific production technology of field pea by front-line demonstration. Recently,

S. Kumar (✉) · S. Pal
Department of Computer Application, V.B.S.P.U, Jaunpur, UP, India
e-mail: jaiswalsachin009@gmail.com

S. Singh
Department of Computer Application, M.G.K.V.P, Varanasi, UP, India

P. Jaiswal
Department of Computer Application, M.C.M.T, Varanasi, UP, India

© The Author(s), under exclusive license to Springer Nature Singapore Pte Ltd. 2023
R. Agrawal et al. (eds.), *International Conference on IoT, Intelligent Computing
and Security*, Lecture Notes in Electrical Engineering 982,
https://doi.org/10.1007/978-981-19-8136-4_37

attempts have been initiated to apprise farmers of prevailing prices by a number of government agencies, and ICAR sponsored projects on market intelligence [1].

In India, the area of green peas has increased continuously from 177.7 thousand ha in 1991–92 to 272.6 thousand ha in 1999–2000. The ratio of the area under cultivation of peas in India to the world area under cultivation of peas increased from 3.2% in 1991–92 to 4.5% in 1999–2000. Green pea production increased from 1.30 million tons in 1991–92 to 3.2 million tons in 2003–2004 [2]. However, the yield of green peas has shown a non-uniform trend, and it has decreased from 14,326 kg/ha in 1991–92 to 10,000 kg/ha in 1997–98 and then to 9143 kg/ha in 1999–00 [2].

The area of green peas in Uttar Pradesh was 218.55 hectare and production 2481.07 Metric Ton in 2015–16 and increased to 220.73 hectare and production 2508.36 Metric Ton in 2016–17 and again increased to 221.00 hectare and production 2511 Metric Ton in 2017–18.

The area of green peas in India was 552 ha and production 5562 MT in 2018–19 and increased to 568 ha and production 5848 MT in 2019–20 and again increased to 573 ha and production 5823 MT in 2020–21.

Cultivation of green peas is labor-intensive like all other vegetable crops [3, 4] and requires high doses of fertilizers. The key component of the cost of growing cod peas is followed by tractors/labor, fertilizers, human/bullock, and chemicals/pesticides. At the same instant of time, the profits per hectare of vegetable crops were almost 4 times, compared to the revenue from pabulum crops [5]. Thus, farmers should have the rigorous to diversify into more profitable cropping patterns such as vegetable growing instead of traditional less profitable patterns (Singh 1995). Similar types of results have been reported by Maurya et al. [6] and Sharma et al. [7]. However, none of the studies has provided an in-depth examination of the economics of cod peas' cultivation in eastern Uttar Pradesh. Therefore, this research is undertaken to

I.   Estimate the economics of the production of green peas in eastern Uttar Pradesh and

II. Study the relative importance of different factors influencing the productivity of green peas (economics of production of green peas (*Pisum sativum L.*) in eastern Uttar Pradesh).

Finally, it is concluded that the marginal and small farmers could reap the highest profit from the cultivation of peas by following scientific procedures. A goal-oriented training program on garden pea production is required to enhance the level of information and adoption of farmers in the study area. Pea growers (especially, the big ones) have been suggested to get rid of traditional methods of marketing in local places and move to telemarket and supermarkets in neighboring towns and add value with standard packing, etc., by forming cooperatives and farmers' interest groups (harnessing high pea secures livelihoods in eastern Uttar Pradesh).

2 Methodology

The forecasted price of cod pea will facilitate farmers of eastern Uttar Pradesh, particularly from Varanasi district, to make a decision when to harvest and produce when to transport it to the marketplace and to which marketplace it should be sold. With this severity, arrival and price data of cod pea were collected from Agricultural Marketing Information Network (AGMARKNET) for the main marketplace of eastern Uttar Pradesh. These markets are Varanasi, Allahabad, Jaunpur, Basti, Faizabad, Ghazipur, Mirzapur, Mau, etc.

2.1 Autoregressive Model

A time-series model named ARIMA is used for short-term predictions. Equation (1) given below represents the ARIMA model:

$$\emptyset_x(b)(1 - A)^D y_t = \beta_q(b) A_t \tag{1}$$

There are two different types of ARIMA, namely stationary ARIMA and non-stationary ARIMA. A stationary time series refers to statistical properties in probability or time distribution within a consistent time. It is identified as follows:

(a) The estimated cost of the time series is not dependent on time.
(b) The autocovariance function is a function of k, where for each k, $S_{y^1} = Cov(y_t, y_{t+k})$.

We have components of autoregressive (AR) and moving ARIMA (MA) in Eq. 1. An ARIMA model can be a combination of an AR (p) model or an MA (q) model or can be an AR model or an MA model.

The autoregressive model (p) is represented by Eq. (2) given below:

$$y_t = \emptyset_1 y_{t-1} + \emptyset_2 y_{t-2} + \ldots + \emptyset_p y_{t-p} + A_t \tag{2}$$

A_t is an error, assumed to be unrelated to each others, where $E\{A_t\} = 0$ and $\text{Var}\{A_t\} = \sigma_A^2$. The coefficients $\emptyset_i$, $i = 1, \ldots, p$ are the factors to be calculated.

The MA model contains the "averages" of the noises of the previous time and the current time. The MA (q) model is represented by Eq. (3).

$$y_t = A_t - \emptyset_1 A_{t-1} - \emptyset_1 A_{t-1} - \cdots - \emptyset_q A_{t-q} \tag{3}$$

The coefficient β_i, $i = 1, p$ are the factors to be determined.

The ARMA (p, q) is a model holding fundamental of AR and MA. ARMA model does not hold the constituent "i" because it is already a stationary model. In another word, the constituent "d" in Eq. (1) is equivalent to 0. We can represent

Table 1 Characteristics of autoregressive (AR), moving average (MA), and autoregressive moving average (ARMA)

	AR (p)	MA (q)	ARMA (p, q)
Model	$y_t = \emptyset_1 y_{t-1} + \emptyset_2 y_{t-2} + \cdots + \emptyset_p y_{t-p} + A_t$	$y_t = A_t - \emptyset_1 A_{t-1} - \emptyset_1 A_{t-1} - \cdots - \emptyset_q A_{t-q}$	$y_t = \emptyset_1 y_{t-1} + \emptyset_2 y_{t-2} + \cdots + \emptyset_p y_{t-p} + A_t A_t$ $- \emptyset_1 A_{t-1} - \emptyset_2 A_{t-2} - \cdots - \emptyset_q A_{t-q}$
PACF	Cutoff after lag p	It reduces sinusoidally or/and exponentially	It reduces sinusoidally or/and exponentially
ACF	It reduces sinusoidally or/and exponentially	Cut off after lag q	It reduces sinusoidally or/and exponentially

the mathematical equation of the ARMA model by Eq. (4) given below:

$$y_t = \emptyset_1 y_{t-1} + \emptyset_2 y_{t-2} + \cdots + \emptyset_p y_{t-p} + A_t A_t - \emptyset_1 A_{t-1} - \emptyset_2 A_{t-2} - \cdots - \emptyset_q A_{t-q} \tag{4}$$

The characteristics of AR, MA, and ARMA procedures are shown in Table 1. This characteristic turns into the guide in calculating the order p and q.

2.2 Forecasting Model of Peas' Modal Prices

The data used for price forecasting of cod peas are the average price of peas from 11 July 2011 to 10 June 2021. The time series model is used to find out the average price of pea in the next time, i.e., autoregressive integrated moving average (ARIMA) on the Google Cloud Platform. The movement or pattern pea price data is shown in Table 1.

3 Result and Discussion

The arrival of a farming commodity in a marketplace refers to the demand for it in the catchment region of a particular marketplace. Access also depends on the ability and the infrastructure to handle the marketplace. It was also indicative of the commodities produced in the region. These indicators must be taken into account by farmers in production and market planning. Therefore, pea access was studied

Table 2 Average monthly arrival of cod pea in important markets of eastern Uttar Pradesh, 2011–21

Month	Important markets in eastern Uttar Pradesh							
	Varanasi	Jaunpur	Allahabad	Basti	Faizabad	Ghazipur	Mirzapur	Mau
November	292.75	N.A	8	42.75	238.9	9.5	N.A	17.6
December	1799.3	385.78	71.6	1,381.00	1029.85	632.6	63.41	209.83
January	1612.625	1434	N.A	N.A	4338.5	466.9	169.63	169.45
February	1257.5	666	517	N.A	3292.25	343	466.58	212.5
March	623.75	604.37	419.6	N.A	627.95	224.6	5.2	94.03

Note N.A is stand for data not available

by compiling monthly data for the main market for the period from August 2011 to August 2021 and is presented in Table 2.

It has been observed that Faizabad (28,110.3 tons) received the biggest volume of pea among the nine markets of eastern Uttar Pradesh and was followed by Varanasi (20,467.5 tons), Basti (7125.1 tons), etc. Hence, it has been stated that the farmers of eastern Uttar Pradesh can take the products in the regional markets like Varanasi, Faizabad, and Allahabad. However, the option of moving products to further markets such as Basti, Deoria, Ghazipur, and Jaunpur can also be explored depending on the price competitiveness.

Seasonal analysis of pea access to key markets in eastern Uttar Pradesh of India revealed that in the Allahabad and Basti market, arrivals of peas were low. It can be said that these are the periods, during which pea must be transported from another district of Eastern UP, especially from Varanasi market to the other market of eastern UP. Allahabad, Siddharthnagar, and Mirzapur markets revealed a low level of supply of pea.

Pea price analysis in eastern Uttar Pradesh markets revealed that the highest price prevails in the Basti market (Table 3). The average price of a pea at Siddharthnagar and Varanasi is comparatively lower than Ghazipur and Jaunpur in the month of January and February and in another month at Basti and Faizabad. Thus, the Siddharthnagar and Varanasi farmers looking for better markets in terms of better prices they should look for Basti, Faizabad, Ghazipur, and Jaunpur markets.

Price behavior based on seasonal indicators revealed that the highest price of pea is in the Basti market in the month of November. The lowest price will prevail in all months in the Siddharthnagar market. And so, it was exposed that Basti market despite getting a low volume of pea compared to Varanasi and Faizabad markets provided a chance to take benefit of the best value existing there. However, far away product markets offered to challenges in terms of low volumes and associated marketing risks faced by farmers. The farmers need to be empowered to collect production to exploit economies of scale and take advantage of recent institutional changes in agriculture marketing. It was possible through the formation of self-help groups, farmers–producers organizations, and the use of the internet e-NAM marketing and connectivity platforms [8].

Table 3 Average monthly price of pea in important markets of eastern Uttar Pradesh, 2011–21

| Month | The average monthly (December–March) price of cod pea in important market of eastern Uttar Pradesh | | | | | | | | |
	Allahabad	Basti	Faizabad	Ghazipur	Jaunpur	Mau	Varanasi	Siddharthnagar
January	1308.42	1400.69	1742.25	2426.29	1777.66	1677.4	1668.19	1229.42
February	1314.56	1363.08	1545.07	1591.07	1601.56	1446.2	1444.14	1144.73
March	2042.5	2175.38	2188.25	1907	1960.96	1695.87	1817.19	1711.15
April	N.A	5620	4416.33	N.A	N.A	1991.66	N.A	N.A
May	N.A	N.A	7000	N.A	4740	N.A	N.A	4675
June	N.A	N.A	N.A	N.A	5020	N.A	N.A	N.A
July	N.A	N.A	6000	N.A	4500	N.A	N.A	N.A
October	N.A	N.A	N.A	N.A	N.A	N.A	N.A	N.A
November	3000	7950	5160	N.A	N.A	5031.25	3172.22	N.A
December	2213.15	2437.69	2364.59	1989.56	2201.87	2339.28	1890.5	1654

Note N.A stands for data not available

Weekly modal prices from pea major markets in eastern Uttar Pradesh of India, we used to forecast prices with the help of autoregressive integrated moving average model as described in the methodology. Using various values for p and q, a set of ARMA and ARIMA models installed for the right choice models. Appropriate models were selected based on the basis of a suitable selection criterion, for example, Schwarz–Bayesian information criteria (SBICs) and Akaike information criteria (AICs). It was observed that:

S. No	District	Best ARIMA model	Total fit time (s)
1	Ghazipur	ARIMA (2, 0, 0)(0, 0, 0) [0] intercept	1.510
2	Faizabad	ARIMA (2, 1, 1)(0, 0, 0) [0]	5.573
3	Siddharthnagar	ARIMA (1, 0, 0)(0, 0, 0) [0] intercept	2.226
4	Basti	ARIMA (2, 1, 1)(0, 0, 0) [0]	5.711
5	Jaunpur	ARIMA (1, 0, 1)(0, 0, 0) [0] intercept	4.979
6	Allahabad	ARIMA (0, 2, 1)(0, 0, 0) [0]	0.656
7	Mirzapur	ARIMA (1, 0, 0)(0, 0, 0) [0] intercept	2.240
8	Varanasi	ARIMA (1, 2, 0)(0, 0, 0) [0]	0.508
9	Mau	ARIMA (0, 1, 1)(0, 0, 0) [0]	2.134

These are found to be the best fitting models. They had the lowest Bayesian information criteria (BICs) and Akaike information criteria (AICs) values. So, a unique model does not apply to different markets of a cod pea. The parameters were estimated through an iterative process by the least square technique which gave the best model as shown in Table 3. The coefficient is found to be statistically considerable; hence, the chosen models were considered highly appropriate and were used for a price forecast. Using the specific model and forecasting prices of cod pea were made for the nine major markets of 2020–21 and compared to actual prices of the same period. Moreover, forecast for the period from December 2021 to March 2022. The cod pea price is developed, and the forecast price was compared to actual market price data for the period from December 20 to February 2021 (Table 4). It was observed the expected price of the main sample. The market was very close to the actual value. The precision of the models has been experimentally verified with the help of concept such as mean and root mean square error (RMSE). The RMSE value ranged from 158.78 for the Allahabad market to 1924.72 for the Varanasi market.

Mean values range from 941.66 for the Mirzapur market to 2265.83 for the Mau market, thus inferring that the chosen ARIMA models were the most appropriate to explain the prices of the cod pea in the respective study markets. These values were in an suitable range.

Different markets were correct and acceptable to a reasonable level. Sankaran [9] and Shukla and Jharkharia [10] note the mean absolute error rate (MAPE) to range from 14 to 20%. Studies have reported that the observed error is acceptable for a new market production where demand and prices are volatile. The study used prediction models that were reached different Pringal markets to predict the future

Table 4 Comparison of forecasted (F) and actual (A) prices of pea for major markets of eastern Uttar Pradesh

Date	Mirzapur		Varanasi		Allahabad		Mau	
	A	F	A	F	A	F	A	F
22 December 2020	900	902.74	1900	1801	1900	1950.99	2050	2643.52
29 December 2020	1500	902.03	1900	1817.80	1900	2333.10	2060	2574.03
4 January 2021	1400	902.97	1580	1784.98	1750	2215.42	2000	2569.40
12 January 2021	800	903.23	1280	1507	1350	1950.99	1375	2578.66
19 January 2021	600	903.30	1150	1393	1200	2273.41	1180	2574.03
26 January 2021	800	901.57	1120	1768.66	1150	2273.41	1100	2574.03
2 February 2021	700	902.74	1200	1621.87	1100	2215.42	1160	2574.03
9 February 2021	900	902.97	1320	1719.72	1300	2273.41	1225	2574.03
23 February 2021	800	903.31	1600	1703.41	1450	2727.15	1500	2606.46
1 March 2021	800	903.30	1750	1703.41	1600	2457.62	1880	2569.40
Accuracy test								
Mean	941.66		1277.83		1436.16		2265.83	
RMSE	327.87		1924.72		158.78		1453.21	

prices of the period from April 29 to July 21, 2020, which is presented in Table 5. The projected prices revealed that the cod pea price was expected to be significantly high in Mau market. Thus, it makes business sense of farmers for production planning and conservation in view of the quality preferences in the Mau market.

Table 5 Forecast price of cod pea for important market of eastern Uttar Pradesh

Date	Mirzapur	Mau	Allahabad	Varanasi
12/1/2021	900.484932	2546.238197	1101.143237	1872.311589
12/7/2021	902.2273022	2574.035422	1119.115211	1768.662198
12/14/2021	902.9712506	2606.465518	1165.189582	1654.490229
12/21/2021	903.2197392	2638.895614	1238.289852	1540.326868
12/28/2021	903.3027377	2671.325709	1338.41602	1426.163511
1/4/2022	903.3304603	2703.755805	1465.568085	982.1929453
1/11/2022	903.33972	2736.185901	1619.746049	957.7065307
1/18/2022	903.3428129	2768.615997	1800.949911	1057.102312
1/25/2022	903.343846	2801.046093	2009.179671	1280.565291
2/1/2022	903.344191	2833.476188	2244.435329	1628.093137
2/7/2022	903.3442965	2861.273413	2467.593444	2024.719677
2/14/2022	903.3443415	2893.703509	2753.040056	2602.653758
2/21/2022	903.3443565	2926.133605	3065.512565	3304.652734
2/28/2022	903.3443616	2958.563701	3405.010973	4130.716605

ALLAHABAD -SARIMAX Results

Dep. Variable:	y	No. Observations:	51
Model:	SARIMAX(0, 2, 1)	Log Likelihood	-286.263
Date:	Sat, 18 Sep 2021	AIC	576.526
Time:	15:38:34	BIC	580.31
Sample:	0	HQIC	577.962
	-51		
Covariance Type:	opg		

	coef	std err	z	P>|z|	[0.025	0.975]
ma.L1	-0.9115	0.085	-10.675	0	-1.079	-0.744
sigma2	6683.867	786.674	8.496	0	5142.013	8225.72

Ljung-Box (L1) (Q):	0.02	Jarque-Bera (JB):	48.7
Prob(Q):	0.9	Prob(JB):	0
Heteroskedasticity (H):	2.93	Skew:	1.17
Prob(H) (two-sided):	0.04	Kurtosis:	7.29

ALLAHABAD-ARIMA Model Results

Dep. Variable:	D2.Modal_Price	No. Observations:	19
Model:	ARIMA(0, 2, 1)	Log Likelihood	-106.043
Method:	css-mle	S.D. of innovations	59.355
Date:	Sat, 18 Sep 2021	AIC	218.087
Time:	15:38:52	BIC	220.92
Sample:	2	HQIC	218.566

	coef	std err	z	P>|z|	[0.02 5	0.975]
const	1.4887	2.301	0.647	0.518	-3.022	6
ma.L1.D2.Modal_Price	-0.9998	0.18	-5.547	0	-1.353	0.647

Roots

	Real	Imaginary	Modulus	Frequency
MA.1	1.0002	+0.0000j	1.0002	0

MIRZAPUR-SARIMAX Results

Dep. Variable:	y	No. Observations:	36
Model:	SARIMAX(1, 0, 0)	Log Likelihood	-227.478
Date:	Sat, 18 Sep 2021	AIC	460.957
Time:	8:51:16	BIC	465.707
Sample:	0	HQIC	462.615

	coef	std err	z	P>|z|	[0.025	0.975]
intercept	122.0687	93.325	1.308	0.191	-60.845	304.983
ar.L1	0.864	0.086	10.005	0	0.695	1.033
sigma2	1.73E+04	3422.448	5.061	0	1.06E+04	2.40E+04

Ljung-Box (L1) (Q):	0.27	Jarque-Bera (JB):	56.65
Prob(Q):	0.6	Prob(JB):	0
Heteroskedasticity (H):	13.22	Skew:	-1.53
Prob(H) (two-sided):	0	Kurtosis:	8.32

MIRZAPUR-ARIMA Model Results

Dep. Variable:	D.Max_Price	No. Observations:	5
Model:	ARIMA(1, 1, 0)	Log Likelihood	-25.514
Method:	css-mle	S.D. of innovations	39.648
Date:	Sat, 18 Sep 2021	AIC	57.029
Time:	8:52:01	BIC	55.857
Sample:	1	HQIC	53.884

	coef	std err	z	P>|z|	[0.025	0.975]
const	17.8962	17.461	1.025	0.305	-16.328	52.12
ar.L1.D.Max_Price	-0.1959	0.869	0.226	0.822	-1.899	1.507

Roots

	Real	Imaginary	Modulus	Frequency
AR.1	-5.1041	+0.0000j	5.1041	0.5

MAU-SARIMAX Results

Dep. Variable:	y		No. Observations:	83		
Model:	SARIMAX(0, 1, 1)		Log Likelihood	-636.727		
Date:	Sat, 18 Sep 2021		AIC	1277.454		
Time:	9:06:22		BIC	1282.267		
Sample:	0		HQIC	1279.386		
	-83					
Covariance Type:	opg					

	coef	std err	z	P>|z|	[0.025	0.975]
ma.L1	-0.7623	0.06	-12.668	0	-0.88	-0.644
sigma2	3.20E+05	9680.592	33.079	0	3.01E+05	3.39E+05

Ljung-Box (L1) (Q):	0.23		Jarque-Bera (JB):	10338.25
Prob(Q):	0.63		Prob(JB):	0
Heteroskedasticity (H):	37.3		Skew:	6.69
Prob(H) (two-sided):	0		Kurtosis:	56.35

MAU-ARIMA Model Results

Dep. Variable:	D.Modal_Price		No. Observations:	52		
Model:	ARIMA(0, 1, 1)		Log Likelihood	-264.366		
Method:	css-mle		S.D. of innovations	39.02		
Date:	Sat, 18 Sep 2021		AIC	534.732		
Time:	9:06:32		BIC	540.586		
Sample:	1		HQIC	536.976		

	coef	std err	z	P>|z|	[0.025	0.975]
const	-15.1035	7.039	-2.146	0.032	-	28.899
ma.L1.D.Modal_Price	0.306	0.127	2.415	0.016	0.058	0.554

Roots

	Real	Imaginary	Modulus	Frequency
MA.1	-3.2685	+0.0000j	3.2685	0.5

VARANASI-SARIMAX Results

Dep. Variable:	y	No. Observations:	55
Model:	SARIMAX(1, 2, 0)	Log Likelihood	-259.664
Date:	Sat, 18 Sep 2021	AIC	523.329
Time:	12:24:00	BIC	527.269
Sample:	0	HQIC	524.844
	- 55		
Covariance Type:	opg		

	coef	std err	z	P>|z|	[0.025	0.975]
ar.L1	-0.6001	0.081	-7.421	0	-0.759	-0.442
sigma2	1049.6099	118.089	8.888	0	818.16	1281.059

Ljung-Box (L1) (Q):	0.02	Jarque-Bera (JB):	72.96
Prob(Q):	0.89	Prob(JB):	0
Heteroskedasticity (H):	0.12	Skew:	1.44
Prob(H) (two-sided):	0	Kurtosis:	7.98

VARANASI-ARIMA Model Results

Dep. Variable:	D2.Modal_Price	No. Observations:	23
Model:	ARIMA(1, 2, 0)	Log Likelihood	-117.917
Method:	css-mle	S.D. of innovations	40.274
Date:	Sat, 18 Sep 2021	AIC	241.834
Time:	12:24:08	BIC	245.241
Sample:	2	HQIC	242.691

	coef	std err	z	P>|z|	[0.025	0.975]
const	7.0358	5.165	1.362	0.173	-3.088	17.159
ar.L1.D2.Modal_Price	-0.655	0.151	-4.348	0	-0.95	-0.36

Roots

	Real	Imaginary	Modulus	Frequency
AR.1	-1.5268	+0.0000j	1.5268	0.5

4 Conclusion

ARIMA is a short-term prediction methodology. Therefore, if we use Eq. (5), then at all times, we had to bring up to date it. There are numerous methods for forecasting time-series data. Perhaps in the future work, we can focus on how to obtain a sound forecasting method. The prediction of pea price for the coming time has been realized. This cost ever realized in the last several intervals. The administration can take an action similar to one in the past. The government gets a picture of the situation of cod pea prices in the state so that they can make a better decision about the price of a cod pea.

The price data and market arrival of nine major marketplaces of eastern Uttar Pradesh were examined. The farmers of Mirzapur and Allahabad were informed to sell their cod pea product to Mau and Varanasi district, where reasonable price prevails. As per the developed forecasting model, we cannot apply a single ARIMA model to all the markets. The ARIMA (2, 1, 1) is suitable for Faizabad and Basti. ARIMA (2, 0, 0) was suitable for the Ghazipur market. ARIMA (1, 0, 0) was suitable for Siddharthnagar and Mirzapur. ARIMA (1, 0, 1) is suitable for Jaunpur. ARIMA (0, 2, 1) is suitable for Allahabad. ARIMA (1, 2, 0) is suitable for Varanasi, and ARIMA (0, 1,1) is suitable for Mau. These models were also found to be predicting prices of cod pea that were almost the same as the actual prices with very good verifications. The predicted price also revealed that the expected prices of cod pea are higher in Mau market. This article has the inference for the farmers of eastern Uttar Pradesh.

References

1. Kumar P, Badal PS, Paul RK, Jha GK, Venkatesh P, Kingsly IT, Anbukkani P et al (2020) Empowering farmers through future price information: a case study of price forecasting of Brinjal in eastern Uttar Pradesh. Indian J Econ Dev 16(4):479–488
2. www.fao.org
3. Rao NS, Tripathi BN (1979) Study of economics of production and marketing of some vegetable crops in Kankipadu Block of Khrishna District, Andhra Pradesh. Allahabad farmer
4. Khunt KA, Desai DB (1996) Economic feasibility and marketing of perennial vegetables in South Gujarat. Finan Agric 28(1):9–14
5. Thakur DS, Sanjay, Thakur DR, Sharma KD (1994) Economics of off-season vegetable production and marketing in hills. Indian J Agric Mark 8:72–82
6. Maurya OP, Singh GN, Kushwaha RKS (2001) An economic analysis of production and marketing of potato in district Varanasi (UP). Encyclopaedia Agric Mark 8:229–238
7. Sharma VK, Sain I, Singh G (2000) Income and employment from summer vegetables vis-à-vis paddy in Punjab. J Agric Dev Policy 12:38–43
8. Kumar P, Manaswi BH, Prakash P, Anbukkani P, Kar A, Jha GK, Lenin V (2019) Impact of farmer producer organization on organic chilli production in Telangana, India. Indian J Trad Knowl (IJTK) 19(1):33–43

9. Sankaran S (2014) Demand forecasting of fresh vegetable product by seasonal ARIMA model. Int J Oper Res 20(3):315–330
10. Shukla M, Jharkharia S (2013) Applicability of ARIMA models in wholesale vegetable market: an investigation. Int J Inform Syst Supp Chain Manage (IJISSCM) 6(3):105–119

Investigation of Micro-Parameters Towards Green Computing in Multi-Core Systems

Surendra Kumar Shukla and Bhaskar Pant

1 Introduction

Multi-core technology has uplifted various multidisciplinary domains like space, automation, robotics etc. Therefore, applications of diverse characteristics have started executing in a time bound manner [1]. Satisfying performance needs of such applications in the presence of energy constraints is a critical concern in the computing domain. New paradigm has made it compulsory to consider energy as a primary criteria before designing the gadgets [2].

Therefore, it does not matter how effective and innovative a hardware design has been proposed by the designer, if the device is not energy savvy it could not be fabricated [3]. The said fact is true for the high performance computing devices as they consume vast amounts of energy. To address the energy consumption issues, DVFS is a widely used approach which tunes the frequency and voltage as per the varying characteristics of the workload [4]. DVFS could be applied at fine grained and coarse grained levels. However, DVFS covers fine grained parameters like ISA (X-86, ARM) in a limited scope.

The energy consumption issues proved to be severe for heterogeneous and cloud computing data centres [5, 6]. To address such issues designs are proposed which are aligned to the thermal constraints [7]. Such techniques are beneficial for the real-time systems and applications. However, an exhaustive study in this area is required to find the latent aspects of the energy.

All the advancements mentioned above are at the hardware side, and little focus has been made at software side for the analysis of energy issues. To find the energy

S. Kumar Shukla (✉) · B. Pant
Department of Computer Science and Engineering, Graphic Era Deemed to be University, Dehradun, India
e-mail: surendrakshukla21@gmail.com

© The Author(s), under exclusive license to Springer Nature Singapore Pte Ltd. 2023
R. Agrawal et al. (eds.), *International Conference on IoT, Intelligent Computing and Security*, Lecture Notes in Electrical Engineering 982,
https://doi.org/10.1007/978-981-19-8136-4_38

effects on the sequential code, an investigation has been done under the Intel processors which opts the turbo boost [8]. In another context, to manage energy consumption, self-aware-based approaches have been proposed. Self-aware-based approaches promote the awareness about the thermal conditions of the applications for themselves and adopting as per the changes received in the environment [9]. Similarly, application-to-core mapping policies have also shown the potential to save energy consumption to a great extent [10].

Considering the hardware and software-based approaches for the green computing, a gap has been identified where to save the energy at the micro-level has not been attempted. Therefore, in this paper, an investigation and analysis of energy saving towards green computing has been performed. To perform the analysis, a Mi-bench benchmark suite has been used [11]. To validate the hypothesis, simulations have been carried out under the state-of-the-art simulator gem5 [12].

Our contributions in this paper are detailed below as

- An in-depth analysis of energy consumption performed under Mi-bench benchmark suite.
- Energy performance trade-off for in-order and out-of-order cores with varying size of cache.
- Exploration of distinct ISAs and their impact on energy consumption.

The research article is organized below as follows—Sect. 2 is devoted for the analysis of Mi-bench benchmark suite characteristics and their impact on performance, Sect. 3 includes the experiments performed on the said benchmarks to find out the energy consumption effects at the micro-level, and Sect. 4 includes the result and analysis and finally the paper has been concluded.

2 Mi-Bench Benchmark Suite Characterization

To perform the analysis of energy consumption, in this paper, Mi-bench benchmarks were used. The general characteristics of Mi-benchmarks are identified and enlisted in Table 1. It could be noted that the benchmarks which involve higher computation like Bitcount are more vulnerable for the energy consumption [14] Similarly, the benchmarks having higher count for the conditional branches are also the candidate for the higher power requirements. Basic block length (BBL) is a parameter which indicates the higher presence of parallelism in the application [15].

Another, important established factor is that, higher the CPU bound instructions indicates the higher computation further higher energy consumption. However, if benchmarks which are CPU bound (Bitcount benchmark: Higher CPU bound instructions) are mapped to the slow processor cores (I/O Bound), they consume a lot of energy [16]. The benchmarks which have poor data locality (Patricia benchmark: random data access pattern) also are energy hungry [17–19]. In Figs. 1 and 2, the general characteristics of the benchmarks were detailed.

Table 1 Mi-bench benchmark characteristics

S. No	Benchmark	Characteristics
1	Dijkstra	Higher number of conditional branch instructions and memory bound instructions
2	Basic Math	Higher branch prediction rate and Basic block length
3	Qsort	Higher number of conditional branches
4	Bitcount	Highest number of CPU bound instructions and low branch prediction rate and BBL size
5	Patricia	Random data access pattern and higher number of memory bound instructions

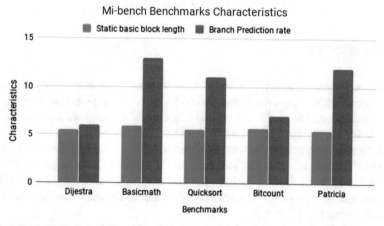

Fig. 1 Mi-bench benchmark characteristics [13]

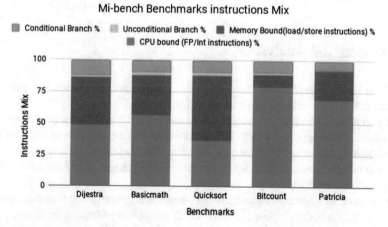

Fig. 2 Mi-bench benchmark instructions mix [13]

Table 2 Script used for the simulation in gem5 simulator

ARM	Inorder L1 = 32 KB L2 = 512 KB	./gem5/build/ARM/gem5.opt -d ./gem5/IOUT /media/shukla/Windows/gem5/configs/example/se.py -c ./gem5/benchmar/MiBench/automotive/basicmath/basicmath_small-arm -o -I=281591461 --cpu-type=MinorCPU --cpu-clock=1ns --caches --l2cache --l1d_size=32kB --l1i_size=32kB --l2_size=512kB --l1d_assoc=2 --l1i_assoc=2 --l2_assoc=2
X86		./gem5/build/X86/gem5.opt -d ./gem5/IOUT /media/shukla/Windows/gem5/configs/example/se.py -c ./gem5/benchmar/MiBench/automotive/basicmath/basicmath_small -o -I=281591461 --cpu-type=MinorCPU --cpu-clock=1ns --caches --l2cache --l1d_size=32kB --l1i_size=32kB --l2_size=512kB --l1d_assoc=2 --l1i_assoc=2 --l2_assoc=2

3 Experiment

To perform the energy analysis, simulations have been performed under the state-of-the-art gem5 simulator. A dedicated python script has been prepared for the gem5, system emulation (SE) mode. The python script used for the simulation is detailed in Table 2. The system configuration used for the simulation is enlisted in Table 3. We have selected five benchmarks from Mi-bench for the simulation; they are Dijkstra, Basic Math, Qsort, Bitcount and Patricia. The L1-I and L1-D cache has been varied from 32 to 64 KB. Cache associativity, peak bandwidth and CPU frequency has been kept constant throughout the simulation. Only three parameters were changed: cache size, CPU model and ISA type.

A dedicated python script has been prepared for the gem5, system emulation (SE) mode. The python script used for the simulation is detailed in Table 3.

4 Results and Analysis

To find out the critical insights on the energy consumption, the simulations were carried out, and the results are detailed in Table 4 and Fig. 3.

It could be noted that the energy consumption for the Dijkstra benchmark has been drastically increased when the benchmarks were executed from in-order core to the out-of-order core. The reason is that, the Dijkstra includes a higher number of conditional branches and memory bound instructions; therefore, it has been reflected in the out-of-order core.

The Basic Math benchmarks have consumed less energy in case of out-of-order core, the reason is that, higher the Basic block length has promoted the higher parallelism which further has reduced the total execution time. Therefore, the Basic Math benchmark has devoted less time in out-of-order core, and energy consumption has been reduced. Energy consumption for the Quicksort benchmark has increased in out-of-order core, because it includes a higher number of instructions, which follow

Table 3 Simulation set-up/system configuration

Benchmark	Applications	L1-D-size (KB)	L1-I size (KB)	L2 size	L1-D asso.	L1-I asso.	CPU model	L2 asso.	Core frequency (GHz)	Total_Inst_num
Mi-bench	Basic Math	32/64	32/64	512 KB/1 MB	2	2	In-order/out-of-order	2	1	281,591,461
	Dijkstra	32/64	32/64	512 KB/1 MB	2	2	In-order/out-of-order	2	1	45,088,085
	Qsort	32/64	32/64	512 KB/1 MB	2	2	In-order/out-of-order	2	1	15,284,694
	Patricia	32/64	32/64	512 KB/1 MB	2	2	In-order/out-of-order	2	1	6012
	Bitcount	32/64	32/64	512 KB/1 MB	2	2	In-order/out-of-order	2	1	5045

Table 4 A comparative analysis of Mi-bench energy consumptions (mJ)

Benchmark	ISA	In-order L1 = 32 KB L2 = 512 KB	In-order L1 = 64 KB L2 = 1 MB	Out-of-order L1 = 32 k L2 = 512 k	Out-of-order L1 = 64 KB L2 = 1 MB
Patricia	ARM	2.853948	2.8560405	1.6032945	1.5993645
	X86	4.099662	4.0863495	2.572257	2.5516215
Bitcount	ARM	8.501617	7.7392455	4.450573	4.331544
	X86	9.273933	9.09624	5.797992	5.616718
Quicksort	ARM	1.431905	1.155978	7.670347	6.125464
	X86	2.457118	2.102839	7.614285	6.185399
Basic Math	ARM	9.421514	8.31398	4.79573	4.452371
	X86	9.15926	8.193406	3.021914	2.451899
Dijkstra	ARM	1.892951	1.880878	7.431331	7.266157
	X86	3.050662	3.02125	7.5111981	7.497335

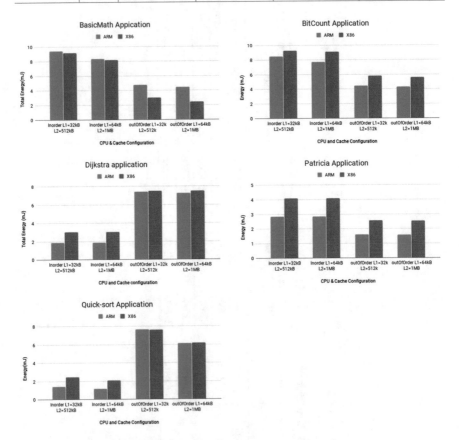

Fig. 3 Analysis of Mi-bench benchmark applications energy consumption

the temporal locality; therefore, out-of-order execution of instructions has increased the miss rate for the last level cache (LLC). And, the potential of the out-of-order has not been utilized properly. Energy consumption for the Bitcount benchmark has been reduced in out-of-order core, as it has a higher Basic block length (BBL) size and higher potential for executing the instructions in parallel. Similar case is true for the Patricia benchmark.

The results could be concluded as follows:

- Increasing the size of the cache slightly reduces the energy consumption. Theoretically, the energy consumption should be increased. The negative results are obtained as the size of cache has not been increased in a higher proportion.
- On the other hand, the energy consumption for the in-order and out-of-order core provides mixed results. For the benchmarks which have higher conditional branches (Dijkstra), the energy consumption is higher, whereas for all the other benchmarks the energy consumption in case of out-of-order core is lower compared to in-order core.
- It indicates that out-of-order core executes the instructions dynamically, and instructions were not in the waiting state, whereas in case of in-order cores the instructions follow the program order; hence, the energy consumption is higher. Theoretically, the energy consumption is higher in case of out-of-order core. However, it also depends on the type of applications that get executed in the system.

5 Conclusion

Green computing is a prime concern of computer architectects. There are various approaches available to address the energy consumption which considers the parameters at the level of hardware. In this paper, we have considered the software aspects, in the micro-level to mitigate the energy consumption of multi-core devices to promote green computing. We could conclude that the application developers need to consider the various aspects related to energy during the designing of the algorithm for the application program. Thus, applications could be executed in the faster hardware efficiently satisfying the power and performance trade-off. Additionally, the conventional ISAs needed the revision to move towards the RISC-V ISA to support the energy constraints of the small devices. In this research article, we have done the analysis of energy with limited parameters (ISA and cache parameters) which could be extended in future work.

References

1. Muttillo V, Tiberi L, Pomante L, Serri P (2019) Benchmarking analysis and characterization of hypervisors for space multicore systems. J Aerospace Inform Syst 16(11):500–511
2. Hamizi I, Kholmatova Z, Succi G (2021) Exploring research hypotheses in green computing. J Smart Environ Green Comput 1:120–130. https://doi.org/10.20517/jsegc.2021.05.
3. Pi Puig M, De Giusti L, Naiouf M (2018) Are GPUs non-green computing devices? J Comput Sci Technol 18(02):e17. https://doi.org/10.24215/16666038.18.e17
4. Hajiamini S, Shirazi BA (2020) Chapter Two—A study of DVFS methodologies for multicore systems with islanding feature. In: Hurson AR (ed) Advances in computers, vol 119. Elsevier, pp 35–71. ISSN 0065-2458, ISBN 9780128203255, https://doi.org/10.1016/bs.adcom.2020.03.005
5. Sheikh SZ, Pasha MA (2022) Energy-efficient cache-aware scheduling on heterogeneous multicore systems. IEEE Trans Parallel Distrib Syst 33(1):206–217. https://doi.org/10.1109/TPDS.2021.3090587
6. Hajiamini S, Shirazi B, Crandall A, Ghasemzadeh H, Cain C (2018) Impact of cache voltage scaling on energy-time pareto frontier in multicore systems. Sustain Comput Inform Syst 18:54–65. ISSN 2210-5379. https://doi.org/10.1016/j.suscom.2018.02.011
7. Sha S et al (2020) On fundamental principles for thermal-aware design on periodic real-time multi-core systems. ACM Trans Des Autom Electron Syst 25(2)
8. Meneses-Viveros A, Paredes-López M, Hernández-Rubio E et al (2021) Energy consumption model in multicore architectures with variable frequency. J Supercomput 77:2458–2485. https://doi.org/10.1007/s11227-020-03349-0
9. Dinakarrao SMP (2021) Self-aware power management for multi-core microprocessors. Sustain Comput Inform Syst 29(Part B):100480. ISSN 2210-5379,https://doi.org/10.1016/j.suscom.2020.100480
10. Gupta M, Bhargava L, Indu S (2021) Mapping techniques in multicore processors: current and future trends. J Supercomput 77:9308–9363. https://doi.org/10.1007/s11227-021-03650-6
11. Guthaus MR, Ringenberg JS, Ernst D, Austin TM, Mudge T, Brown RB (2001) MiBench: a free, commercially representative embedded benchmark suite. In: Proceedings of the fourth annual IEEE international workshop on workload characterization. WWC-4 (Cat. No.01EX538), pp 3–14. https://doi.org/10.1109/WWC.2001.990739
12. Binkert N, Beckmann B, Black G, Reinhardt SK, Saidi A, Basu A, Hestness J, Hower DR, Krishna T, Sardashti S, Sen R, Sewell K, Shoaib M, Vaish N, Hill MD, Wood DA (2011) The gem5 simulator. SIGARCH Comput Archit News 39(2):1–7. https://doi.org/10.1145/2024716.2024718
13. Shukla SK, Chande PK (2019) Parameter analysis of interfering applications in multi-core environment for throughput enhancement. Int J Eng Adv Technol (IJEAT) 9(2):1272–1286
14. Khokhriakov S, Manumachu RR, Lastovetsky A (2020) Multicore processor computing is not energy proportional: an opportunity for bi-objective optimization for energy and performance. Appl Energ 268:114957. ISSN 0306-2619,https://doi.org/10.1016/j.apenergy.2020.114957
15. Kambadur M, Tang K, Kim MA (2014) ParaShares: finding the important basic blocks in multithreaded programs. In: Silva F, Dutra I, Santos Costa V (eds) Euro-Par 2014 parallel processing. Euro-Par 2014. Lecture notes in computer science, vol 8632. Springer, Cham. https://doi.org/10.1007/978-3-319-09873-9_7
16. Butko A et al (2016) Full-system simulation of big.LITTLE multicore architecture for performance and energy exploration. In: 2016 IEEE 10th international symposium on embedded multicore/many-core systems-on-chip (MCSOC), pp 201–208. https://doi.org/10.1109/MCSoC.2016.20
17. Atachiants R, Doherty G, Gregg D (2016) Parallel performance problems on shared-memory multicore systems: taxonomy and observation. IEEE Trans Softw Eng 42(8):764–785. https://doi.org/10.1109/TSE.2016.2519346

18. Gangodkar D, Gupta S, Gill GS, Kumar P, Mittal A (2012) efficient variable size template matching using fast normalized cross correlation on multicore processors. In: Thilagam PS, Pais AR, Chandrasekaran K, Balakrishnan N (eds) Advanced computing, networking and security. ADCONS 2011. Lecture notes in computer science, vol 7135. Springer, Berlin, Heidelberg. https://doi.org/10.1007/978-3-642-29280-4_26

19. Chandran SN, Gangodkar D, Mittal A (2015) Parallel implementation of local derivative pattern algorithm for fast image retrieval. In: International conference on computing, communication & automation, pp 1132–1137. https://doi.org/10.1109/CCAA.2015.7148545

Hope Project: Development of Mobile Applications with Augmented Reality to Teach Dance to Children with ASD

Mónica R. Romero⬤, Ivana Harari⬤, Javier Diaz⬤, and Estela Macas⬤

1 Introduction

One of the great challenges facing developing countries is the search for equity to educate in diversity [1]; currently, there are a large number of children who are diagnosed with developmental disorders, and these children generally need particular attention and the implementation of strategies to improve the education they receive [4, 5], since it is directly proportional to benefit their environment and therefore the quality of life [6].

Autism spectrum disorder, henceforth ASD, can be defined as a complex neurodevelopmental disorder [8–10], which is detected in the early years and lasts a lifetime [12]. The NICT, considering it as a facilitator of the decoding of information, is logical, concrete, located in a space, and not the verbal language that is invisible, temporary, and abstract [17, 18]. Thus, the research first proposes an intervention plan that uses NTIC that can be used by educators, psych pedagogues, therapists, parents who work daily with ASD children [15, 16], making use of augmented reality [17]; for this purpose, the software called Hope (Hoope) is used, in order that children can develop certain skills and abilities.

M. R. Romero (✉) · I. Harari · J. Diaz
Faculty of Informatics, Research Laboratory in New Information Technologies (LINTI), National University of La Plata, Calle 50 y 120, 1900 La Plata, Buenos Aires, Mexico
e-mail: monica.romerop@info.unlp.edu.ar

I. Harari
e-mail: iharari@info.unlp.edu.ar

J. Diaz
e-mail: jdiaz@info.unlp.edu.ar

E. Macas
International Ibero-American University—UNINI MX, Mexico City, Mexico
e-mail: estela.macas@doctorado.unini.edu.mxc

© The Author(s), under exclusive license to Springer Nature Singapore Pte Ltd. 2023   473
R. Agrawal et al. (eds.), *International Conference on IoT, Intelligent Computing and Security*, Lecture Notes in Electrical Engineering 982,
https://doi.org/10.1007/978-981-19-8136-4_40

The study structured is as follows: Sect. 2 explains the methodology for the investigation. Section 3 presents the results of the study, Sect. 4 presents a reflection of the results found in the application of the Hoope software in the experimental study, and Sect. 5 presents the conclusions, recommendations, limitations, and future lines of research.

2 Material and Method

The research addresses a mixed approach supported by the qualitative and quantitative method, additionally the study is of the type: descriptive, exploratory, because it seeks to know in a detailed way the relationship between pedagogical practice through technological innovation mediated using new emerging technologies and the benefits of the application of Hoope software in the teaching–learning processes of children with ASD.

The modality used is documentary and field research [18, 19]. This is because the experimentation is conducted on a specific software called Hoope created in the Laboratory for Research of New Computer Technologies LINTI, of the National University of La Plata in Argentina, and documentary because the process and the results are supported in a methodology and in the theoretical support of previously conducted research.

The field work conducted in Ecuador, specifically in the city of Quito at the Ludic Place Therapeutic Center, which welcomes ASD children offers support for the prevention and addressing of specific learning needs associated with the presence and risk of spectrum disorders. In this area, various methodologies, programs, techniques, and instruments are used to be able to support the children who attend consultations. The procedure for data collection will be through scheduled sessions where a multidisciplinary team intervenes through an interview, deep observation.

2.1 Population

The Director of the Ludic Place Center, destined for three professionals (teacher, psychologist, and psych pedagogue), participated in this process. These professionals were receptive to new and innovative strategies that include the use of new information and communication technologies. Additionally, parents, supported the proposal and signed the informed consent for their children to participate in the intervention. Children with ASD, Matias, who from now on will be identified with the letter M, is 4 years old and has high-functioning ASD, while Eidan, who from now on will be identified with the letter E, is 5 years old and has medium-functioning ASD, have been evaluated and diagnosed in the Ecuadorian Institute of Social Security of Ecuador (IESS).

2.2   Work Plan

The work plan was developed for a period of six months, from February to June 2021, where several activities were planned: the conceptualization of the project, the bibliographic review, the viability of the project, validation of the current situation of infants, contextualization, needs analysis, development of the intervention plan: diagnostic phase, intervention phase, evaluation phase. For the intervention plan, the sessions were designed to work for 20–25 min, twice a week, for a period of several months.

2.3   Phases of the Intervention

Phase I: Diagnostic or initial evaluation. Diagnostic and detailed evaluation of the student, before starting the intervention, we conduct a complete and in-depth evaluation of M and E, to approach the intervention process in an individualized way. It is necessary to emphasize that the center has the diagnosis of children.

Phase II: During the intervention. For the intervention phase, strategies were proposed to conduct a playful activity mediated through technology using the Hoope system. This system allowed the child to interact alone or with the help of the professional who guides the session. The activities that were conducted have a defined order, each session seeks a purpose, and previously, some aspects considered essential have been considered, such as the organization of spaces, the time of the sessions, the necessary materials; and, the collaboration of the center team is counted on therapeutic and with parents.

Phase III: Final evaluation—psych pedagogical. The purpose is to contrast the results obtained in the diagnostic evaluation with those that will be obtained after the process. Using the interviews, it is possible to obtain the necessary information to capture the results of the intervention of children with ASD with the Hoope software.

2.4   Resources Used in the Intervention Plan

To conduct this research, some resources were used, which are indicated below.

Human resources: Multidisciplinary group made up of teachers, psychologists, educational psychologists, doctoral students, systems engineers, parents, and children with ASD.

TIC resource: In relation to technological resources, the Hoope system created in the Research Laboratory of New Computer Technologies LINTI of the National University of La Plata, Argentina, was used a Kinect device and a laptop. The Hoope system is a system that is based on augmented reality, and it is focused on children with ASD from 3 to 14 years old. This software allows the participant to choose options that allow reinforcing teaching learning areas. Next, Fig. 1 shows main menu software and the resources used for the process is shown in Fig. 2.

Fig. 1 Main menu of the software Hoope. Capture made of the software used for pedagogical intervention

Fig. 2 Resources used in pedagogical intervention

2.5 Activities Designed to Reinforce Teaching–Learning Processes

Next, the activities planned for the teaching–learning processes presented; for the intervention plan, we choose to reinforce several processes perception, imitation, fine motor skills, gross motor skills, and visual motor skills, the same ones that are presented in Table a planning temporary activities (Table 1).

Figure 3 shows the activities proposed in the software called Hoope to work the space of perception, imitation, fine and gross motor skills, and visual motor coordination in children with ASD.

Table 1 Curricular content planning intervention project

Area: education	Directed to: children with autistic disorder ASD	Time: 25 min per session
Theme:	Learn by dancing—playful activity. use of Hoope system	
Objectives of the intervention plan		
Perception	Recognition, awareness, and playful experimentation of the body. Visual perception (fundamental to the basis of cognitive processing and reasoning) is the ability to recognize and interpret different visual material correctly and transform this information into an adapted motor response. Therefore, it is an important skill, indispensable for school success	
Imitation	Recognition, awareness, and playful experimentation of the body as an expressive medium with the elements that make up the language of dance, space, time, and energy	
Fine motor	Recognition, awareness, and playful experimentation of the body	
Gross motricity	Recognition, awareness, and playful experimentation of the body	
Visio coordination driving	Recognition, awareness, and playful experimentation of the body	
Name of the activity		
Contact points/touch point	They are interactive zones that appear randomly around the upper part of the user and are activated by being touched with the hands	
Kick points/kick points	They are interactive zones that appear randomly around the bottom of the user and are activated when touched with the feet	
Route tracking/tracking match	They are a set of strokes that appear randomly around the top of the user, and they are activated by touching the starting point and dragging the hand along the entire path to the end point	
Avatar pose/match poses	They are a set of poses that appear randomly at each end of the user and are activated when he manages to imitate the pose by more than 80%	
Mix of poses and exercises	They are a set of poses that appear randomly	
Skills with performance criteria	Learning activities: perception, imitation, fine motor skills, gross motor skills	

Fig. 3 Activity proposed to work imitation, perception, fine and gross motor skills, and visual motor coordination. Software Hoope—playful game for children with ASD

3   Results

Once the application of the intervention project is concluded, the purpose is to contrast the results obtained in the diagnostic evaluation with the results after the intervention process using the Hoope application. We focus on determining if the intervention plan generated favorable results and if there is evidence of any progress in the teaching–learning processes of children M and E.

For the evaluation of the proposed curricular activities, a scale of three possible options is used. The multidisciplinary team that accompanied the development of the pedagogical intervention plan was asked: if the child with ASD is this M or E carried out the proposed activity, it is classified as passed and it is scored as 3; if the child tries to carry out the activity, it is determined that the activity is emergent and is scored with 2; and if on the contrary, the child fails in the process, the activity is scored with 1. The results of the proposed curricular activities are shown in Table 2.

In the following image, we can observe E using the Hoope software during a scheduled session. Figure 4 describes the activity proposed to work imitation: software Hoope—playful game for children with ASD.

Next, the results presented to show the progress of each participant after complying with the proposed work schedule, through the designed phases and after the proposed

Table 2 Results of the proposed curricular activities processes before and after AR

Results of the proposed curricular activities post-AR use				
Actions	Before RA		After RA	
Child with ASD	M	AND	M	AND
Activity to work imitation	1	1	3	3
Activity to work perception	2	2	3	2
Activity to work fine motor skills	1	2	2	3
Activity to work fine motor skills	1	2	2	3
Activity to work visual motor coordination	2	1	2	2

Fig. 4 Activity proposed to work imitation, software Hoope—playful game for children with ASD

sessions. Table 3 shows the comparison of M results, and an analysis of the activities is conducted at the beginning or diagnostic phase and after the use of the Hope System that includes several activities. The interview is conducted in the third phase of this intervention plan, it is conducted in the educational center, and they were informed in advance of the day and time where the meeting was to take place.

Imitation activities: Question: Do you consider that the child's ability to imitate has improved after the use of augmented reality applications, specifically through the Hoope software? Analysis: When asking the multidisciplinary team (psychologist, teacher, and psych pedagogue), they totally agreed that the children reinforced the imitation process, developing the proposed exercises in an easier and more intuitive way with the application of augmented reality, using the option 1 of the proposed Hoope game.

Perception activities: Question: Do you consider that children's perception has improved after using the Hope software application that includes a natural interface with augmented reality? Analysis: The child's perception has improved, after the use

Table 3 Representative graphs of data analysis

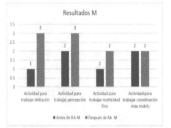	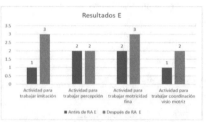
Results of M after the intervention with AR	Results of M after the intervention with AR

Proposed activity Interview results after the intervention process Imitation process	Perception process interview results
	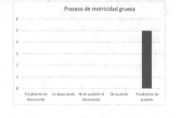
Results of the fine motor process interview	Gross motor process results .

of the Hoope software, people from the multidisciplinary team indicated that they fully agree, and others agree.

Fine motor activities: Question: Do you think that the child's fine motor skills have improved after using the Hoope software? Analysis: The fine motor skills of the child has improved after the use of applications with augmented reality, some people from the multidisciplinary team indicated that they were in complete agreement, and others indicated that they agreed, as shown in figure.

Gross motor activities: Question: Do you consider that gross motor skills on the part of the child have improved after the use of applications with augmented reality? Analysis: The gross motor skills of the child has improved after the use of applications with augmented reality, and imitations of movements of the robot from the Hoope game were carried out by the children during the sessions.

Visual motor coordination activities: Question: Do you think that the child's ability to associate animals with colors has improved after the use of augmented reality applications? Analysis: The visual motor coordination capacity of the children has improved after the use of applications with augmented reality, and all the people on the team indicated that they were in complete agreement.

4 Discussion

It is necessary to review the fulfillment of the proposed objectives of the intervention project and determine to what extent these developed and fulfilled.

In relation to: Review and analyze updated bibliography in relation to the teaching–learning process of children with ASD that favor the approach, concretion and deepening of the proposal, it conducted considering different theories and research that exist in this regard.

As for carrying out an analysis of educational needs that allows knowing the incidence of the difficulty of certain teaching–learning processes of children with ASD who attend the Ludic Place Therapeutic Center, it was achieved by conducting interviews with the treating psychologist, who can I obtain relevant information, taking into account that they are the ones who share directly with children with ASD and know what the needs of each one of them are.

About: Designing a plan for intervention mediated by information and communication technology, in particular augmented reality, was carried out taking into account the needs of analysis, since the idea is precisely to cover the deficiencies that exist, once this aspect has been analyzed. It helped a lot to take into account the general issues that were wanted to be addressed and then to determine which were the areas that would need to be worked to obtain the desired results and the time in which those changes are expected to be seen.

To select those activities that are the most appropriate for learning such as the processes of imitation, perception, fine and gross motor skills, and visual motor coordination, taking into consideration the context first; after that, the bibliographic review was taken into consideration, to finally plan those activities that could be more suitable according to the augmented reality software application called Hoope.

5 Conclusions

The intervention plan allowed to include emerging technologies in our case the use of software Hoope, helping the children who participated in reinforcing teaching–learning processes as perception, imitation, fine, and gross motor skills and visual motor coordination are essential to reduce the existing gap and inequality to which they exposed daily. Regarding the work in the field, direct contact with the community was established through the Ludic Place Therapeutic Center, where they worked with children with ASD, parents, and support professionals (psychologist, educational psychologists, teachers, information, and communication technology professionals).

These studies, which include experimentation as a fundamental basis, are important since they not only come to verify theories, concepts, and information from similar works, but also serve to develop new teaching processes, and hence the importance of being able to identify to personalize teaching. The incorporation of models, methodologies, and strategies, especially with children with autism, is a fundamental requirement, understanding that everyone has their own learning process. For the development of the intervention program, as well as for its monitoring, evaluation, and joint decision-making, it is necessary to work in a comprehensive and multidisciplinary way to obtain better results, given the multitude of professionals involved, the intervention approached from an interdisciplinary approach unifying goals, objectives, and methodology used with the child.

Using innovative resources in the classroom, it seeks to stimulate the teaching–learning processes, it is essential to offer children with ASD during their school stage an adequate teaching–learning process that allows them to strengthen their skills, and this intervention is an alternative in the educational process, to work in coordination toward joint goals and priorities (parents, professionals who accompany the child with ASD, psych pedagogues among others). New ICT information and communication technologies, and specifically augmented reality, are providing teachers with new and effective strategies that allow them to be more effective in education, generating significant interest in learning in children. The intervention proposal included two children with ASD diagnosed with moderate ASD (requires notable help) and severe ASD (they require a lot of help); however, it would be opportune to carry out the intervention in children with mild ASD, and it is possible that this plan and its results are better received and that this intervention based on information and communication technology is of relevant help in these cases.

It is important to note that this intervention plan can be applied and reinforced if applicable, and these adaptations related to content have been proposed as an orientation and exemplification; however, it will be the teacher in collaboration with his team of treating professionals, who will specify the elaboration of the individualized plan based on this intervention program, as well as on the orientations offered by the pedagogical counselor. Autism is a complex disorder that has characteristics of one child with another; therefore, the interventions must be different. Individuality is precisely the factor that should never lost sight of when planning an intervention and what works for one case may not receive in the same way with another child; however, with the help and patience of the professionals in charge, adaptations of the plans can be made to individualize them and achieve better results.

In the case of children with ASD, their development is not stable or predictable; therefore, this plan must evaluate regularly, and it should modify and perfected as many times as necessary. Computer applications in the field of education provide important advantages since they are means that tend to generate intrinsic motivation, being attractive and stimulating. We look at M and E, who like games as well as the music and sound effects provided by the Hoope software, as well as animated characters. Regarding future lines of research, the application of this Hoope program would be of great interest not only in the therapeutic center but also in the home of children with ASD or during schooling to reinforce the processes. We are grateful to the LINTI New Computer Technologies Research Laboratory of the National University of La Plata, Argentina, and the National Secretary of Higher Education, Science and Technology SENESCYT, Ecuador, as well as the Ludic Place Therapeutic Center where this project was conducted.

References

1. Troya I, Lalama ADR, Pacheco M, Yépez M (2018) Los retos de la docencia, frente a la educación inclusiva en el Ecuador. Espirales Rev. Multidiscip. Investig. 2(14):61–70. [Online]. Available: http://www.revistaespirales.com/index.php/es/article/view/190/131
2. Pantoja A (2014) El modelo tecnológico de intervención psicopedagógica. REOP—Rev. Española Orientación y Psicopedag. 13(2):189. https://doi.org/10.5944/reop.vol.13.num.2. 2002.11595
3. Tipo CR, Quir C (2019) Propuesta de intervención psicopedagógica para el refuerzo de la lectura en el tercer año de primaria
4. Guerrero Barona E, Blanco Nieto LJ (2004) Diseño de un programa psicopedagógico para la intervención en los trastornos emocionales en la enseñanza y aprendizaje de las matemáticas. Rev. Iberoam. Educ. 34(2):1–14. https://doi.org/10.35362/rie3422990
5. Romero M, Harari I (2017) Uso de nuevas tecnologías TICS-realidad aumentada para tratamiento de niños TEA un diagnóstico inicial. CienciAmérica 6(3):131–137. [Online]. Available: https://dialnet.unirioja.es/descarga/articulo/6163694.pdf
6. Marín FA, Esteban YA, Iturralde SM (2016) Prevalence of autism spectrum disorders: data review. Siglo Cero 47(4):7–26. https://doi.org/10.14201/scero2016474726
7. García-Primo P (2014) La detección precoz de trastornos del espectro autista (TEA). El programa de cribado con M-CHAT en España y revisión de otros programas en Europa, p 252
8. Málaga I, Lago RB, Hedrera-Fernández A, Álvarez-álvarez N, Oreña-Ansonera VA, Baeza-Velasco M (2019) Prevalence of autism spectrum disorders in USA, Europe and Spain: coincidences and discrepancies. Medicina (B. Aires). 79(1):4–9
9. Diagn C (1923) Am Psychiat Assoc 9(5)
10. Rodríguez Medina J (2019) Mediacion entre iguales, competencia social y percepcion interpersonal de los ninos con TEA en el entorno escolar. https://doi.org/10.35376/10324/ 39475.
11. Reaño E (2015) La Tríada de Wing y los vectores de la Electronalidad: hacia una nueva concepción sobre el Autismo, April 2015, pp 0–13. [Online]. Available: https://www.resear chgate.net/publication/274510152
12. Ventoso R, Brioso Á (2007) Ángel Rivière: La búsqueda del sentido en la clínica del autismo. Infanc. y Aprendiz. 30(3):413–437. https://doi.org/10.1174/021037007781787444
13. Corbellini S, Real LC, Silveira N (2016) Intervenções Psicopedagógicas e Tecnologias Digitais na Contemporaneidade. In: An. dos Work. do V Congr. Bras. Informática na Educ. (CBIE 2016), vol. 1(Cbie), p 1394. https://doi.org/10.5753/cbie.wcbie.2016.1394.
14. Sobrado L, Ceinos C, García R (2012) Utilización de las TIC en orientación profesional: Experiencias innovadoras. Rev. Mex. Orientación Educ. 9(23):2–10
15. Romero M, Macas E, Harari I, Diaz J (2019) Eje integrador educativo de las TICS : Caso de Estudio Niños con trastorno del espectro autista. In: SAEI, Simp. Argentino Educ. en Informática Eje, pp 171–188
16. Romero MR, Diaz J, Harari I (2017) Impact of information and communication technologies on teaching-learning processes in children with special needs autism spectrum disorder, pp 342–353

17. Romero MR, Macas E, Harari I (2020) Is it possible to improve the learning of children with ASD through augmented reality mobile applications? Springer, Cham
18. Oliva CR (2015) The use of ICT in educational guidance: an exploratory study on the current situation of use and training among educational guidance professionals. Rev. Esp. Orientac. y Psicopedag. 26(3):78–95. https://doi.org/10.5944/reop.vol.26.num.3.2015.16402
19. Monica R, Ivana H, Javier D, Jorge R (2020, June) Augmented reality for children with Autism Spectrum Disorder. A systematic review. In: 2020 international conference on intelligent systems and computer vision (ISCV). IEEE Computer Society, pp 1–7

Printed in the United States
by Baker & Taylor Publisher Services